Standardization of Epidemiologic Studies of Host Susceptibility

NATO ASI Series

Advanced Science Institutes Series

A series presenting the results of activities sponsored by the NATO Science Committee, which aims at the dissemination of advanced scientific and technological knowledge, with a view to strengthening links between scientific communities.

The series is published by an international board of publishers in conjunction with the NATO Scientific Affairs Division

A	**Life Sciences**	Plenum Publishing Corporation
B	**Physics**	New York and London
C	**Mathematical and Physical Sciences**	Kluwer Academic Publishers
D	**Behavioral and Social Sciences**	Dordrecht, Boston, and London
E	**Applied Sciences**	
F	**Computer and Systems Sciences**	Springer-Verlag
G	**Ecological Sciences**	Berlin, Heidelberg, New York, London,
H	**Cell Biology**	Paris, Tokyo, Hong Kong, and Barcelona
I	**Global Environmental Change**	

Recent Volumes in this Series

Volume 267 — Bioelectrochemistry IV: Nerve Muscle Function—Bioelectrochemistry, Mechanisms, Bioenergetics, and Control
edited by Bruno Andrea Melandri, Giulio Milazzo, and Martin Blank

Volume 268 — Advances in Molecular Plant Nematology
edited by F. Lamberti, C. De Giorgi, and David McK. Bird

Volume 269 — Ascomycete Systematics: Problems and Perspectives in the Nineties
edited by David L. Hawksworth

Volume 270 — Standardization of Epidemiologic Studies of Host Susceptibility
edited by Janice S. Dorman

Volume 271 — Oscillatory Event-Related Brain Dynamics
edited by Christo Pantev, Thomas Elbert, and Bernd Lütkenhöner

Volume 272 — Photobiology in Medicine
edited by Giulio Jori, Roy H. Pottier, Michael A. J. Rodgers, and T. George Truscott

Volume 273 — Targeting of Drugs 4: Advances in System Constructs
edited by Gregory Gregoriadis and Brenda McCormack

Series A: Life Sciences

Standardization of Epidemiologic Studies of Host Susceptibility

Edited by

Janice S. Dorman

Graduate School of Public Health
University of Pittsburgh
Pittsburgh, Pennsylvania

Springer Science+Business Media, LLC

Proceedings of the NATO Advanced Research Workshop on
Standardization of Epidemiologic Studies of Host Susceptibility,
held June 23–27, 1992,
in Pittsburgh, Pennsylvania

NATO-PCO-DATA BASE

The electronic index to the NATO ASI Series provides full bibliographical references (with keywords and/or abstracts) to more than 30,000 contributions from international scientists published in all sections of the NATO ASI Series. Access to the NATO-PCO-DATA BASE is possible in two ways:

—via online FILE 128 (NATO-PCO-DATA BASE) hosted by ESRIN, Via Galileo Galilei, I-00044 Frascati, Italy

—via CD-ROM "NATO Science and Technology Disk" with user-friendly retrieval software in English, French, and German (©WTV GmbH and DATAWARE Technologies, Inc. 1989). The CD-ROM also contains the AGARD Aerospace Database.

The CD-ROM can be ordered through any member of the Board of Publishers or through NATO-PCO, Overijse, Belgium.

Library of Congress Cataloging-in-Publication Data

On file

ISBN 978-1-4899-1215-2 ISBN 978-1-4899-1213-8 (eBook)
DOI 10.1007/978-1-4899-1213-8

Originally published by Plenum Press, New York in 1994
Softcover reprint of the hardcover 1st edition 1994

PREFACE

The incidence of insulin-dependent diabetes mellitus (IDDM) varies dramatically across racial groups and countries, with annual age-adjusted rates of approximately 40/100,000 per year in Finland, but only 0.5/100,000 per year in China. Although reasons for these marked geographic differences are unknown, it is likely that genetic variations across populations play a major role. To determine the contribution of genetic factors to the global patterns of IDDM incidence, international comparative studies are now being undertaken as part of the WHO Multinational Project for Childhood Diabetes, known as the DIAMOND Project. It is, therefore, necessary to develop and implement epidemiologic standards for these investigations which can be applied across populations. This will ensure that comparable data are obtained in all countries, and that relevant scientific questions can be properly addressed.

The development of standards for molecular epidemiologic studies of IDDM is the objective of the NATO Advanced Research Workshop. During this meeting at the University of Pittsburgh, scientists from across the world convened to discuss issues relating to the standardization of:

1. the collection of family history data to assess the risk of IDDM in first degree relatives,

2. case-control molecular epidemiology studies of IDDM susceptibility,

3. HLA family studies,

4. laboratory methods and DNA technology transfer for genetic marker evaluations.

International comparative research based upon these standards will permit an accurate evaluation of the contribution of genetic factors to the global patterns of IDDM incidence. By identifying factors responsible for the worldwide differences in IDDM incidence, it will be possible to begin to focus on preventive strategies, which is the ultimate objective of the DIAMOND Project.

ACKNOWLEDGEMENTS

* North Atlantic Treaty Organization
* National Institute of Diabetes and Digestive and Kidney Diseases/
National Institutes of Health Grant R13 DK45393
* Juvenile Diabetes Foundation International/
Greater Pittsburgh Chapter
* University of Pittsburgh Medical Center
Department of Conference Management
* Boehringer Mannheim

CONTENTS

EPIDEMIOLOGIC STANDARDS FOR DIAMOND

FAMILIAL IDDM EPIDEMIOLOGY

METHODS FOR EPIDEMIOLOGIC STUDIES OF HOST SUSCEPTIBILITY

POPULATION, CASE-CONTROL AND FAMILY STUDIES OF HOST SUSCEPTIBILITY

LABORATORY STANDARDS FOR EPIDEMIOLOGIC RESEARCH

POSTERS

PREVENTING DIABETES - A HEALTH PARADIGM FOR THE NEXT CENTURY: PERSPECTIVES FROM THE INTERNATIONAL DIABETES FEDERATION (IDF)

Shigeaki Baba

Chairman, The 15th IDF Congress, 1994
Professor Emeritus, Kobe University
Hyogo Medical Center for Adults
13-70 Kitaoji-cho
Akashi, Hyogo 673
JAPAN

Dr. Janice S. Dorman, NATO Course Director, Ladies and Gentlemen, I am very honored to attend this NATO Advanced Research Workshop and to be given the opportunity to address the opening session. Since I am not an expert in the field of epidemiology, many of you may think it somewhat strange that I should be here. I, like all of you, think that for diabetes research and clinical practice, epidemiological studies are most important and are indispensable for the control of diabetes. I believe this research group will make a large contribution in this field.

Epidemiological research, along with public hygiene, has undergone dramatic developments and is now expanding with new objectives, new viewpoints, and new techniques. This field may now be termed clinical epidemiology, and especially for the practicing clinician, modern epidemiology is one of the most important medical sciences. Why is this so?

Clinical epidemiology is able to offer a wealth of valuable information regarding diabetes and its treatment and prevention. Moreover, modern epidemiology is now not only the basis for medical decisions, but it also provides many invaluable suggestions for future research and targets in diabetology. For example, it can indicate the risk of various complications in diabetics at different stages of the disease.

It can be said that we have advanced from epidemiology centering on mortality rates and prevalence in population bases to incidence studies. This means that by using cohort studies, we can evaluate treatment methods and their risks, as well as make future predictions. Through registration we can further increase accuracy, and we may consider that annual research results on incidence will extend public health not only to treatment, but also to prevention by improvement in lifestyles. Furthermore, it is possible to determine the most effective means of limiting economic costs and provide data on health care utilization.

I believe that epidemiology has entered a new era. In particular, advances in genetics and molecular epidemiology, as well as the development of new techniques will most likely give birth to new research fields and powerful ideas on prevention. This NATO Advanced Research Workshop will focus on the development and implementation of epidemiologic standards for these investigations and, thus, is a very timely meeting.

If we look back on the history of diabetic therapy, 73 years have passed since the monumental discovery of insulin. Along with advancements in science, we have seen advances in the pathogenesis and treatment of diabetes. However, the diabetic population is still continuing to increase; blindness, cardiovascular disease, foot amputations, diabetic nephropathy and neuropathy are also increasing. These severe complications are causing pain and suffering in daily life. Considering the increase in the number of patients suffering from non-communicable diseases and the critical nature of such disorders, we can easily foresee that the current methods of medical investigation, health education, and new knowledge on diseases will not be adequate to cope with the future urbanized aging society. In order to achieve a satisfactory lifestyle in this society and to ensure the health of the population, we must examine the quality of life with the improvements in medical treatments and prevention of disease.

I feel that new research programs shall function according to health promotion, including the task of environmental control, and the development of a system or network of specific epidemiologic activities. These include: 1) a registration system for disease on a population basis, 2) annual reporting of epidemiological data, 3) long-term cohort studies, 4) studies of the economics of health with simulations of an experimental model area, and 5) standardization of epidemiologic studies worldwide.

In 1994, the 15th IDF Congress will be held in Kobe, Japan. As chairman, I am already making preparations. I have been considering the manner, the objectives, and the results that will make an impact at this IDF Congress. The mid-1990s will be the foothold for the 21st century, and the starting point for the construction of a new, healthier world. Therefore, in considering both past results, as well as a new age and new approaches towards diabetes, I have chosen the following slogan for this 15th IDF Congress:

"Preventing Diabetes: A Health Paradigm for the Next Century"

I now hope that all of you understand the reason I am here. The purpose of this Workshop is "the development and implementation of epidemiologic standards for these investigations". I am greatly pleased at the timeliness of this project. This action plan which focuses on specific issues, will be expanded. I firmly believe without a doubt that along with comparative studies of populations throughout the world, significant results will be achieved.

THE CONTRIBUTION OF DIAMOND AND MULTINATIONAL STUDIES TO EPIDEMIOLOGICAL RESEARCH

Lewis H. Kuller

Department of Epidemiology
Graduate School of Public Health
University of Pittsburgh
Pittsburgh, PA 15261
USA

INTRODUCTION

Epidemiology is, by definition, the study of epidemics. We study the differences in the frequency of a disease in a defined population in relationship to time, place, geography and the characteristics of the population. Identification of an epidemic then leads to specific analytical studies to determine the "causes of the epidemic" and, subsequently, and most importantly, the best methods for prevention of disease.

Epidemiological research differs from many types of laboratory studies, such as in physiology, biochemistry or molecular biology. The tools of epidemiology work best in a changing environment, with sharp contrasts in terms of disease frequency in comparison to another environment or to previous experiences in the same environment. The epidemiologist thrives on a chaotic situation that creates natural experiments among humans. The bench laboratory scientists have the luxury of tightly controlled experiments in which specific variables can be manipulated in a very well defined environment. Furthermore, the laboratory bench scientist can often carefully control the host, i.e., the genetic characteristics of the animal model. The human is not just a 70 kilogram adaptation of a mouse or even a human primate with just an enlarged cerebral cortex.

The ultimate success of understanding disease pathogenesis and control is based on studies in human populations. The understanding of the disease process in man depends on the interrelationships between the laboratory, animal experimental studies and epidemiological research. None of these approaches has a unique advantage in solving the numerous puzzles related to disease pathogenesis in human beings.

Standardization of Epidemiologic Studies of Host Susceptibility
Edited by J. Dorman, Plenum Press, New York, 1994

INSULIN-DEPENDENT DIABETES MELLITUS

Our interest in insulin-dependent diabetes mellitus (IDDM) in the Department of Epidemiology at the Graduate School of Public Health at the University of Pittsburgh began in the 1970's. Three key observations lead to our initial interest: 1) there was a substantial seasonal variation in the apparent onset of insulin-dependent diabetic cases, with a higher frequency in winter than in summer, 2) there appeared to be a racial and socioeconomic gradient, with lower rates in African Americans and in the lower socioeconomic classes, and 3) there was a higher risk in family members. These initial observations suggested to us that insulin-dependent diabetes might be caused by an infectious agent that was transmitted in families and that risk was primarily age dependent. Moreover, the higher rates within families suggested the possibility of an important host or genetic component. The possibility of a respiratory virus or at least some type of person-to-person transmission was also a reasonable hypothesis. Our initial research efforts began with a doctoral student who played an important role in developing a registry for insulin-dependent diabetes in Allegheny County.[1]

The development of this registry was feasible because of the strong clinical program in insulin-dependent diabetes at Children's Hospital of Pittsburgh directed by Dr. Allan Drash. Dr. Ronald LaPorte joined the diabetes epidemiology program during its early developmental stages. Our initial research focused on host susceptibility, HLA typing and risks of diabetes. Estimations of risk within families, including the secondary attack rate, defined as the risk of diabetes among first degree relatives, were made; the time interval between the initial index case in the family and secondary cases among the family members was also examined. We also began studies of various immunological markers, especially islet cell antibodies, and attempted to determine if we could identify an infectious etiology to the disease.[1]

Graduate students from several countries joined our research program and brought with them information about the apparent differences in the incidence of diabetes among countries (i.e., the apparent epidemics in certain countries compared to others). The strong association of insulin-dependent diabetes with HLA antigens suggested that genetic, as well as environmental factors, might account for the geographic variations. An equally likely hypothesis, however, was that errors in measuring the incidence of disease across populations accounted for the substantial geographic differences in insulin-dependent diabetes mellitus.[2]

Dr. LaPorte was instrumental in successfully developing the WHO DIAMOND Project and the collaborative multinational studies.[2] Dr. Janice Dorman and Dr. Massimo Trucco collaborated to successfully bring the new techniques of molecular biology, especially in studying the major histocompatibility genes, from the laboratory to the epidemiological studies in Pittsburgh and then to many other countries. Their paper in the Proceedings of the National Academy of Sciences[3] was a pioneering report on the ability to bring molecular biology, epidemiology and international studies successfully together to examine the molecular epidemiology of a major disease like insulin-dependent diabetes mellitus.

The multinational studies probably are one of the most powerful tools in epidemiology to evaluate the geographic variations and temporal trends in the incidence of disease. The large variation in insulin-dependent diabetes mellitus incidence across countries provides a rich natural experiment.[2] Why are the rates of insulin-dependent diabetes mellitus so much higher in Finland than in France, Japan, or China? The two major hypotheses are related to differences in host susceptibility or to variations in exposure to an environmental agent (or agents) that cause diabetes, which differ in prevalence among countries. For example, the distribution of the frequency of alleles coding for an amino acid other than aspartic acid at position 57 of the HLA-DQ β chain is correlated with the incidence of insulin-dependent diabetes mellitus among populations.[3]

It has been suggested that cow's milk protein may be an environmental cause of diabetes mellitus.[4] Could variations in milk consumption during early life among countries be a reason for the differences in the risk of IDDM? Perhaps the higher wine consumption, as compared to milk in France, or the avoidance of milk and meat at the same time in Israel has something to do with the relatively low rates of IDDM in these countries. Certainly at the present time this is a far-fetched idea, but easily testable in multinational studies.

Epidemiological research measures the interrelationships between the host, agent and environment. The response of the host to an agent that causes a specific disease or diseases is referred to as susceptibility. Basically, if an individual is exposed to some agent, be it a virus, chemical, food protein, or specific nutrient, he or she will respond with no disease, complete immunity, subclinical disease or a spectrum of clinical disease. For example, approximately one in a thousand individuals who were exposed to the poliomyelitis virus developed clinical poliomyelitis; the rest had an immune response in terms of development of antibodies proving that they had been exposed to the agent, but developed no clinical disease. Is there a specific agent(s) that causes insulin-dependent diabetes mellitus only in susceptible individuals? Antibodies to glutamic acid decarboxylase (GAD), for example, may be markers of subclinical disease following exposure to an environmental agent.[5]

In the past, susceptibility was very difficult to measure prior to exposure to the agent. We usually defined susceptibility in relationship to demographic variables, such as age, sex, race, nutritional status, or the presence of other diseases. The study of insulin-dependent diabetes has provided the first large scale attempt to identify the molecular genetic basis of susceptibility, and as such is a unique and important contribution. We have successfully related molecular polymorphisms of both the HLA-DQ β and α chain genes to the incidence of insulin-dependent diabetes in individuals, families, and among countries.

These genetic markers are very powerful predictors of susceptibility. Whether there is a specific environmental agent or small group of agents that cause the disease among the susceptible individuals is still unknown. It is rather presumptuous to assume that insulin-dependent diabetes is "an autoimmune disease," with no specific environmental cause or causes. "Autoimmunity" is sometimes used synonymously with disease etiology, but in reality is a general term for diseases of unknown etiology in which the immune response plays an important role.

We should avoid getting into the trap of labeling insulin-dependent diabetes mellitus as an autoimmune disease without careful, rigorous approaches to try to identify the interaction between the susceptibility genes and possible environmental agents. The population studies in different countries offer a unique opportunity in epidemiology to study the interrelationship between susceptibility at the molecular level and the environmental agents. This has never been possible before because of our inability to define susceptibility at the individual level. There may be a few important agents that "cause" insulin-dependent diabetes or there may be many non-specific agents that cause the disease in a susceptible host.

Viruses or other living organisms may "cause" insulin-dependent diabetes. There may be a common virus in a very susceptible host that leads to the disease. A search for secondary cases within families, especially subclinical disease, may be a fruitful research project. There may be a mutation of a common virus that has changed its pathogenicity and virulence to cause insulin-dependent diabetes. New molecular biology methods may help identify the agents similar to the success in identifying HIV. There may be unidentified types of viruses, or other living organisms that contribute to the development of insulin-dependent diabetes. New molecular biology methods in combination with epidemiology and molecular immunology may be useful in trying to unravel the problem. A key question is, what is the incubation period of insulin-dependent diabetes? For example, how long after exposure to an agent does the disease develop, especially the first "subclinical stigmata." The monitoring of very susceptible individuals, such as siblings who have similar HLA-DQ α and β susceptibility genes may provide a measure of the incubation period. Differences in the secondary attack rates in families in both high and low incidence populations may provide further clues to environmental etiology.

The search for "non-living" agents, such as milk proteins, may be more complex. The initial studies identified milk proteins[4] in the serum of diabetics as compared to controls. Longitudinal studies of high risk populations, combined with a challenge or experimental epidemiological approaches may be of great importance. The most powerful tools, however, may be to compare the lifestyles and exposure to specific environmental agents with risk of disease among populations and within families. Because of the substantial large variations in nutrition, occupation and other lifestyles among countries, the approach of comparing heterogeneous rather than homogeneous populations is more likely to lead to the identification of specific environmental agents which contribute to the disease.

Epidemiology would be a rather sterile discipline without a primary focus on disease prevention. The prevention of insulin-dependent diabetes will depend on the identification of susceptible individuals and, hopefully, the modifications of the environment to reduce the risk of disease. Obviously, a major question at present is what treatment, if any, should be provided for the very high risk susceptible individuals.

CONCLUSIONS

In summary, the search for the cause or causes of IDDM is still an important issue. The success in molecular epidemiology has made it possible to identify very susceptible populations, and within these populations, families and individuals at very high risk of developing IDDM. We can utilize the known variations in susceptibility within and across populations to test for specific environmental agents. The study of very low risk populations, as compared to those with very high risk, as well as studies of new epidemics of IDDM, and evaluations of the development of disease within families, offer opportunities to identify the causes and hopefully the prevention of insulin-dependent diabetes.

REFERENCES

1. L.H. Kuller, D.J. Becker, K.J. Cruickshanks, et al., Evolution of the Pittsburgh studies of the epidemiology of insulin-dependent diabetes mellitus, *Genet Epidemiol.* 7:105 (1990).
2. J.S. Dorman, R.E. LaPorte, M. Trucco, Genes and environment, *Baillere's Clin Endocrinol Metab.* 5:229 (1991).
3. J.S. Dorman, R.E. LaPorte, R.A. Stone, M. Trucco, Worldwide differences in the incidence of Type I diabetes are associated with amino acid variation at position 57 of the HLA-DQ-β chain, *Proc Natl Acad Sci. USA* 87:7370 (1990).
4. J. Karjalainen, J.M. Martin, M. Knip, et al., A bovine albumin peptide as a possible trigger of insulin-dependent diabetes mellitus, *N Engl J Med.* 327:302 (1992).
5. M. Rowley, I.R. Mackay, Q.Y. Chen, et al., Antibodies to glutamic acid decarboxylase discriminate major types of diabetes mellitus, *Diabetes* 41:548 (1992).

BEYOND INSULIN-DEPENDENT DIABETES MELLITUS REGISTRIES: CAPTURE-RECAPTURE APPROACHES FOR MONITORING INCIDENCE

Ronald E. LaPorte,[1] Daniel McCarty,[1] Eugene Tull,[1] Ingrid Libman[1] and Masato Matsushima[2]

[1]Department of Epidemiology
Graduate School of Public Health
University of Pittsburgh
Pittsburgh, PA 15261
USA

[2]Third Department of Medicine
Jikei University School of Medicine
3-25-8 Nishishinbashi, Minato-Ku, Tokyo 105
JAPAN

INTRODUCTION

Since 1980, the most important development in the area of childhood diabetes epidemiology has been the establishment of standardized registries for monitoring the incidence of the disease. In this paper, the evolution of the registries will be discussed, as well as concepts as to where the field should progress.

In laboratory science, technological advances have often driven rapid improvements in our knowledge and understanding of the world around us. Microscopes, telescopes and stethoscopes have allowed man to see and hear better. Typically, such advancements have meant the development of new instruments, or techniques that facilitate and standardize measurement. However, diabetes epidemiologists have the population as their laboratory. We believe that there are new methods available that can aid diabetes epidemiologists to count incident cases of IDDM.

To investigate diseases such as childhood diabetes, the fundamental measurement unit is that of incidence rates. The better we are able to measure incidence rates, the better we are able to understand disease on a population basis. In diabetes epidemiology, we make our living calculating incidence by counting new cases of IDDM and comparing rates across time and place to determine the risk factors for disease. Moreover, the ultimate goal is the prevention of IDDM. Prevention can be defined as the reduction of the incidence of disease. Thus, to eradicate IDDM, we need to know its incidence. Measurement of incidence is the core and heart of the WHO DIAMOND Project. Without incidence, there would be no DIAMOND Project. Without incidence, there would be no epidemiology.[1] We propose that it is time to modernize traditional approaches of counting new onset cases of IDDM for the 21st century. The technology is available to transform IDDM epidemiology.

A fundamental goal for incidence rates is the identification of new cases. Traditional methods of surveillance in public health fit into two broad categories, the first of which is a registry.[2] Standardized incidence registries have rapidly emerged in IDDM epidemiology. We had a dream in 1983 that incidence registries would be developed in many countries and that we could begin to map the incidence of IDDM across the world.[3] At that time, there were few registries, none was using comparable approaches, and none was collaborating. Moreover, the utility of IDDM registries was questioned by many basic scientists who thought that the cure for diabetes was right around the corner. Also, there were no international resources for registry development. While at first the dream of standardized IDDM registries did not seem to be immediately feasible, people soon began to realize that the frequency with which the disease occurs in their country was a fundamental piece of knowledge. It appeared that the country in which a child lived was the most potent determinant of risk of disease. Therefore, there was an explosion of interest in IDDM registries in a very short period of time with rapid registry development using standardized approaches on almost all continents.

As presented in Table 1, a simple set of criteria was set forth for registries.[3,4] These consisted of the definition of a case, and the description of the population at-risk. It also emphasized for the first time in any disease registry, the need of a secondary source of case ascertainment which would provide a check on the degree of ascertainment. It was agreed that all registries should strive for at least a 90% ascertainment for the primary source. These simple criteria have served us well for the past nine years. The registry development has led to many important findings concerning the etiology of IDDM.

Table 1. Criteria for IDDM registries.

1.	Age at onset 0-14 years
2.	Diagnosed with diabetes and placed on insulin
3.	Diagnosed within the defined time period
4.	A member of the defined community at the time of diagnosis
5.	The establishment of a secondary source of case ascertainment
6.	Ascertainment of at least 90% in the primary source.

Despite the enormity of the effort and the remarkable findings from the registry work, it is important to weigh the pluses and minuses of registries based upon our current understanding of the etiology of diabetes. As the registries evolved and epidemics were found, the limitations became evident. The major difficulty is that registries have proven to be too limiting for the identification of rapidly changing patterns of disease. It is in the epidemics that the causes of disease will be found, and the prevention of epidemics has been fundamental to the prevention of disease. However, registries are ill suited for epidemic identification and prevention as they do not monitor sufficiently large populations, and the latency between changing patterns of incidence and the identification of these changes is too long to be in "hot pursuit" of the factors causing the epidemics. Thus, approaches need to be developed to rapidly obtain accurate IDDM incidence data to complement current registry approaches.[5,6]

The fundamental principle of IDDM registries is that they are active surveillance systems where the research is led by a small group of interested investigators, who actively go to hospitals, pharmacies, etc. to identify cases. The goal is to identify as close to 100%

of the cases as possible. As presented, a goal of the registries has been to achieve at least a 90% ascertainment from the primary source. As we in DIAMOND have found, this has been extremely important, yet at the same time it is also very costly.

The goal of "near-perfect" ascertainment is exactly the same philosophy which has been used in cancer, myocardial infarction and multiple sclerosis incidence studies, for example, to establish registries with a high degree of ascertainment. Identifying everyone, though, is costly and thus leads to limited geographic coverage, and a long latency between the development of the disease and registration. IDDM epidemiology, however, is different from other chronic disease systems in that we required a systematic check on the degree of ascertainment with a secondary source of case identification. The secondary checks have been infrequent with cancer registration, for example, and not standardized across registries.

Registry work for non-communicable diseases and, in fact, chronic disease epidemiology can be contrasted with the second approach for the identification of new cases of disease which has been established for communicable diseases. The infectious disease system consists of passive disease surveillance. These systems usually are in public health agencies and are voluntary. They evaluate multiple diseases, rapidly identify cases and have low cost per identification of cases. Using the U.S. as an example, less than 5% of the U.S. is being monitored for IDDM, however, almost the whole country is being monitored for serious communicable diseases. Thus, the communicable disease system has the advantages of broad, efficient, and timely coverage compared to the registry based system we have been using for the past nine years. However, the fundamental drawback of the communicable disease model is that the incidence rates derived are low and variable and thus very inaccurate. Evaluation of passive systems reveal that frequently <20% of the cases are identified.[7] The imprecision of the incidence rate estimates is not a major problem for communicable disease research where, with "spiking" incidence rates, a very imprecise measure can identify a 10-100 fold increase in the number of cases. However, with IDDM and chronic diseases in general, the monthly or yearly change in incidence is of a much lower magnitude. Therefore, to identify changes, one needs to have much more precise measurement tools. Hence, the methods of registries were established with the primary concept that the only way to achieve precise incidence rates is to achieve 100% ascertainment of cases. However, even with registries in general, 100% ascertainment is frequently not achieved and the costs preclude broad monitoring.

In the 21st century, we need to modernize our approach towards determining incidence and to move to the next generation of incidence technology to complement existing registry systems. Recently we have argued that the technology is available in an article titled "Counting Birds, Bees, and NCDs".[8] To paraphrase the first line of the paper, "Why is it that we know more about the number of vampire bats, sperm whales, and Ulysses butterflies there are in the world than we know about the number of new cases of childhood diabetes?" It is because we are better able to count wildlife than we are to count incident cases of disease. We believe that the bird, fish and insect literature provides the methods to revolutionize our approaches towards counting diabetes, and in fact any non-communicable disease. It is from the annual ecology literature that the next generation of IDDM incidence systems will evolve.

The approach towards the counting of new cases and, thus, the determination of incidence can be made through the use of capture-recapture. Capture-recapture approaches are the fundamental method by which the population and range of wildlife are determined.[9-12] To illustrate the approach, suppose we wanted to determine the number of trout in Lake Geneva. The animal ecologists would not try and capture all the trout in the lake, as we try and capture all the new onset cases of diabetes. The bird, fish and mosquito people realized that it would be almost impossible to identify all cases, yet still they obtain accurate incidence rates. They accomplish this through the simple capture-recapture procedure where they first would go to the Lake and net fish. The fish would then be tagged, and released, then the following day the fish would be recaptured, the proportion of cases in common is then used to estimate the number of fish in the total lake with 95% confidence intervals. The bird and fish scientists rarely have over 20% of the cases in each sample, yet obtain very accurate estimates of the number of cases. In contrast, we constantly strive for over 90% ascertainment in our primary sources, although this is truly not needed.

The primary assumption of the two sample case is one of independence of sources. In health, this is a difficult assumption to achieve. For example, pediatricians will refer to specific hospitals, and insulin prescriptions may come from a certain set of physicians. However, in an elegant method, Feinberg and Bishop[9] showed that if we use more than two sources for the identification of cases, then source dependencies can be modelled and taken into consideration when estimating the number of new cases. In IDDM epidemiology, typically, there are many possible incomplete sources of ascertainment including hospitals, physicians, diabetes associations, laboratories, newspaper advertisements, etc. What is critical is not the exact nature of the log linear modelling, but rather what this means for estimating incidence. The modelling provides a means for accurately estimating incidence rates with even a high and variable degree of undercount so that inexpensive sources of ascertainment are extremely valuable. Moreover, more than two sources of identification of cases are needed. It typically is much more costly to identify all cases of IDDM coming from hospitals through active reporting systems than it would be to obtain a mix of active and passive systems employing multiple sources of ascertainment.

To understand how these approaches towards counting animals can be employed to count new cases of IDDM, it is best to contrast them with the traditional approaches of registries and passive surveillance methods. When we go out to identify new cases of disease, frequently we will go to multiple sources with registries, or with passive systems, the data will be reported to us from multiple sources. To determine the numerator (e.g., the number of new cases), the sources are aggregated eliminating the duplicates. It is this total that is used for determining the incidence rates. This approach assumes that the aggregate of the rosters represents all (or nearly all) of the cases in the population. It assumes that the number of missing cases is zero, and thus any possible undercount is ignored.[13]

The traditional methods for calculating incidence rates can be contrasted with the ascertainment corrected rate approach.[13] In this method, instead of eliminating the duplicates, attention is paid to the duplicates as this is important information for determining the degree to which cases may have been missed. Thus, the degree of undercount is estimated, and the rates of disease are calculated using capture-mark-recapture.

We, in IDDM epidemiology, were actually the first group to recognize that we should standardize all of our incidence rates to assure compatibility. Two researchers from NATO countries, Dr. de Beaufort[14] from Luxembourg, and Dr. Vaandrager[15] from the Netherlands were the first to use these approaches. The importance of determining and controlling for undercount in a standardized manner has not as yet been recognized by most cancer registries, where among the thousands of cancer epidemiology papers, few have evaluated the degree of ascertainment, let alone established standardized procedures for adjustment as what we are using in diabetes.

HOW CAN THIS APPROACH REVOLUTIONIZE DETERMINATION OF THE INCIDENCE OF IDDM?

With our registries, we spend a considerable amount of time attempting to identify all cases occurring within a community. This is very burdensome and costly. We were being like fishermen trying to catch all the fish in Lake Geneva to count them. We can obtain almost the same degree of precision for considerably lower cost using multiple incomplete sources of case ascertainment.

In our multinational DIAMOND Project,[1] capture-recapture procedures serve as a "check" on our registries. It has thus been viewed as a secondary procedure with registries that were "near-perfect", i.e.,had a high degree of ascertainment. We have still had the goal of trying to find all the cases. However, this goal is very expensive and thus cannot be used across broad areas or over time. However, we have not as yet recognized in IDDM epidemiology that we can break out of the concept that we need to achieve close to 100% ascertainment in our sources to identify cases and still achieve very accurate incidence rates. If very accurate estimates of wildlife can be produced with less than 20% ascertainment per sample, then certainly the same would apply to IDDM and all chronic diseases. We thus can even consider that cheap sources of ascertainment which only identify a small proportion of cases can yield very precise, standardized estimates of new cases.

IMPLEMENTATION

Recently, many of the WHO Directors of Collaborating Centers and Recognized Health Officials have begun to argue that there is a global need for the monitoring the incidence of non-communicable diseases in the 21st century. IDDM can take the lead as a model disorder for all diseases. It has certain advantages in that it is ideally suited for being captured because it is a severe disease, but not too severe; thus the children are seen by multiple health care officials, and are tagged. Moreover, it is a disease with a classic set of symptoms. Therefore, there are few misdiagnosed cases, and a simple, inexpensive blood test, that of a blood sugar, rapidly diagnoses the disease. In addition, it is a frequent disease, but not too frequent. Therefore, the establishment of the monitoring system would not be too burdensome.

We have argued that the monitoring of non-communicable diseases should move into the public health system, as it is the health departments that have the responsibility and resources for monitoring of diseases. Moreover, by having IDDM into the public health system, we have an advantage that governments must pay attention to it, as by law, they would be required to collect the data. National health systems, however, such as the Centers for Disease Control, have shied away from the inclusion of chronic diseases into their monitoring efforts. With the simple introduction of making IDDM in children a reportable disease, and with the introduction of capture-recapture, the local health departments, thus, become the sentinels to obtain accurate national incidence data on the incidence of IDDM.

We need to begin to push governments into making IDDM a reportable disease.[5,6] It is also our responsibility to assist local health departments in developing methods of operation so that we can be able to directly compare the national incidence of IDDM in the United States, with that of other NATO countries, and the world.

If we can demonstrate the feasibility of national monitoring of the incidence of IDDM, this will serve as a model for the national monitoring of the incidence of communicable and non-communicable diseases alike. NATO countries can take this lead in the vision for the 21st century. All new cases of diseases will be reported to local health departments. We recognize that the data, as such, would not be directly comparable across countries. However, for each disease we can easily determine the degree of undercount, and use this to ascertainment-adjust the rates so that we can not only compare the incidence of IDDM across countries, but all diseases. The measurement of the health of a nation will thus be determined through the comparisons of overall incidence of all diseases.

Acknowledgements

Research supported by NIH grant #R01 DK24021.

REFERENCES

1. WHO DIAMOND Project Group, The WHO Multinational Project for Childhood Diabetes, *Diabetes Care* 13(10):1062 (1990).
2. J. Goldberg, G.M. Belfand, P.S. Levy, Registry evaluation methods: A review and case study, *Epidemiol Rev.* 2:210 (1980).
3. R.E. LaPorte, N. Tajima, H.R. Akerblom, N. Berlin, J. Brosseau, M. Christy, A.L. Drash, H. Fishbein, A. Green, R. Hammon, M. Harris, H. King, Z. Laron, A. Neil, Geographic differences in the risk of IDDM: The importance of registries, *Diabetes Care* 8(suppl.1):101 (1985).
4. Diabetes Epidemiology Research International Registry Group, Diabetes registries in Asia, *Bull WHO* 65(6):897 (1987).
5. WHO DIAMOND Project Group on Epidemics, Childhood diabetes, epidemics and epidemiology: An approach for controlling diabetes, *Am J Epidemiol.* 135:803 (1992).
6. R.E. LaPorte, S. Baba, Magic bullets, reportable disease, and prevention of childhood diabetes, *Diabetes Care* 15(1):128 (1992).
7. S.B. Thacker, R.C. Berkelman, Public health surveillance in the United States, *Epidemiol Rev.* 10:164 (1988).
8. R.E. LaPorte, D.J. McCarty, E.S. Tull, N. Tajima, Counting birds, bees, and NCDs, *Lancet* 339:494 (1992).
9. Y.M.M. Bishop, S.E. Feinberg, P.W. Holland, Discrete Multivariate Analysis: Theory and practice, Cambridge, Mass: MIT Press, pp. 229-56 (1975).
10. D.G. Chapman, The estimation of biological populations, *Ann Math Stat.* 25:1 (1954).
11. G.F. Seber, The Estimation of Animal Abundance and Related Parameters, 2nd ed., New York: MacMillan.
12. R.E. LaPorte, E.S. Tull, D. McCarty, Monitoring the incidence of myocardial infarctions: Applications of capture-mark-recapture technology, *Int J Epidemiol.* 21(2):258 (1992).
13. D.J. McCarty, E.S. Tull, C.S. Moy, C.K. Kwoh, R.E. LaPorte, Ascertainment corrected rates: Applications of capture-recapture methods, *Int J Epidemiol.* 22(3):559 (1993).
14. C.E. de Beaufort, G. Michel, G. Glaesener, The incidence of Type I (insulin-dependent) diabetes mellitus in subjects aged 0-19 years in Luxembourg: A retrospective study from 1977 to 1986, *Diabetologia* 31:758 (1988).
15. G.J. Vaandrager, G.J. Bruining, F.H. Veenhot, N.M. Drayer, Incidence of childhood diabetes in the Netherlands: A decrease from north to south over North-Western Europe?, *Diabetologia* 27:203 (1984).
16. NCD Directors of Collaborating Centers, Needed: Universal monitoring of all serious diseases of global importance, *Am J Public Health* 83:941 (1993).

EPIDEMIOLOGY OF IDDM IN CHILDHOOD IN THE BALTIC SEA REGION

Jaakko Tuomilehto,[1] Toomas Podar,[2] Bela Adojaan,[2] Irina Kalits,[2] Eva Tuomilehto-Wolf,[1] Marjatta Karvonen,[1] Girts Brigis,[3] Brone Urbonaite,[4] Zilvinas Padaiga,[4] Zygimantas Cepaitis,[1] Vilius Grabauskas,[4] Marian Rewers,[5] Mieczyslaw Walczak,[6] Evgene Shubnikov,[7] Juri Nikitin,[7] and Ronald LaPorte[8]

[1]Diabetes Research Centre
Department of Epidemiology and Health Promotion
National Public Health Institute
SF-00510 Helsinki
FINLAND

[2]Republic Endocrinology Centre
Tartu University
EE-2400 Tartu
ESTONIA

[3]Latvian Medical Academy
Department of Public Health and Epidemiology
LV-1007 Riga
LATVIA

[4]Kaunas Medical Academy
3000 Kaunas
LITHUANIA

[5]Department of Preventive Medicine
University of Colorado School of Medicine
Denver, Colorado
USA

[6]Institut Pediatrii
Academy of Medicine,Szpitalna 27133
60-572 Poznan
POLAND

[7]Institute of Internal Medicine
Academy of Medical Sciences, Siberian Branch
630090 Novosibirsk
RUSSIA

Standardization of Epidemiologic Studies of Host Susceptibility
Edited by J. Dorman, Plenum Press, New York, 1994

[8]Department of Epidemiology
Graduate School of Public Health
University of Pittsburgh
Pittsburgh, PA 15261
USA

INTRODUCTION

Insulin-dependent diabetes mellitus (IDDM) is a chronic disease with unknown aetiology and, therefore, without effective measures for its prevention. Both genetic and environmental factors play a role in the development of the disease. Comparative studies between different populations of varying genetic and environmental backgrounds are needed for a better understanding of the aetiology of IDDM.

Previous data have demonstrated a very large geographic variation in the incidence of childhood IDDM. In Finland, the incidence of IDDM is the highest in the world and it also has increased markedly during the last decades.[1-4] In Scandinavian countries (Sweden, Norway and Denmark), IDDM is also common and has been increasing. [1,3,5-12] In marked contrast, Poland, which is located south of the Baltic Sea, has one of the lowest incidence rates of IDDM reported within Europe.[1,5,6,13] Comparative data have now become available for the three Baltic Republics of Estonia, Latvia and Lithuania on the eastern coast of the Baltic Sea.[14-19] Since no other information on incidence in the Russian population were available, we used data from Novosibirsk.[20] This report describes the central epidemiologic findings from the active research carried out in this region.

DIAGNOSTIC CRITERIA

The diagnostic criteria applied were those used for IDDM registers.[5-7] All cases were diagnosed by a doctor and permanent insulin therapy was initiated. The age at diagnosis of the children was less than 15 years. Details of the collection of data used in this paper have been reported in separate publications.[1,2,4,13-21]

OUTBREAKS

The incidence of IDDM varies considerably from year to year. The identification of years with "outbreaks" of IDDM in five populations have been evaluated applying the following method:[22]

$$\frac{(\text{Number of observed - number of expected cases})^2}{\text{Number of expected cases}} > 3.84$$

This model means that if the product exceeds 3.84, the number of cases per year is more than twice the standard deviation.

VARIATION AMONG COUNTRIES

The age standardized incidence of IDDM in Finland was higher than in other populations around the Baltic Sea (Table 1). There was little variation in the risk of IDDM among the populations living on the southeastern side of the Baltic Sea; the incidence was approximately 6-7/100,000 per year. On the western and northern side of the Baltic Sea, the incidence was clearly higher. It varied from approximately 10/100,000 per year in Estonia to 34/100,000 per year in Finland. The lowest incidence was reported for the district of Novosibirsk, Russia, 4.6/100 000 per year.

The relative risk of IDDM in Sweden was about 70% of that in Finland, whereas in Poland it was less than 20% of that in Finland. The relative risk in the incidence compared with Finland was similar for both sexes, but the differential between Finland and other populations, except Norway, was slightly smaller in females than males.

Table 1. The risk of IDDM in children aged 14 years and under in the populations of the Baltic Sea region.

Population	Period	Incidence/100,000		Relative Risk¶	
		Males	Females	Males	Females
Finland	1983-88	36.9	31.6	1.00	1.00
Sweden	1978-87	25.1	23.8	0.68	0.75
Norway	1978-82	24.1	21.2	0.58	0.52
Denmark	1970-76	14.5	13.4	0.39	0.42
Estonia	1983-88	10.7	10.0	0.29	0.32
Germany	1982-84	7.0	7.0	0.19	0.22
Lithuania	1983-88	6.5	7.0	0.17	0.22
Latvia	1983-88	6.3	6.5	0.18	0.21
Poland	1983-88	6.0	6.4	0.16	0.20
Russia	1983-89	4.4	4.7	0.12	0.15

¶Compared with Finland

In the youngest age group (0-4 years), the annual incidence in Finland reached approximately 20/100,000 per year (Table 2). In most other countries, the incidence was six times lower, approximately 3/100,000 per year. In each country the incidence rose with age. The steepest rise was in Estonia and Russia, where the incidence rates in the 10-14 year age group were 4-6 times higher than for those age 0-4 years. The rate ratio in Estonia was significantly higher than in Latvia, Lithuania or Poland. Among 0-4 year old children, the incidence in Estonia did not differ from that in the other three Baltic populations. In Sweden and Norway, this rate ratio was similar to that for Finland (approximately 2), whereas in Denmark it was slightly higher.

Table 2. The incidence of IDDM per 100,000 per year during childhood in the Baltic populations.

	Boys' Age Group			Girls' Age Group		
Country	0-4 yrs	10-14 yrs	Rate Ratio*	0-4 yrs	10-14 yrs	Rate Ratio*
Finland	23.1	46.3	2.0	18.4	37.7	2.0
Sweden	16.8	31.6	1.9	14.5	29.4	2.1
Norway	11.9	32.6	2.7	13.1	27.9	2.1
Denmark	7.7	22.7	2.9	7.4	20.1	2.7
Estonia	3.1	18.1	5.8	3.2	13.3	4.2
Latvia	3.2	10.5	3.3	3.7	10.0	2.7
Lithuania	4.1	8.6	2.1	2.9	9.6	3.3
Poland	2.9	9.5	3.3	3.2	9.2	2.9
Germany	2.9	10.2	3.5	3.2	11.1	3.5
Russia	2.3	8.7	3.7	1.9	7.8	4.1

*Ratio: 10-14 years compared with 0-4 years

In populations where the incidence was high, there was a male excess in incidence (Figure 1). The higher the incidence, the higher was the male/female (M/F) ratio. This increase was exponential. In Finland, the sex ratio was 1.17. The lowest ratio was found in Poland and Lithuania (0.94), approximately 20% lower than that of Finland. The relationship between the incidence and the M/F ratio followed the regression equation: M/F ratio = 0.880 + (0.0151) (incidence) - (0.0002) (incidence)^2. The variation in incidence explained 77% of the variation in the M/F ratio ($p=0.0025$). According to this model, the M/F ratio is 1.0 when the incidence is 9.0/100,000 per year.

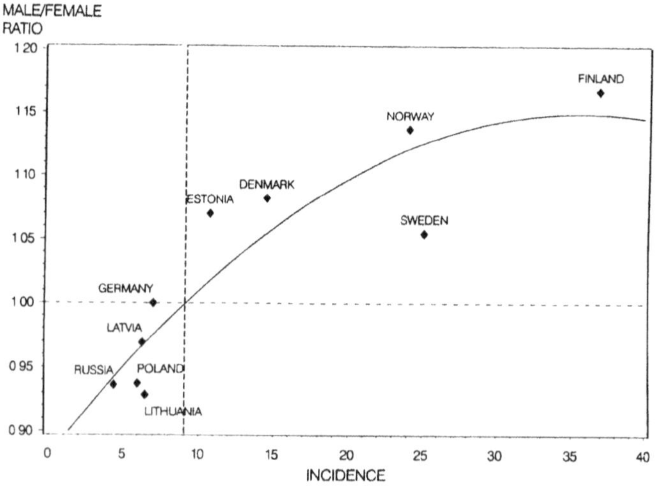

Figure 1. The male/female ratio in incidence of childhood diabetes as a function of incidence in males.

VARIATION WITHIN COUNTRIES - GEOGRAPHIC AREAS

Within-country regional variation in incidence has been studied in Finland, Sweden and Norway. The earlier data from Finland suggested that in the 1970's, the incidence was lower in the northern than in the southern parts of the country.[23] More recent findings, however, reveal little variation in incidence of IDDM among the 12 provinces of Finland.[21]

In Sweden, significant differences among the 24 counties have been found.[10] There was no systematic geographic pattern, but the highest incidence was reported for the southeastern part of the country near or at the Baltic Sea. In Norway, rather pronounced geographic variation has been documented, with higher incidence in the southeastern parts of the country and lower rates in the western and northern regions.[11]

Overall, it appears that the variation in incidence of IDDM within Finland, Sweden and Norway does not follow the clear geographic pattern which is seen for several other diseases such as cancer and cardiovascular disease. The variation in IDDM is much larger between countries than within a country. This could be attributed to genetic factors. On the other hand, the abrupt change in incidence from one country or population to another that share a certain genetic background cannot be explained by genetic variation alone.

The children living in the City of Novosibirsk had a higher risk of IDDM of 5.7/100,000 per year (95% CI = 4.8 - 6.8) than children living in the surrounding counties where the rate is 3.6/100,000 per year (95% CI = 2.9 - 4.5).[20] The children of the City of Novosibirsk had a peak in incidence in 1985-1986 when the rate reached 8.8/100,000 per year. No increase in incidence was observed for children living in the surrounding counties. The relatively small area surrounding the Baltic Sea is inhabitated by several distinct ethnic groups, although they are mixed to some degree (Table 3).

Table 3. Ethnicity of the populations living in the countries around the Baltic Sea.

Ethnic Group	Country
Scandinavians	Sweden, Norway, Denmark, Finland
Finno-Ugrians	Finland, Estonia, Sweden
East-Baltic	Latvia, Lithuania
Polish	Poland, Lithuania, Germany
Germans	Germany, Denmark, Poland
Russians	Estonia, Latvia, Lithuania

VARIATION WITHIN COUNTRIES - ETHNIC GROUPS

The within-country variation can also be studied by comparing the risk of the disease among different ethnic groups living in one country. Such an analysis was done in Estonia where the incidence of IDDM in native Estonians was compared with immigrant non-Estonians during 1980-89 (Table 4).[18] Average annual incidence was higher in Estonians 11.8/100,000 per year (95% CI = 10.4 - 13.3) than in non-Estonians (7.8/100,000 per year; 95% CI = 6.3 - 9.6). These data indicate that immigrant populations need not acquire the same risk of IDDM as the native population. Genetic factors are likely to be the main reason for this differential. There are, however, data from a few small migrant studies to suggest that the incidence of IDDM in migrants tends to resemble that of the host populations rather than the rates for native locations.[24] However, due to methodological differences, these results have not been conclusive.

VARIATION WITHIN COUNTRIES - AGE, SEX AND TIME

Of the populations living near the Baltic Sea, the incidence of IDDM has increased in Finland,[1,3,4] Sweden,[12] Norway,[11] and Poland,[13] but not in Estonia, Latvia and Lithuania.[19] The analysis of the Danish birth cohorts according to the National Conscript Registry revealed that the cumulative incidence of IDDM during the first 20 years of life increased from 2.37 to 2.90 per 1000 (22.4%) for the first eight and last eight birth cohorts, respectively.[25] The data from Novosibirsk also suggest that the incidence may be increasing among Russians. Current data on trends are not available for Denmark and Germany. Incidence trends for Finland, Sweden and Poland were analyzed using the Poisson regression model. This method permits the evaluation of the effect of age, sex, birth cohort and calendar time period on the increase in incidence. The results from these three separate analyses showed that the temporal increase was mainly due to the calendar time effect, not the birth cohort effect.

Table 4. Incidence of IDDM per 100,000 per year by ethnic group in Estonia and in the district of Novosibirsk in Western Siberia.

		Estonia*		Novosibirsk°†
		Estonians	Non-Estonians †	Russia
Males:	Incidence	12.1	8.5	4.6
	95% CI	9.0-16.3	6.4-11.2	3.8-5.6
Females:	Incidence	11.4	7.2	4.9
	95% CI	8.3-15.6	5.2-9.7	4.0-6.0
Total:	Incidence	11.8	7.8	4.7
	95% CI	10.4-13.3	6.3-9.6	4.1-5.4

*Period 1980-1989, °Period 1983-1989, †Ethnicity mainly Russians

In Finland[4] and Poland,[13] the temporal variation did not differ between the two sexes and the three five-year age groups. In Sweden,[12] however, the increase in incidence of IDDM was found only for females, not for males. In all three countries, the regression analyses suggested that there was a linear increase, as well as a non-linear component which contributed to the temporal variation in the incidence of IDDM. These changes in incidence over time have been very drastic in Finland, Sweden, Norway and Poland (Table 5).

The long-term trend in Finland shows an approximate 300% increase in incidence during 30 years since the early 1950's.[1,4,26] During the 1970's and early 1980's, the relative increase in incidence was similar, approximately 2.5% per year, in Finland, Sweden and Norway[4,11,12] and approximately 6.5% per year in Poland.[13] Such a rapid increase is unlikely to be due to genetic reasons alone. It is likely that some environmental exposures have become stronger or more prevalent in these countries. The nature of these environmental factors are unknown and have been the subject of many speculations. Dietary factors are likely to be involved in this increased environmental exposure, but few studies give insight to this question and most of them are only descriptive or speculative.

During 1983-1988, the yearly variation in incidence of IDDM was compared in five Baltic countries. The ratio lowest-to-highest yearly incidence was 80% in Finland,

Table 5. The average increase in incidence of childhood IDDM in the Baltic populations.

Country	Change/year	Period
Finland	2.5%	1965-84
Sweden	2.6%	1978-87
Norway	2.2%	1973-82
Russia	2.1%	1983-89
Estonia	no change	1980-88
Latvia	no change	1983-88
Lithuania	no change	1983-88
Poland	6.5%	1970-85
Denmark	no recent data	
Germany	no recent data	

78% in Estonia, 58% in Latvia, 68% in Lithuania, and 77% in Poland.[19] In spite of this large variation, no clear time trend was found except in Finland. The year-to-year variation in incidence was more strongly expressed among males than females (Table 6).

In 1986, there was a peak in incidence among males in all countries except Poland, and an excess of female IDDM cases in Finland, Latvia and Lithuania. The evaluation of possible outbreaks also revealed another pattern among males; the yearly peak in incidence seemed to move from the north (Finland in 1983) towards the south year by year (Table 7).

Table 6. Evaluation of possible yearly outbreaks* in incidence of IDDM in five countries around the Baltic Sea.

			Males		
Year	Finland	Estonia	Latvia	Lithuania	Poland
1983	6.86*				
1984		5.24*		13.02*	0.90
1985			10.21*		0.90
1986	6.07*	9.50*	4.41*	10.8*	
1987		0.94			
1988	1.49			1.51	0.02

			Females		
Year	Finland	Estonia	Latvia	Lithuania	Poland
1983	1.36				
1984			2.90		
1985		2.94		2.56	2.78
1986	4.80*	0.01	2.90	3.24*	
1987	1.79		0.88	0.64	6.26*
1988		6.38*			0.04

*An outbreak may be suspected if $\frac{(\text{observed-expected})^2}{\text{expected}} > 3.84$

Table 7. The peak years in incidence of IDDM in the Baltic Sea region during the 1980's.

Country	Peak years					
Finland			1983			1986	
Sweden			1983			1986	
Norway		1982					
Estonia				1984		1986	
Latvia					1985	1986	
Lithuania				1984		1986	
Poland	1981			1984	1985		1987
Russia					1985	1986	

GENETIC SUSCEPTIBILITY

The major genetic susceptibility to IDDM is conferred by genes in the HLA region of chromosome 6. This susceptibility is necessary but not sufficient to produce the disease. Due to the lack of population-based and standardized data on full HLA haplotypes among diabetic subjects, it is not possible to make proper comparisons among the populations. In Finland, such data have become available recently,[27] and a collaborative research work is in progress with a number of other countries in this region.

Approximately 85-90% of cases of childhood IDDM are "sporadic", i.e., there is no earlier case of IDDM among the first degree relatives. At present, data are not sufficient to determine whether the familial aggregation of IDDM is greater in high incidence populations because standardized survey methods have not been applied.[28] One way to estimate the familial aggregation of IDDM is the assessment of the prevalence of IDDM in relatives at the time of the diagnosis of the child. Table 8 illustrates such findings from different studies. It is not known whether this variation in familial aggregation is real or due to a bias in study methods.

Table 8. Proportion of IDDM parents (%) among parents of children with IDDM.*

Reference	Population	Number of Children	Either Parent	IDDM Fathers	IDDM Mothers
West et al. 1979	Canada	518	3.2		
Chern et al. 1982	USA, MN	493	3.0		
Gorsuch et al. 1982	UK	160	3.5	4.5	1.3
Wagener et al. 1982	USA, PA	1128	2.6	3.7	1.4
Tajima et al. 1985	Japan	1572	0.4		
Tillil, Kobberling 1987	Germany	554	2.5	4.1	1.7
Dahlquist et al. 1989	Sweden	3503	4.3	6.2	2.3
Tuomilehto-Wolf et al. 1989	Finland	446	3.9	5.4	2.4

*From Reference 29.

The studies where parental IDDM has been assessed for fathers and mothers separately have consistently shown that the prevalence of IDDM is higher in fathers than in mothers (Table 8). The data from Finland[21,27] and Sweden[10] are fully comparable and the results are similar; approximately 5-6% of the fathers and 2.5% of the mothers of the newly diagnosed children have IDDM.

In Finland, Sweden and Norway, infant mortality has been the lowest in the world during the last decades.[30] It is possible that, in these countries, males with childhood IDDM have survived better than in other countries, as suggested by the recent mortality evaluation.[31] This could have resulted in a larger pool of genetically high-risk subjects in Scandinavia than in other countries, especially among males with IDDM who have been reproducing more than females with IDDM. However, this issue is very poorly understood and further studies are warranted.

It is very likely that certain genetic traits are responsible for part of the incidence variation among the populations living in the Baltic Sea region. The HLA haplotype A2, Cw1, Bw56, DR4, DQw8, that has not been found in diabetic patients in any other Baltic population or elsewhere, contributes approximately 11% to the observed genetic susceptibility among the Finns.[27] This HLA haplotype was also found among patients whose parents had IDDM and, similarly, among mothers and fathers. It was the most commonly transmitted HLA haplotype in Finland. It has not as yet been found in Estonia, Sweden, Norway and Lithuania. This suggests that there may be a genetic reason for the high incidence of IDDM in Finland.

CONCLUSIONS

Epidemiological studies of IDDM in the populations living at the Baltic Sea are very important because of the large variation in incidence recorded over the region. The Finns, who live north of the Baltic Sea, have the highest incidence of IDDM in the world, followed by Sardinia[32] and Sweden. Within Europe, Poland has reported one of the lowest incidence rates. There are no other data from the Russian population living near the Baltic Sea except those from Estonia.[18,20] They suggest that the incidence may be higher in Estonia than in Eastern parts of Russia (Siberia). Comparative analyses, including several populations different in many respects from this relatively small geographic area, may offer an effective way to obtain new knowledge about the occurrence and aetiology of childhood IDDM.

The recent studies show that the incidence of IDDM in Finland is at least three times higher than in Estonia, and approximately five times higher than in Latvia, Lithuania and Poland. The incidence in Latvia and Lithuania resembles that in Poland rather than that in Estonia. The incidence among the Russian population of 4-6/100,000 per year[20] appears to resemble closely that of Latvia and Lithuania. In Estonia, the incidence of IDDM among the native population has been found to be 12/100,000 per year.[18]

It has been proposed that the incidence of childhood IDDM varies as a function of the distance of the population from the equator.[5-7] The data from the Baltic Sea region do not show such a gradient in the incidence. There are several exceptions, and there is little evidence that latitude per se would be the causal determinant for this variation. High incidence of IDDM has been reported also in Sweden and Norway, although large areas of their territories are at the same latitude as Estonia, Latvia and Lithuania where the incidence is lower. The present data would point towards genetic contributions to the population difference in incidence.

The data from the Baltic Sea region confirm earlier findings that in high incidence populations, a male excess in the risk of IDDM exists, whereas in low incidence populations, there is often a slight female preponderance. The reasons for this diverging sex differential in incidence of IDDM are unknown.

In Finland, and in a number of other countries around the Baltic Sea, the trend has been increasing and it has persisted for several decades. In Estonia, Latvia, Lithuania, Poland and Russia the incidence has not increased since 1983.[19] However, there seems to be a non-linear, epidemic-type variation in incidence in every population. Such a variation was more clearly expressed in males than in females. We have recently pointed out that IDDM is a rather atypical chronic disease in that apparent "epidemics" are relatively common.[22]

An interesting observation in our present data was that we found the highest incidence in 1986 in most of the Baltic countries and also in Novosibirsk, Western Siberia. A recent report from Hungary[33] has also shown an increase in incidence during 1986. We could speculate that the reason must be a major environmental triggering event. Since there was no known major viral epidemic in Northern Europe around

1985-86, another unspecific exposure could be the reason for this peak. Another "epidemic wave" took place among males during 1983 to 1987, moving from the north towards the south. Additional elevations of IDDM incidence rates in Finland and Sweden (1983), Estonia (1984) and Poland (1982), together with the present data, suggest that during these years some environmental triggering influence has been particularly strong. It seems to operate similarly in high and low incidence populations and the effect is slightly stronger in males than females.

Given that the Finns and Estonians are genetically close, factors other than genetics are needed to result in such a large difference in the incidence of IDDM between these two populations. Actually, the incidence in Estonia was closer to that in Latvia, Lithuania and Poland, despite obvious major genetic differences. The increase in incidence in Finland, Norway and Sweden did not match that observed in the other Baltic populations, and may suggest that some environmental factors that operate in the before-mentioned "high-risk" countries are responsible for the observed excess risk of IDDM. It is unlikely that those countries would be free of viral infections that operate in Finland. There are certainly distinct differences in dietary habits between countries. However, all of them are relatively high milk consumers. Although the milk protein theory[34,35] cannot be fully excluded, it is unlikely to be the reason for the observed incidence differences. There is little doubt that genetic factors are probably responsible for a considerable proportion of the reported differences in incidence of IDDM between the Baltic countries.

Even though these populations have been mixed to some extent during the past centuries, each of them have their own distinct genetic and cultural patterns. It seems obvious that the genetic susceptibility to IDDM is one of the major determinants of this large variability in incidence. We can speculate that the susceptibility genes for IDDM are likely to be those that are particularly prevalent in the Finnish, Swedish and Norwegian populations. Since the incidence in Estonians, who are genetically close to Finns, is not particularly high, it might suggest that the high-risk genotypes are probably of Scandinavian origin rather than of Finno-Ugrian origin. The excess in the risk of IDDM in Finland over Sweden and Norway might, at least partially, be explained by the presence of one single high risk HLA haplotype, A2, Cw1, Bw56, DR4, DQw8.[27]

The increasing time trends and the relative lack of within-country variation in incidence in Finland, Sweden and Norway suggest that a major environmental factor may play a role in unmasking the underlying genetic susceptibitility in these countries. At the moment, too little is known about the putative environmental risk factors of IDDM, but evidence is accumulating that these factors may not necessarily be such unspecific triggering factors for IDDM as virus infection. Given the large variation in the risk of IDDM within the Baltic Sea region, this part of the world provides an excellent opportunity to investigate the genetic-environmental interaction in the aetiology of IDDM and the reasons for obvious outbreaks in incidence of IDDM.

Acknowledgements

This study was partly supported by the grants from the Academy of Finland, National Institutes of Health (DK-37957) and the Nordisk Insulinfond.

REFERENCES

1. H.K. Åkerblom, A. Reunanen, The epidemiology of insulin-dependent diabetes mellitus (IDDM) in Finland and in Northern Europe, *Diabetes Care* 8 (suppl. 1):10 (1985).

2. A. Reunanen, H.K. Åkerblom, J. Tuomilehto, High incidence of insulin-dependent diabetes mellitus (IDDM) in children in Finland, *Arctic Med Res.* 47 (suppl. 1):535 (1988).
3. Diabetes Epidemiology Research International Group, Secular trends in incidence of childhood IDDM in 10 countries, *Diabetes* 39:858 (1990).
4. J. Tuomilehto, M. Rewers, A. Reunanen, et al., Increasing trend in Type I (insulin-dependent) diabetes mellitus in childhood in Finland: Analysis of age, calendar time and birth cohort effects during 1965 to 1984, *Diabetologia* 34:282 (1991).
5. R.E. LaPorte, N. Tajima, H.K. Åkerblom, et al., Geographic differences in the risk of insulin-dependent diabetes mellitus: The importance of registries, *Diabetes Care* 8:101 (1985).
6. Diabetes Epidemiology Research International Group, Geographic patterns of childhood insulin-dependent diabetes mellitus, *Diabetes* 37:1113 (1988).
7. M. Rewers, R.E. LaPorte, H. King, J. Tuomilehto, Trends in the prevalence and incidence of diabetes: Insulin-dependent diabetes mellitus in childhood, *Wld Hlth Statist Quart.* 41:179 (1988).
8. H.J. Ustvedt, E. Olsen, Incidence of diabetes mellitus in Oslo, Norway 1956-65, *Br J Prev Soc Med.* 31:251 (1977).
9. B. Christau, H. Kromann, O.O. Andersen, et al., Incidence, seasonal and geographical patterns of juvenile-onset insulin-dependent diabetes mellitus in Denmark, *Diabetologia* 13:281 (1977).
10. G. Dahlqvist G, L. Blom, G. Holmgren, et al., The epidemiology of diabetes in Swedish children 0-14-years: A six-year prospective study, *Diabetologia* 28:802 (1985).
11. G. Joner, O. Søvik, Increasing incidence of diabetes mellitus in Norwegian children 0-14 years of age 1973-1982, *Diabetologia* 32:79 (1989).
12. L. Nyström, G. Dahlquist, M. Rewers, S. Wall, The Swedish childhood diabetes study. An analysis of the temporal variation in diabetes incidence 1978-1987, *Int J Epidemiol.* 19:141 (1990).
13. M. Rewers, R.E. LaPorte, M. Walczak, et al., Apparent epidemic of insulin-dependent diabetes mellitus in Midwestern Poland, *Diabetes* 36:106 (1987).
14. I. Kalits, T. Podar, Incidence and prevalence of Type I (insulin-dependent) diabetes in 1988, *Diabetologia* 33:346 (1990).
15. J. Tuomilehto, T. Podar, A. Reunanen, et al., Comparison of the incidence of insulin-dependent diabetes mellitus in childhood between Estonia and Finland, 1980-1988, *Diabetes Care* 34:982 (1991).
16. G. Brigis, Incidence rate of childhood insulin-dependent diabetes mellitus (IDDM) in Latvia, *in*: "Epidemiology and Etiology of IDDM in the Young", P. Czernichow, ed., *Pediatr Adolesc Endocrinol.*, Basel: Karger, 21:72 (1992).
17. V. Grabauskas, B. Urbonaite, Z. Padaiga, Incidence of childhood insulin-dependent diabetes mellitus in Lithuania 1983-88, *Acta Paediatr Scand.* 80:718 (1991).
18. T. Podar, E. Tuomilehto-Wolf, J. Tuomilehto, R.E. LaPorte, B. Adojaan, Insulin-dependent diabetes mellitus in native Estonians and immigrants to Estonia, *Am J Epidemiol.* 135:1231 (1992).
19. J. Tuomilehto, T. Podar, G. Brigis et al., Comparison of the incidence of insulin-dependent diabetes mellitus in childhood among five Baltic populations during 1983-1988, *Int J Epidemiol.* 21:518 (1992).
20. E. Shubnikov, T. Podar, J. Tuomilehto, Y. Nitikin, Low incidence of childhood IDDM in district of Novosibirsk (Russia), *Diabetes Care* 15:915 (1992).
21. J. Tuomilehto, R. Lounamaa, E. Tuomilehto-Wolf, et al., Epidemiology of childhood diabetes in Finland: Background of a nationwide study of Type I (insulin-dependent) diabetes mellitus, *Diabetologia* 35:70 (1991).

22. World Health Orgnization DIAMOND Project Group on Epidemics, Childhood diabetes, epidemics, and epidemiology: An approach for controlling diabetes, *Am J Epidemiol.* 135:803 (1992).
23. A. Reunanen, H.K. Åkerblom, M.L. Käär, Prevalence and ten-year (1970-1979) incidence of insulin-dependent diabetes in children and adolescents in Finland, *Acta Paediatr Scand.* 71:893 (1982).
24. J. Siemiatycki, E. Colle, S. Campbell, R. Dewar, et al., Incidence of IDDM in Montreal by ethnic group and by social class and comparisons with ethnic groups living elsewhere, *Diabetes* 37:1096 (1988).
25. A. Green, P.K. Andersen, A.J. Svendsen, K. Mortensen, Increasing incidence of early onset Type I (insulin-dependent) diabetes mellitus: A study of Danish male birth cohorts, *Diabetologia* 35:178 (1992).
26. O. Somersalo, Studies of childhood diabetes incidence in Finland, *Ann Paediatr Fenn.* 1:239 (1955).
27. E. Tuomilehto-Wolf, J. Tuomilehto, Z. Cepaitis, R. Lounamaa, DIME Study Group. New susceptibility haplotype for Type I diabetes, *Lancet* ii:299 (1989).
28. The WHO Multinational Project for Childhood Diabetes Group, Familial insulin-dependent diabetes mellitus (IDDM) epidemiology: Standardization of data for the DIAMOND Project, *Bull WHO* 69:767 (1991).
29. T. Podar, J. Tuomilehto, E. Tuomilehto-Wolf, R.E. LaPorte, J.S. Dorman, The risk of insulin-dependent diabetes in offspring of parents with insulin-dependent diabetes, *in:* "Standardization of Epidemiologic Studies of Host Susceptibility", NATO ASI Series, J.S. Dorman, ed., Plenum Press, 1994.
30. J. Tuomilehto, Health status in Finland and other Nordic Countries with special reference to chronic non-communicable diseases, *Ann Nutr Metab.* 435(suppl.1):41 (1991).
31. Diabetes Epidemiology Research International Mortality Study Group, International evaluation of cause-specific mortality and IDDM, *Diabetes Care* 14:1:55 (1991).
32. M. Songini, Incidence of Type I diabetes mellitus in Sardinia, *Medicographia* 13:10 (1992).
33. G. Soltesz, L. Madacsy, D. Bekefi, I. Danko, The Hungarian Childhood Diabetes Epidemiology Group, Rising incidence of Type I diabetes in Hungarian children (1978-1987), *Diabetic Med.* 7:111 (1990).
34. R.B. Elliott, J.M. Martin, Dietary protein: A trigger of insulin-dependent diabetes in the BB rat? *Diabetologia* 24:297 (1984).
35. J. Karjalainen, M.M. Julio, M. Knip, J. Ilonen, B.R. Robinson, E. Savilahti, H.K. Åkerblom, H.M. Dosch, A bovine albumin peptide as a possible trigger of insulin-dependent diabetes mellitus, *New Eng J Med.* 327:302 (1992).

STANDARDS FOR IDDM EPIDEMIOLOGY IN ASIA

Naoko Tajima and Masato Matsushima

Third Department of Medicine
Jikei University School of Medicine
3-25-8 Nishishinbashi, Minato-ku
Tokyo 105
JAPAN

INTRODUCTION

The risk for individuals to develop IDDM varies remarkably based on the country of residence and race, with Orientals having lower incidence figures, almost one twentieth of that seen in Finland.[1,2] There is no question that to find the factors related to this global pattern is the key to untangling the etiology of IDDM. However, we first should confirm that the data describing the epidemiology of IDDM were obtained in a standardized manner and are accurate.

Standardization is extremely important, especially for low risk countries. In these areas, IDDM is not a common disease and the public awareness has been low, potentially resulting in many missing cases. Secondly, in many low risk areas, population statistics are not easy to obtain. Thirdly, in most of the areas, cases have not been identified through ongoing population-based surveillance, but through the retrospective review of hospital records without secondary validation. In this case, the degree of case ascertainment may not be high and often times is not even estimated. Using two data sets from Japan, the importance of standardization for IDDM incidence registries is discussed in this report.

REPORTED IDDM INCIDENCE DATA WITHOUT STANDARDIZATION

Japanese IDDM Registry System

In Japan, two major population-based systems were developed to detect childhood diabetes during the mid 1970's. One was the central registry for free medical care for both

IDDM and NIDDM patients who were under age 18 years, and the other was urine glucose screening for school children.[3] The establishment of these systems made it possible to accumulate IDDM incidence data rather rapidly during the past 10 years in the various parts of Japan. To set up IDDM registries, investigators typically have employed the central registry data and hospital records as primary sources of case ascertainment, and the list of diabetes summer campers and information from schools as secondary sources. Urine screening tests, however, are more sensitive for detecting NIDDM than IDDM.

IDDM Incidence Rates and Secular Trends

Table 1 summarizes reported non-standardized IDDM incidence data from seven areas in Japan.[4] There was a 2-3 fold intra-country difference with apparent rising incidence. However, each study used different procedures, data sources, age groups and time periods. Moreover, the degree of case ascertainment was not clearly described. Therefore, we cannot simply compare the data even if they are all from one country.

Table 1. Reported IDDM incidence rates by prefecture (per 100,000).

Prefecture	Period	Age (years)	Rate
Hokkaido	1979-81	0-15	1.7
Tokyo	1979-80	0-15	0.9
Chiba	1974-83	6-15	1.5
Osaka	1981-87	0-17	2.3
Tottori	1973-77	0-14	1.4

As for the secular trends, there is an impression that IDDM has been increasing over time in Japan. This is based on clinical opinions, as well as the report by the Ministry of Health and Welfare regarding the total number of the new applicants for free medical care for diabetic children under 18 years (Figure 1).[5] However, this represents only the number of case applicants from all over Japan, not the incidence of disease. We could speculate that during the early time period, physicians may not have been aware of the program. Thus, there may have been considerable under-ascertainment. As awareness of the system grew, potentially more and more people applied, providing evidence for an artificial rise in incidence. Moreover, the applicants included youth onset NIDDM. Therefore, it is difficult to assess the trends in IDDM incidence from these data.

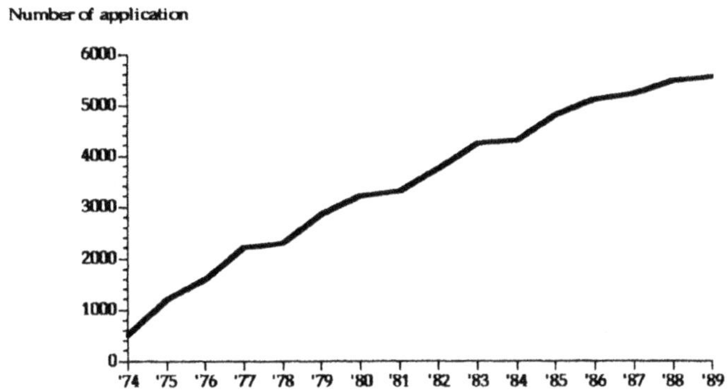

Figure 1. Number of applicants for free medical care.

Discrimination of NIDDM From IDDM

The proportion of NIDDM patients included in the registry was determined by reviewing the original file at the Ministry of Health and Welfare. It was found that 100% of the cases diagnosed under age 5 years were IDDM. However, NIDDM, both chemical and clinical diabetes, began to appear after age 6 years and comprised almost half the cases diagnosed after the age 10 years.[6] These NIDDM cases have slightly different genetic backgrounds than IDDM patients and they often use insulin to control diabetes. This seems to be the case with other Oriental populations. We potentially need to be more careful in standardizing the case definition for IDDM in Asia.

Increasing Incidence of NIDDM Among Children in Japan

NIDDM incident cases are most effectively identified by urine screening for school children. Kitagawa et al. has been conducting these screening tests in the Tokyo metropolitan area since 1975 and demonstrated that the number of NIDDM cases has increased from 1.5/100,000 (1975-1977) to 3.8/100,000 (1984-1986) over the past 10 years, in parallel with the increase in the number of obese children.[7] Thus, the secular trend observed in Figure 1 is thought to reflect the increase in NIDDM, and not in IDDM, among children.

IDDM INCIDENCE USING STANDARDIZED PROCEDURES

Annual IDDM Incidence Rates After 1985 in Japan

We learned a lesson from the previous studies of IDDM epidemiology in Japan. Without standardization, no reliable comparative studies can be performed. Therefore, the most accurate existing data from three areas of Japan (Hokkaido, Tokyo and Kagoshima) were re-analyzed using the same procedures.[8] The three areas were selected for their high case ascertainment and the geographical distribution. The total population examined was 3.6 million, estimated from the Japan Census Bureau, 1985. Case ascertainment was over 95% in each registry.

From the standardized procedures, the overall incidence by area for the period of 1985 to 1989 for children age 0-14 years was approximately 2/100,000 per year, ranging from 1.65 to 2.07, with no north-south gradient. The age-specific incidence rates are amazingly similar across very different areas (Table 2). In accordance with the previous reports in Caucasian populations, the peak onset age was 10 to 14 years. What is of particular importance is the extremely low variation in the incidence rates and the "tight" 95% CI surrounding the variation. Thus, unlike the 2.5-fold variation with the non-standardized data, there was only 1.2-fold difference in the standardized incidence rates.

Table 2. Annual IDDM incidence rates per 100,000 and 95% CI.

Age	Hokkaido	Tokyo	Kagoshima
0-4 yr	0.74 (0.39-1.26)	1.16 (0.72-1.78)	1.48 (0.88-2.33)
5-9 yr	1.43 (0.96-2.06)	1.68 (1.14-2.37)	1.76 (1.11-2.63)
10-14 yr	3.65 (3.29-4.56)	2.01 (1.46-2.67)	2.07 (1.39-2.98)
0-14 yr	2.07 (1.73-2.48)	1.65 (1.35-2.01)	1.78 (1.40-2.26)

Hokkaido: 1985-89, Tokyo: 1987-89, Kagoshima: 1980-89

Secular Trends of IDDM in Hokkaido and Kagoshima, 1980-89

Temporal trends in IDDM incidence were examined for Hokkaido and Kagoshima, which have ongoing population-based registries that have been active for more than 10 years. The incidence rates for both sexes were quite constant during the past decade, with a tendency towards a decline after 1987. No epidemic was detected (Figure 2).

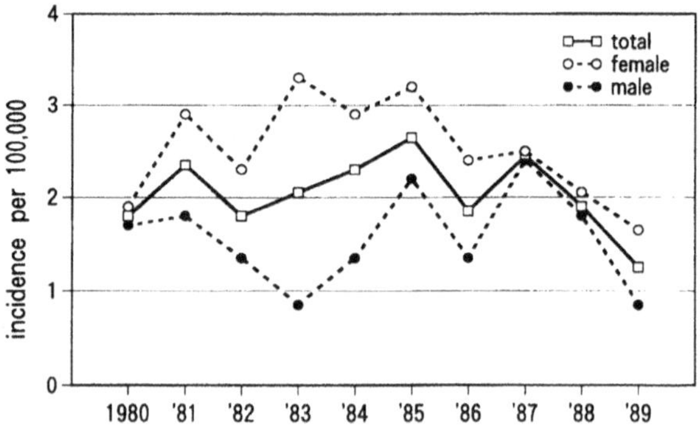

Figure 2. Secular trends of IDDM incidence in Japan (data are from Hokkaido and Kagoshima).

MIGRANT STUDIES

Importance of Migrant Studies

Once we obtain accurate incidence data from a low risk country, such as Japan, the next step is to evaluate IDDM incidence rates in migrant populations. Migrant studies are one of the best approaches for evaluating the effect of environmental factors on the occurrence of IDDM.

It is extremely important to identify whether the IDDM incidence rate among the second generation of Japanese immigrants, without interracial marriage, in high risk areas such as Hawaii or California, is increased. The Japanese represent a very homogeneous population[9,10] which appears to be resistent to IDDM. Therefore, only very strong environmental factors could potentially increase the incidence of IDDM among Japanese immigrants. This is the area of research where standardized protocols are most needed.

China Heritage Studies

China is an exciting place to explore the relationships between the genetic and environmental factors in IDDM. It is one of the lowest risk countries,[11] having a huge population with diversity in ethnicity, geography and culture. The genetic background of Chinese people was investigated by examining the distribution of immunoglobulin Gm and Km allotypes in 74 Chinese geographically diverse populations.[12] Heterogeneity was demonstrated, with striking differences between the northern Mongoloid cluster and the southern Mongoloid cluster, with a boundary near the latitude of 30 degrees north along Yangze River. The Northern Mongoloid group was associated with haplotype Gm 1, 21, while the Southern group had Gm 1, 3, 5. The presence of Gm 3,5 phenotype among this population appears to be the result of admixture from Caucasoid-derived population along the Silk Road, beginning from the current city of Xi'an.

Ethnic Diversity in China

Another important component is ethnic diversity within China. China is a nation with 56 different ethnic groups. The Han ethnic group is the largest, consisting of 93.3% of the total population. The minority groups include Koreans, Mongols, Uygurs, Zhuang, Yao, Dong, Miao, Li and others. Most of the minority groups live in remote mountainous regions and have their own customs and cultural characteristics. However, Hans live in geographically and culturally diverse areas, scattered all over China. The genetic background of the Chinese people appears to have been determined by a mixture of geographic and ethnic components. Another complicated issue is that in China, the ethnic background of individuals in the census is determined by the person's surname, which is unlike the situation in the U.S., where it is determined by what they declare. There is great genetic diversity among Hans.

Studies in China will include the identification of ethnic background and examination of the possible environmental factors. Because of the genetic variation, we would predict that despite the low incidence, much larger intra-country variation would be seen compared to that in Japan.

FUTURE DIRECTION FOR THE STANDARDIZATION OF IDDM INCIDENCE STUDIES

WHO DIAMOND Project Approach

One of the strategies used by the WHO DIAMOND Project Group is to educate young researchers concerning the importance of standardization of international epidemiological studies and to make friends.[13] The first Japan-U.S. Diabetes Epidemiology Training Course, very similar to the course in Cambridge, was held in Japan in the summer of 1991.[14] Twenty-eight participants from 11 different countries in Asia, the U.S., Canada and Australia, and similar number of international faculty gathered and spent eight days at the guest house of Jichi Medical School, about 100 km away from Tokyo.

Diabetes Epidemiology Training Course in Beijing

In the spring of 1992, a Diabetes Epidemiology Training Course was held in Beijing, organized by Dr. Yang Ze, from the Institute of Occupational Medicine, Chinese Academy of Preventive Medicine. It was a five day course with 38 participants from 15 centers from various parts of China. Twelve centers brought local Methods of Operation Manuals (MOOs) for incidence studies which the faculty reviewed. Potential problems were discussed during the course. The procedures for establishing the IDDM registries in China will be identical to those employed by the WHO DIAMOND Project so that the data from China will be directly comparable with the data from across the world. The preliminary data were really intriguing. For example, the Shanghai registry demonstrated a very low incidence rate of 0.5 per 100,000 for 1980 to 1991, with an epidemic in the year 1986. There appeared to be a low prevalence of IDDM among minority ethnic groups, and a relatively high prevalence in Xi'an city, which is the start of the Silk Road.

Standardization is required for every aspect of multinational studies and are critical for incidence registries. Incorrect conclusions can be reached when incidence studies are not standardized. After a standardized surveillance is in place, the next step will be to conduct genetic and other studies. The genetic component of the DIAMOND Project is already making progress as outlined in this Proceedings. With standards for host-susceptibility studies, we will be able to identify ethnic-specific and genotype-specific IDDM incidence rates, which will be a major cornerstone to identify the risk factors for IDDM.

REFERENCES

1. R.E. LaPorte, N. Tajima, H.K. Akerblom, et al., Geographic differences in the risk of insulin-dependent diabetes mellitus: The importance of registries, *Diabetes Care* 8 (suppl.1): 101 (1985).
2. M. Rewers, R.E. LaPorte, H. King, J. Tuomilehto, Trends in the prevalence and incidence of diabetes: Insulin-dependent diabetes mellitus in childhood, *World Health Stat Quart.* 41:178 (1988).
3. N. Tajima, R.E. LaPorte, I. Hibi, et al., A comparison of the epidemiology of youth-onset insulin-dependent diabetes mellitus between Japan and the United States (Allegheny County, Pennsylvania), *Diabetes Care* 8 (suppl.1):17 (1990).
4. T. Kuzuya, C. Ito, A. Sasaki, et al., Prevalence and incidence of diabetes in Japanese people compiled from the literature - A report of the epidemiology data committee, The Japan Diabetes Society, *J Japan Diabetes Soc.* 35:173 (1992).
5. Y. Goto, Epidemiology of diabetes mellitus in Japan, *Nippon Rinsho*. 614:222 (1990).
6. T. Kitagawa, M. Ohwada, N. Tajima, A. Sugaya, Incidence of childhood diabetes in Tokyo metropolitan area during 1988-1989, *J Japan Diabetes Soc*. 34:262 (1991).

7. M. Ohwada, Y. Hanaoka, Y. Tanimoto, et al., Descriptive epidemiology of non-insulin-dependent diabetes mellitus detected by urine glucose screening in school children in Japan, *Acta Paediatr Japan* 32:716 (1990).
8. Japan IDDM Epidemiology Group, Lack of regional variation in IDDM risk in Japan, *Diabetes Care* 16:796 (1993).
9. H. Matsumoto, On the origin of the Japanese race: Studies of genetic markers of the immunoglobulins, *Proc Japan Acad.* 60:211 (1984).
10. H. Matsumoto, Characteristics of mongoloid and neighboring populations based on the genetic markers of human immunoglobulins, *Hum Genet.* 80:207 (1988).
11. M.Z. Bao, J.X. Wang, J.S. Dorman, et al., HLA-DQß, non-Asp-57 allele and incidence of diabetes in China and the USA, *Lancet* 2:497 (1989).
12. Z. Tongmao and D.L. Tsung, Gm and Km allotypes in 74 Chinese populations: A hypothesis of the origin of the Chinese nation, *Hum Genet.* 83:101 (1989).
13. WHO DIAMOND Project Group, WHO Multinational Project for Childhood Diabetes, *Diabetes Care* 13:1062 (1990).
14. Japan-U.S. Diabetes Epidemiology Training Course - Educational component of WHO multinational project for childhood diabetes "DIAbetes MONDiale" (DIAMOND), T. Kuzuya, Y. Kanazawa, N. Tajima, eds., Shinohara Publisher Inc. Tokyo (1992).

STANDARDS FOR FAMILY HISTORY DATA COLLECTION AND ANALYSES FOR THE WHO DIAMOND PROJECT

Maria T. Tenconi

Università Degli Studi di Pavia
Dipartimento di Medicina Preventiva
Occupazionale e di Comunità
Sezione di Igiene
Viale Forlanini n. 2
27100 Pavia
ITALY

INTRODUCTION

The need to standardize the collection of family history data for IDDM is due to the great importance of these data in research regarding the etiology of this disease. The establishment of standardized IDDM registries has revealed great geographical variability in IDDM incidence; marked differences among countries and ethnic groups indicate variation in the prevalence of common etiological factors, either genetic or environmental. Recent studies have proposed that the primary determinant of the worldwide patterns of disease could be population variation in the prevalence of susceptibility genes,[1] with environmental factors playing a secondary role. The investigation of familial cases of disease is a starting point for testing etiological hypotheses, which is a primary objective of the WHO Multinational Project for Childhood Diabetes, known as the DIAMOND Project.

Although family aggregation of IDDM has been described, attempts to define a specific mode of inheritance have been unsuccessful. The presence of susceptibility genes and potential environmental factors are likely to cluster in the families of affected individuals and potentially increase their risk of developing the disease. Presently, familial IDDM epidemiology is expanding beyond descriptive evaluations to include molecular assessments of potential genetic, environmental and other epidemiological risk factors in first degree relatives. These analytical studies permit the testing of specific hypotheses regarding the causes of IDDM clustering within families. When conducted in incidence cohorts, such investigations provide estimates of the absolute risk of disease for relatives with specific host-environmental exposures.[2]

The establishment of IDDM incidence registries in many different countries through the DIAMOND Project provides an excellent foundation for the development of research

Standardization of Epidemiologic Studies of Host Susceptibility
Edited by J. Dorman, Plenum Press, New York, 1994

on family IDDM epidemiology worldwide. Family history information will permit: 1) the estimation of IDDM recurrence risks in families, 2) the assessment of familial aggregation of the disease, 3) evaluations of the epidemiology of IDDM in family members, and 4) the identification of probands and families for genetic marker studies.[2]

IDDM RECURRENCE RISK IN FAMILIES

Although several studies have reported risk percentages of developing IDDM for relatives of an IDDM proband, the data are not directly comparable because of methodological differences.[2] Most of the study populations consisted of patients identified from hospitals or diabetes clinics. Owing to a potential referral bias, multiple case families were likely to be over-represented in these cohorts. This may have increased the IDDM risk estimates for family members to above those obtained from more representative groups. Furthermore, the definition of diabetes among family members varied considerably. Several studies did not distinguish IDDM from NIDDM. By not reporting the ages of siblings, the population at-risk could not be defined accurately. Moreover, sources of data and data collection methods were not uniform, prohibiting accurate comparisons of IDDM risks between populations.

FAMILIAL AGGREGATION OF IDDM

One of the most characteristic patterns of the familial aggregation of IDDM is the sex difference in the prevalence of parental diabetes. Diabetic children from families with an IDDM parent are significantly more likely to have an affected father than an affected mother.[2] This pattern has been observed in different countries, but the mechanism by which paternal IDDM is more frequently transmitted remains poorly understood. Possible explanations include: 1) an increase in miscarriages of susceptible fetuses by IDDM mothers, 2) maternal immunological tolerance of the fetus to diabetic autoantigens, 3) an increase in paternal transmission of susceptibility genes, and 4) genomic imprinting. Still unknown is the pattern of sex differences in parental IDDM in non-Caucasian populations.

EVALUATION OF THE EPIDEMIOLOGY OF IDDM IN FAMILY MEMBERS

Investigations performed in several populations have revealed particular characteristics associated with an increased IDDM risk among siblings.[2] These are: 1) high birth order, 2) older maternal age, 3) IDDM in fathers versus mothers. However, the consistency of these relationships across populations and racial groups has not been well-established.

Epidemics of IDDM were reported in several countries in the 1980s. Whether similar temporal increases in risk occurred among family members is not known. An evaluation of the temporal changes in familial IDDM incidence would provide important insights as to the nature of the environmental determinants of the disease, which tend to cluster in families, and their interaction with host susceptibility.

In order to evaluate the absolute risk of the disease for relatives with specific risk factors, analytical studies among family members are needed.[2] At a multinational level, these studies are very important to estimate the recurrence risk of the disease in families in different populations; at a national level they would allow us to calculate specific risks associated with genetic markers or environmental risk factors. This also has clinical implications, as one can provide more accurate risk estimates for families if children are tested for susceptibility genes, islet cell antibodies, etc. However, this is only possible if the overall recurrence risk in families is known.

IDENTIFICATION OF FAMILIES FOR GENETIC MARKER STUDIES

The collection of family history data in a standardized manner is also important in order to identify probands and informative simplex and multiplex families for serological and molecular genetic studies, which are comparable across populations.

FAMILY HISTORY DATA COLLECTION

The collection of family history data for epidemiological purposes, although relatively easy and inexpensive, has been criticized in the past because of the unreliability of the information. In the Tecumseh Community Health Study,[3] a longitudinal investigation performed in a defined community, the agreement between the results obtained from interviews performed on 9,159 subjects regarding family history of major diseases in first degree relatives and those from direct interviews of siblings was poor, reaching only 35% for diabetes. This finding justified the conclusion that both mortality and morbidity data obtained by survey should be used very cautiously. However, a more recent analysis from the San Louis Valley Diabetes Study[4] found that there was complete agreement between proband and family history reports from first degree relatives, although only a moderate percentage of the family members (48%) reported ever having been tested for diabetes. Thus, the actual prevalence of undiagnosed NIDDM in families cannot be ascertained without individual screening of all family members.

When considering the reliability of family history data in IDDM probands, especially those under 15 years of age, it is important to note that the interviewed person is usually a parent. Today, parents are very concerned about their children's diseases and are often well-informed about inheritance of the condition. Consequently, under-reporting in first degree relatives can be considered unusual.

The collection of family data for a multinational study, such as the WHO DIAMOND Project, requires standardized forms. As for IDDM incidence registries, the core information should be minimal and easy to ascertain so that the standards will be applicable in developed and developing countries across the world. To achieve these goals, standards are needed for: 1) definition of the family structure, 2) definition of core variables to determine the presence of diabetes in all family members, 3) methods of family history data collection, 4) data storage and analyses.[2]

Definition Of The Family Structure

The risk of developing IDDM in a specified geographical area is computed by dividing the number of ascertained IDDM cases (numerator) by the population at-risk (denominator). The denominator can be expressed as cumulative risk or in terms of person-years at-risk. The enumeration of IDDM cases in first degree relatives at-risk requires the determination of the structure of the nuclear family and the definition of IDDM (versus NIDDM), in order to obtain a full family pedigree for the proband. Potential barriers to accurately determining family structure, such as divorce/remarriage, lack of contact with relatives, death of family members, adoption, differences in cultural practices and family roles (especially in developing countries), may lead to biased evaluation of the family structure. Other obstacles may include inadequate data collection instruments and poor interviewing skills.

Definition Of Core Variables

The definition of IDDM among probands, whose family history data will be collected, is the same as that adopted for population incidence registries and consists of: 1) a confirmed diagnosis by a physician, 2) age less than 15 years at disease onset, 3) on insulin at hospital discharge, and 4) residency in the area of target population at diagnosis.

The core data which must be collected for each first degree relative of the IDDM proband include sex, living status, date of birth, diabetes status, and, if deceased, the cause, date and place of death. Special attention should be given to the type of diabetes which can be summarized in two questions concerning the age at onset of diabetes and the treatment or age at which continuous insulin therapy was started. Family members who were less than age 35 years at disease onset and were placed on insulin treatment at diagnosis are most likely to be insulin-dependent. The classification of type of diabetes would be more accurate if determined by direct measurements (C-peptide, etc.). However, this is only possible for small registries with a major interest in evaluating IDDM in families. Obviously, due to the need to collect accurate data for the DIAMOND Project, information from first degree relatives only is requested. Extended family history data on the occurrence of diabetes in second degree relatives may be obtained at individual centers if desired.

Methods Of Collection Of Family History Data

Family history information can be collected by survey and by clinical examination. Both methods used together would provide the best results in terms of reliability. However, because not all centers have the possibility of examining all the family members, the questionnaire is the least expensive and the most acceptable option for multinational studies. A standardized form has been developed for the DIAMOND Project[2]. The form was designed so that it could be completed by IDDM cases who were aged 15 years and over, or by their parents if the proband was less than age 15 years. This form can be mailed to each registered case, completed at home and returned in a stamped envelope. Advantages of a mail survey include the possibility of obtaining more accurate information about other family members. However, mailed surveys have a low response unless there are several reminders from the research center. The response rate is higher from telephone interviews with the proband or his/her first degree relative. In this case, the data may be less accurate due to recall bias. A combination of these methods will facilitate complete data collection in developed countries. Family history data may also be obtained from short personal interviews at the time of first hospital admission or during out-patient follow-up. This method is good for developing countries and for prospective registries.

Data Storage And Analysis

The data collected for comparative analysis must be computerized in a standardized manner. The core data for each first degree relative must be entered into a computer file on a separate record. Identification numbers must be defined to distinguish between families (family identification numbers), diabetic relatives, (case identification numbers) and family relationships (member identification numbers). Standards for data storage will also facilitate the sharing of computer analyses programs, designed to estimate the recurrence risks to relatives. Such programs are being developed for the DIAMOND Project using the computer software dBase IV. A special session during the NATO Workshop focused on the use of the programs. They will permit the identification of single versus multiple IDDM case families and distinguish between those with affected siblings and parents. Programs for the analyses of familial IDDM aggregation will permit a characterization of unaffected and affected relatives by age at onset of diabetes, sex, birth order, living status of relatives, etc. Furthermore, life-table analyses programs will estimate IDDM recurrence risks in siblings, parents and children. These programs can be linked to those being established for DIAMOND incidence studies, providing comparisons between the IDDM incidence rates among relatives with the risk estimates for the general population.

THE PAVIA IDDM REGISTRY FAMILY STUDY

In northern Italy, within the Pavia Province Registry which is part of the Lombardy Eurodiab Registry, family history data have been collected to ascertain the presence of diabetes in family members and to collect sera for HLA typing and virological analyses and DNA for molecular studies in IDDM probands diagnosed under 15 years of age and their relatives. The Pavia IDDM Registry is a prospective population registry established in 1988 in the Pavia Province (population at-risk age 0-29 years = 164,766). From 1988 to 1991, 50 new cases of IDDM were registered corresponding to an incidence of 7.59/100,000 per year in the population for those under 30 years (Table 1).

Table 1. IDDM incidence rate per 100,000 per year and 95% confidence intervals (CI) in Pavia province in 1988-1991.

	Males			Females			Total		
Age	n	Rate	(C.I.)	n	Rate	(C.I)	n	Rate	(C.I.)
0-14 years	13	10.3	(5.5-17.6)	10	8.3	(4.0-45.3)	23	9.3	(5.9-14.0)
15-29 years	12	5.7	(2.9-9.9)	15	7.4	(4.1-18.2)	27	6.5	(4.3-9.5)
0-29 years	25	7.4	(4.8-10.9)	25	7.7	(5.0-11.5)	50	7.6	(5.6-10.0)

Population at-risk 0-29 = 164,766.

The family data were collected using the standard DIAMOND survey[2] by direct interview with patients or their parents if the IDDM cases were less than 15 years. A comparison between these data and those obtained from medical records at the time of the probands' hospitalization was made retrospectively. The reliability of the questionnaire compared to medical records was equal to 94%, and the completeness was 92%.

The risk of developing IDDM for first degree relatives of probands was 1% for parents; only one father had IDDM (Table 2). Among the siblings, the only two who had IDDM came from a multiplex family resident in Pavia Province and were of Sardinian origin. The number of children of probands was equal to 5 and none had IDDM. Two percent of the parents had NIDDM.

Table 2. Pavia IDDM Registry Family Study: Risk of developing IDDM and NIDDM for relatives of IDDM probands.

	Cumulative Risk	
Relationship to Probands	IDDM	NIDDM
Parents	1%	2%
Siblings	3.1%	-

Risk for parents to age of 69 years, siblings to age of 32 years.

Table 3. Pavia IDDM Registry Family Study: Age at birth of mothers of IDDM probands.

Mothers Age at Birth of IDDM Proband	N	%
< 26 years	13	26.5
26-29 years	21	42.9
30 + years	15	30.6

The proportion of mothers aged more than 30 years at the time of birth of the proband was higher than expected (Table 3).

The proportion of children with IDDM also increased with birth order (Table 4). Interestingly, the proportion of families of IDDM probands with three or more children was 22%, while in the Pavia Province general population it is less than 10%. This finding has already been described by different authors[5] who observed a high reproductive performance in parents of IDDM cases. A case-control study will give more satisfactory results in terms of relative risks.

Table 4. Pavia IDDM Registry Family Study: Birth order of IDDM proband by number of children in the family.

Number of Children		Birth Order of IDDM Proband			
in the Family	N	1st	2nd	3rd	4th +
1	14	100%	-	-	-
2	25	46%	54%	-	-
3	5	20%	40%	40%	-
4+	6	-	33%	16%	50%

In conclusion, the development of standards for family history data collection will greatly facilitate international collaborative research in areas where incidence registries are established and will allow investigators to develop hypotheses regarding possible environmental factors and even preventive strategies for IDDM in family members.

REFERENCES

1. J.S. Dorman, R.E. LaPorte, R.A. Store, M. Trucco, Worldwide differences in the incidence of Type I diabetes are associated with aminoacid variation at position 57 of the HLA-DQ beta chain, *Proc Natl Acad Sci. USA* 87:7370 (1990).
2. The WHO Multinational Project for Childhood Diabetes Group, Familial insulin-dependent diabetes mellitus (IDDM) epidemiology: Standardization of data for the DIAMOND Project, *Bull WHO* 69:767 (1991).

3. J.A. Napier, M.A. Metzner, B.C. Johnson, Limitations of morbidity and mortality data obtained from family histories - A report from the Tecumseh Community Health Study, *Am J Public Health* 62:30 (1972).
4. L.B. Kahn, J.A. Marshall, J.Baxter, S.M. Shetterley, R.F.Hamman, Accuracy of reported family history of diabetes mellitus. Results from San Louis Valley Diabetes Study, *Diabetes Care* 13:796 (1990).
5. J. Diamond, Our phantom children, *Natural History* 5:18 (1992).

SIBERIAN IDDM REGISTER AND THE COLLECTION OF STANDARDIZED FAMILY HISTORY INFORMATION

Evgene V. Shubnikov and Lada A. Kalashnikova

Siberian Branch of Russian Academy of Medical Sciences
Zhemchuzhnaia 22-11
630090 Novosibirsk
RUSSIA

INTRODUCTION

The first diabetes registry in Novosibirsk was created in 1985, according to WHO recommendations.[1] This registry is focused on adults over the age of 20 and covers approximately 100,000 individuals in one district of the City of Novosibirsk. It was developed using the same methods employed for the first registry of diabetes mellitus in Kaunas, Lithuania in the former Soviet Union.[2] However, the IDDM registry was not initiated at the same time due to insufficient scientific information, as well as financial and administrative difficulties. In 1988, a visit by Professor Jaakko Tuomilehto to Novosibirsk led to the beginning of work in this direction, based on the experience of other IDDM registries[3,4,5] and recommendations by the Diabetes Epidemiology Research International Group.[6] Our IDDM registry will be a base for future investigations of etiological factors associated with IDDM among children from across the world. The geographical position of Novosibirsk also has played an important role, since it was the first IDDM registry to be developed in Russia.

METHODOLOGY

The Siberian IDDM Registry covers the territory of the Novosibirsk Region, which is an administrative unit of Russia. Most of the territory is situated on the West Siberian plane. Only in the eastern part can one see mountains. The main river is the Ob. Annual precipitation is approximately 300-450 mm. The annual mean air temperature is close to zero Celsius. The Novosibirsk Region contains 178,300 square kilometers. The capital of the Novosibirsk Region is Novosibirsk City, covering 477 square kilometers. Novosibirsk City is a large industrial center on the bank of the Ob river and Ob artificial sea.

A population census was conducted in the Novosibirsk Region in 1979 and 1989. Based on the census and the annual registration of people (born, dead and migrants), the Novosibirsk Department of the Government Statistical Office has defined the regional population of Novosibirsk City from other parts of the Novosibirsk Region.

In 1989, the population of the Region was 2,778,724; 645,355 of whom were under 15 years of age. The Novosibirsk City population in 1989 was 1,435,889; 305,508 of whom were under age 15 years. The Novosibirsk Region is 98% Russian, 2.0% Ukrainian and 1.3% German.

IDDM case identification is based on two sources. The primary source is a retrospective review of medical documentation from two endocrinology departments in Novosibirsk's hospitals, where all children from the Region with suspected IDDM are admitted for initial diagnosis and treatment. The secondary source is a retrospective review of pediatric endocrinologists' records of children with IDDM or suspected IDDM.

The minimum data required for registration included name, sex, nationality, date of birth, date of diagnosis, date of first insulin administration, the address at diagnosis, current address, and sources of identification.

We used current address information to contact families and collect standardized family history information. We conducted interviews with all families identified during 1983-1990 by mail using the questionnaire specially developed for the WHO DIAMOND Project.[7]

RESULTS

From 1983 to 1990, 245 children, 0-14 years of age, were diagnosed with IDDM in the Novosibirsk Region. The average annual incidence was relatively low, approximately 4.4/100,000 among boys and 4.7/100,000 among girls. The total incidence was approximately 4.6/100,000. In both sexes, the highest incidence of IDDM was in the 10-14 year old age group. We compared the incidence among children living in Novosibirsk City to that for children living in the Novosibirsk Region. The risk of IDDM in children living in the Novosibirsk City was 5.7/100,000, significantly higher than for children living in the Novosibirsk Region, 3.6/100,000.[8]

The families of these 245 children were interviewed. One hundred eleven families completed our questionnaire, including 8 of 12 multiplex families, where both a parent and a child had IDDM. From these 12 multiplex families, there were 8 fathers with IDDM and 4 mothers.

In addition, by their own report, 3 fathers had NIDDM, 2 sibs IDDM, 3 grandmothers IDDM, 25 grandmothers NIDDM, and 7 grandfathers NIDDM. Thirteen cases had other relatives with IDDM and 7 had other relatives with NIDDM.

CONCLUSIONS

The standardized collection of family history information within the framework of the WHO DIAMOND Project will lead to other standardized investigations of the pathogenesis and etiology of IDDM. We hope the Siberian IDDM Registry will participate in these remarkable studies.

Acknowledgements

Many thanks to Dr. J. Dorman for initiating the standardized collection of family history information among IDDM families in Siberia and for assistance in preparation of this article.

REFERENCES

1. WHO Expert Committee on Diabetes Mellitus: Second Report, WHO Technical Report Series N646, Geneva (1980).
2. R. Ostrauskas, L. Butkune, V. Dargis, The revealing of diabetes mellitus and impaired glucose tolerance according to data of City's polyclinica/Aktualnie voproci issledovania rosta i reparaii tkanei: Tez.recp.nauch.konf.26 aprela, Kaunas, 34 (1985).
3. R.E. LaPorte, H.A. Fishbein, A.L. Drash, et al., The Pittsburgh IDDM registry: The incidence of IDDM in Allegheny County, PA (1965-1976), *Diabetes* 30:279 (1981).
4. A. Green, P.K. Anderson, Epidemiology and studies of diabetes mellitus in Denmark. Clinical characteristics and incidence of diabetes among males aged 0-19 years, *Diabetologia* 25:226 (1983).
5. M. Rewers, R.E. LaPorte, M. Walcak, et al., Apparent "epidemic" of insulin-dependent diabetes mellitus in Mid-western Poland, *Diabetes* 36:106 (1987).
6. Diabetes Epidemiology Research International Study Group, Diabetes Registries in Asia, *Bull WHO* 65:897 (1987).
7. The WHO Multinational Project for Childhood Diabetes Group, Familial insulin-dependent diabetes mellitus (IDDM) epidemiology: Standardization of data for the DIAMOND Project, *Bull WHO* 69:767 (1991).
8. E. Shubnikov, T. Podar, J. Tuomilehto, Y. Nikitin, Low incidence of Childhood IDDM in District of Novosibirsk (Russia), *Diabetes Care* 15:915 (1992).

THE EPIDEMIOLOGY OF FAMILIAL INSULIN-DEPENDENT DIABETES MELLITUS (IDDM) IN HUNGARY

Gyula Soltész[1] and the Hungarian Childhood Diabetes Epidemiology Study Group*

[1]Department of Pediatrics
University of Pécs
H-7623 Pécs
HUNGARY

INTRODUCTION

The establishment and the international coordination of standardized population-based registries in the last decade have provided reliable incidence data for childhood insulin-dependent diabetes mellitus (IDDM) in many parts of the world.[1,2] A number of descriptive studies have shown that parents and siblings of diabetic children, and children of a diabetic proband, have a much higher IDDM risk than unrelated individuals in the background population.[3] In most of these studies of familial IDDM, however, the patients were identified from hospitals and diabetes clinics and not from population-based registries. The definition of diabetes among family members was not uniform and there were inconsistencies in the methods used for data collection. Furthermore, most of these studies represented only Northern European and North American populations with high incidence rates.[3]

Our study, based on a nationwide population-based registry with a high degree of ascertainment,[1,4,5] is one of the first to provide reliable descriptive epidemiological data of familial IDDM for a low incidence Eastern European population (incidence rate in 1989-90 was 7.9 per 100,000).

MATERIAL AND METHODS

The Hungarian Childhood Diabetes Registry has recorded all children with IDDM in Hungary since 1978, with an ascertainment rate of < 96%.[4] The Hungarian Childhood Diabetes Study Group is one of the participating centers in the WHO DIAMOND Project and the EURODIAB ACE Study, two large international collaborative epidemiological studies of childhood IDDM.[1,2] Childhood IDDM in affected probands was defined on the basis of the following criteria: 1) insulin therapy at discharge, 2) insulin treatment started before the 15th birthday, and 3) permanent address within the study region at the time of diagnosis. Data from patients diagnosed between 1978 and 1991 were used for this analysis. IDDM in first degree relatives (parents and siblings) was defined as insulin therapy for treatment of diabetes and < 36 years of age at the time of diagnosis. The core data used in this study included: 1) for all first degree relatives: date of birth, sex, living status and diabetes status, 2) for diabetic first degree relatives: exact date of diagnosis and initiation of insulin therapy.

For statistical analysis, the chi square test, Wilcoxon's rank sum test and Spearman's rank correlation were used.

RESULTS

Between 1978 and 1991, 1,816 new cases of childhood IDDM were registered in Hungary. Among the parents of these children, 1.6% of the mothers and 3.03% of the fathers were reported to have IDDM, and 2.2% of the diabetic children had a sibling with IDDM. Information was available for 34 father-offspring, 26 mother-offspring and 25 sibling-sibling families.

As shown in Table 1, the family size in the parent-offspring families and in the families of non-familial diabetic children was similar to that of the background population. Sibling-sibling families, however, tended to be larger; the mean number of children was 2.5.

Table 1. Family size and male/female ratio (mean) in familial IDDM.

	Number of Children	Male/Female Ratio
Background population	1.9	1.05
Non-familial IDDM	2.2	1.06*
Father-offspring families		
All children	1.9	1.13
IDDM children		1.27
Mother-offspring families		
All children	1.7	1.14
IDDM children		1.00
Sibling-sibling families		
All children	2.5	1.63
IDDM siblings		1.32

*See Reference 1.

As compared to "non-familial" IDDM, which almost equally affects boys and girls, there was a sex difference in the incidence of IDDM in children of multiple case families. Except for IDDM children of diabetic mothers (male/female ratio = 1.0), considerably more sons compared to daughters were affected among children of diabetic fathers (Table 1). Similarly, there was a male excess for IDDM children in the sibling-sibling families. Interestingly, a higher than expected male/female ratio was also found among all children, including diabetics and non-diabetics, in multiple IDDM families.

As is shown in Figure 1, there was a striking difference in the age-structure at diagnosis between "non-familial" diabetic children and diabetic children of parent-offspring families. Diabetic children in the parent-offspring families tended to be younger at diagnosis; 32% of them were younger than 5 years of age at onset as compared to 16% in non-familial IDDM. If the children were divided into age groups, 0-4 years and 5-14 years at diagnosis, a particularly high proportion of the children in the age group 0-4 years had parents with IDDM ($\chi^2 = 8.24, p < 0.01$). Children of diabetic fathers and diabetic mothers developed IDDM at similar ages, and there was no significant difference between the ages at diagnosis of boys and girls of diabetic fathers and mothers.

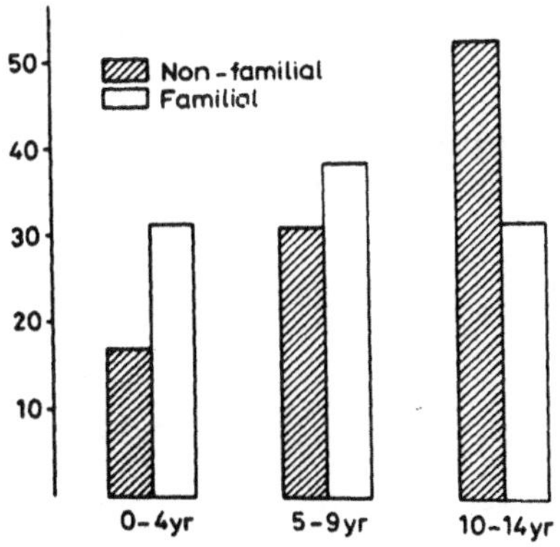

Figure 1. Percentage of children aged 0-4 years, 5-9 years and 10-14 years at diagnosis in parent-offspring and single case families.

Figure 2 shows that there was a significant positive correlation between the age at onset of affected parents and the age at onset of their diabetic children (Spearman's rank correlation coefficient $r = 0.45, p < 0.01$).IDDM children in parent-offspring families were, on average, 16 years younger at diagnosis as compared to the age at onset of their affected parents.

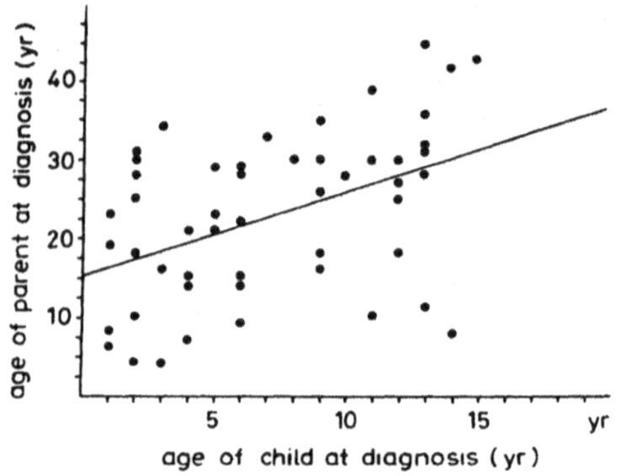

Figure 2. Relationship between age at diagnosis of diabetic children and that of their similarly affected parents ($r=0.45$, $p<0.01$).

In the sibling-sibling families, the mean ages at diagnosis for the index case and the IDDM sibling were similar (9.8 years versus 10.1 years), with no significant relationship between the ages of the two groups at disease onset. The mean time interval between the occurrence of disease in the index cases and the siblings was 4.4 years, with a wide range (1-18 years).

Finally, Figure 3 shows that there were temporal increases in the number of newly diagnosed diabetic children in the multiple IDDM families each year. These increases in risk tended to occur during periods of increasing population incidence.

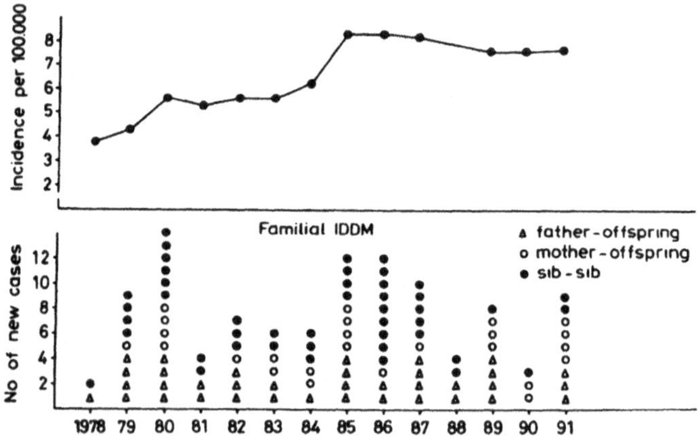

Figure 3. Incidence of childhood IDDM in Hungary (1978-1991) (Upper-panel shows national incidence data. See Reference 4).

CONCLUSIONS

Family history data from several studies representing mainly North American and Northern European populations have shown that the prevalence of paternal diabetes in IDDM families is significantly higher than the prevalence of maternal diabetes.[6-11] Several theories have been proposed to explain the differential transmission of IDDM from diabetic fathers as opposed to diabetic mothers, i.e., selective loss of the fetus that carries IDDM susceptibility genes,[6-9] and preferential transmission by the father of the HLA-DR4 linked diabetes predisposing allele.[10] Recently, an alternative mechanism, genomic imprinting, has been proposed and tested by the Pittsburgh group.[11] In agreement with previous publications,[6-11] we have also found that in our low incidence East European population, the prevalence of IDDM among the fathers of diabetic children was approximately twice the prevalence among mothers. Furthermore, in accordance with the recent data of McCarthy et. al.,[11] we have also shown that sons of diabetic fathers were more often affected than daughters. This sex-difference was not found in the diabetic children of IDDM mothers. A male excess for IDDM children was also observed in the multiple sibling case families in our study. Our data, however, should be interpreted cautiously, since in our cohort there was an unexpectedly high male/female ratio among all children born to diabetic mothers and diabetic fathers.

One of the most striking findings of our study was the significantly younger age at diagnosis of diabetic children of affected parents compared to children with non-familial IDDM. A similar observation based on a single year survey of new IDDM cases has recently been made in the United Kingdom.[12] Another interesting observation of our study was the significant relationship between the age at onset of diabetic children and that of their affected parents. Similar positive correlations were found between the age at diagnosis of the index cases and that of their diabetic fathers in the British study.[12] They were also able to demonstrate a positive correlation between the age of diagnosis of the index cases and that of their diabetic siblings, a relationship not found in our cohort of children.[12] It should be noted, however, that the correlation between the age at diagnosis of the diabetic sibling pairs in Metcalfe and Baum's study was not a strong one. Finally, we were able to demonstrate a tendency for increased numbers of familial IDDM cases during periods of increasing population incidence, an observation which has not been made before.

In summary, our study described the epidemiologic characteristics of familial IDDM in a low incidence Eastern European population. Our data have shown that the risk of developing diabetes for children of fathers with IDDM was almost twice that for children of mothers with IDDM. Children with familial IDDM were younger at diagnosis with an increased male/female ratio (except for children of IDDM mothers). There was a significant positive relationship between the age at diagnosis of children with IDDM and the age at diagnosis of their affected parents; the proportion of familial IDDM cases increased during periods of increasing population incidence. We are currently conducting an HLA study in some of these multiple IDDM families, but clearly large and standardized international studies are needed for the assessment of potential genetic, environmental and other putative risk factors in familial IDDM.

Acknowledgements

This study was supported by the Hungarian Ministry of Welfare (Grant No. T-02 664/93).

*Members of the Hungarian Childhood Diabetes Epidemiology Study Group: Acsai L., Ambró I., Barkai L., Blatniczky L., Bódis J.-né, Békefi D., Bökönyi Zs., Buzogány M., Dósa M., Gajzer É., Godó B., Györkó B.-né, Hosszú É., Jány A., Karádi Zs., Karsten J., Károlyi Gy., Kiss I., Korányi J., Kozári A., Keszei K., Körner A., Kürthy K., Madácsy L., Molnár M.,

Niederland T., Petrásy K., Petheö I., Retkes T., Rippl I., Román F., Soltész Gy., Sótér M., Szabó L., Tóth Gy., Tóth P., Tóth T.

REFERENCES

1. A. Green, E.A.M. Gale, C. Patterson, for the EURODIAB ACE Study Group, Incidence of childhood onset insulin-dependent diabetes mellitus: The EURODIAB ACE Study, *Lancet* 339:905 (1992).
2. WHO DIAMOND Project Group, WHO Multinational Project for Childhood Diabetes, *Diabetes Care* 13:1062 (1990).
3. The WHO Multinational Project for Childhood Diabetes Group; Familial IDDM epidemiology: The importance of standardization for the DIAMOND Project, *Bull WHO* 69:767 (1991).
4. G. Soltész, L. Madácsy, D. Békefi, I. Dankó and the Hungarian Childhood Diabetes Epidemiology Study Group, Rising incidence of Type I diabetes in Hungarian Children, *Diabetes Med.* 7:111 (1990).
5. G. Soltész and the Hungarian Childhood Diabetes Epidemiology Study Group, IDDM in Hungarian children: Population-based clinical characteristics and their possible implication for diabetic health care, *Padiatr Padol.* 27: 63 (1992).
6. J.H. Warram, A.S. Krolewski, M.S. Gottlieb, C.R. Kann, Differences in risk of insulin-dependent diabetes in offspring of diabetic mothers and diabetic fathers, *N Engl J Med.* 311:149 (1984).
7. G. Dahlquist, L. Blom, T. Tuvemo, et al., The Swedish childhood diabetes study - Results from a nine year case register and a one year case-reference study indicating that Type I (insulin-dependent) diabetes mellitus is associated with both Type II (non-insulin-dependent) diabetes mellitus and autoimmune disorders, *Diabetologia* 32:2 (1989).
8. J.A. Gavard, J.S. Dorman, R.E. LaPorte, An increased secondary attack rate of insulin-dependent diabetes mellitus for children of diabetic fathers compared with children of diabetic mothers, *Am J Epidemiol.* 130:822 (1989).
9. J.H. Warram, A.S. Krolewski, C.R. Kahn, Determinants of IDDM and perinatal mortality in children of diabetic mothers, *Diabetes* 37:1328 (1988).
10. C.M. Vadheim, J.I. Rotter, N.K. Maclaren, et. al., Preferential transmission of diabetic alleles within the HLA gene complex, *N Engl J Med.* 315:1314 (1986).
11. B.J. McCarthy, J.S. Dorman, C.E. Aston, Investigating genomic imprinting and susceptibility to insulin-dependent diabetes mellitus: An epidemiologic approach *Genet Epidemiol.* 8:177 (1991).
12. M.A. Metcalfe, J.D. Baum, Family characteristics and insulin-dependent diabetes, *Arch Dis Child.* 67:731 (1992).

THE IDDM REGISTRY IN THE PROVINCE OF TURIN (ITALY): PRELIMINARY DATA OF THE FAMILIAL STUDY

Graziella Bruno, Mauro Giorio, and Gianfranco Pagano

Piedmont Study Group for Diabetes Epidemiology
Institute of Internal Medicine
University of Turin
ITALY

INTRODUCTION

In recent years, the increasing interest of diabetologists towards the epidemiology of IDDM has resulted in the worldwide institution of population-based registries. Incidence data are now available in many geographic areas, as part of either the EURODIAB[1] or DIAMOND Projects,[2] showing striking variation in the distribution of the disease. Particularly in Italy, a 4-fold difference between Northern Italy and Sardinia has been recently observed.[1]

In spite of the increasing availability of data worldwide on the incidence of IDDM, little population-based data on familial risk are available. Comparative studies of descriptive and analytical familial IDDM epidemiology may reveal less variability than the differences observed in general population incidence rates.[3] Therefore, as part of WHO DIAMOND Project, we have undertaken this study to evaluate familial aggregation of cases in the IDDM Registry of the Province of Turin, Italy.

METHODS

The registry of IDDM in the Province of Turin, Italy (1 million inhabitants under age 29 years) has retrospectively collected incident cases diagnosed in the age group 0-29 years since 1984.[4] The identification of cases is based on the list of patients attending diabetic clinics, as well as discharge diagnoses from public and private hospitals in the Piedmont Region, with an estimated ascertainment rate of 97%.

In the Province of Turin, family history of diabetes data are not available in most medical records, and no attempt has been made before now to standardize routine collection of this information. Since the identification of incident cases for the period 1989-91 has already begun, family history data have been included in the core information collected for each case. We are currently focusing on the collection of family history data for the cohort of the 298 incident cases for the period 1984-88, according to the standardized method suggested by the DIAMOND Project.[3]

A questionnaire was administered to probands and/or parents during their periodic diabetes clinic appointments. Living status, age and development of IDDM were obtained for all first degree relatives as of December 31, 1991. A sibling was considered to have IDDM if he/she were on continuous insulin therapy from time of onset and were diagnosed < 30 years of age. A small proportion of cases (10%) have moved from the Province of Turin, and only 23% seldom visit the diabetic clinics.

In order to obtain complete family information, these demographic data are now being collected by the registry office for all probands not yet contacted through diabetic clinics. Family physicians of every family member will be contacted in order to assess whether he/she is affected by diabetes and to collect onset data.

RESULTS

Table 1 shows age-specific IDDM incidence rates for the Province of Turin during the period 1984-88; the incidence of IDDM was 8.18/100,000 for the age group 0-14 years, 5.82/100,000 for the age group 15-29 years, and 6.79/100,000 for the age group 0-29 years. Significant sex differences were evident, with higher risk in males for the age group 15-29 years.

Table 1. Age-specific IDDM incidence rates/100,000 for residents of the Province of Turin for 1984 to 1988.

	Males		Females		Total	
Age (yr)	Cases	Incidence (95% CI)	Cases	Incidence (95% CI)	Cases	Incidence (95% CI)
0-14	81	8.75 (6.98-10.94)	67	7.58 (5.87-9.74)	148	8.18 (6.94-9.63)
15-29	97	7.36 (6.02-8.98)	53	4.21 (3.12-5.56)	150	5.82 (4.94-6.85)
0-29	178	7.93 (6.84-9.20)	120	5.60 (4.66-6.72)	298	6.79 (6.06-7.61)

At the present time, preliminary familial data are available for 169 of 298 families (57%). However, as shown in Table 2, most of our family data were available for families in which probands were younger at onset of the disease (68% vs. 46%).

Table 2. Representatives of the sub-group of incident cases from the IDDM Registry in the Province of Turin for whom family data have been collected.

	Family Data Collected			Family Data Not Yet Collected			
Age (Yr)	Cases	%	Age at Onset (mean±sd)	Cases	%	Age at Onset (mean±sd)	p Value
0-14	100	68	8.48±4.1	48	32	10.94±3.0	<0.001
15-29	69	46	20.33±3.9	81	54	21.89±4.2	0.02
0-29	169	57	13.32±7.0	129	43	17.81±6.54	<0.001

Table 3 shows the prevalence of diabetes in family members of incident cases. No significant difference was evident between mothers and fathers. However, the prevalence of IDDM in siblings was higher than in parents.

Table 3. Prevalence of diabetes among the family members of IDDM incidence cases for the period 1984-88 in the Province of Turin.

Family Members	Diabetic Status	n	%
Fathers	Non-Diabetic	159	94.6
	IDDM	3	1.8
	NIDDM	6	3.6
Mothers	Non-Diabetic	161	95.2
	IDDM	4	2.4
	NIDDM	4	2.4
Siblings	Non-Diabetic	191	93.6
	IDDM	13	6.4

In Table 4, characteristics of probands and siblings are compared. Of 204 biologic siblings, only 13 (6.4%) were IDDM; 12 were IDDM at the time of diagnosis of the proband. No significant differences were evident with regard to sex, age at onset or current age.

Table 4. The IDDM Registry in the Province of Turin: Preliminary data on familial aggregation and the characteristic of probands.

	IDDM Probands		IDDM Siblings		Non Diabetic Siblings	
	#	%	#	%	#	%
Males	99	59	7	54	94	49
Females	70	41	6	46	97	51
Age at Onset (yr) (mean ± sd)	13.32 ± 7.09		13.15 ± 6.94		---	---
Current Age (yr) (mean ± sd)	18.8 ± 7.30		18.23 ± 6.94		21.18 ± 9.28	

CONCLUSIONS

The familial aggregation of a disease is a good indicator that genetic factors are important in determining disease risk. Although family history is among the strongest known determinants of IDDM, the predictive power of a positive family history is not very high. In this survey, the majority of cases were sporadic, with no history of IDDM among first degree relatives. No significant difference was evident between the prevalence of IDDM among fathers and mothers, whereas the prevalence of IDDM in siblings was higher than in parents. However, our preliminary results may be biased by the younger age at onset of probands recruited until now. In fact, in the light of the hypothesis that the effect of a positive family history for diabetes may differ with age, the subgroup of families we have studied may not be representative of the whole cohort of incident cases, since both the mean current age and the mean age at onset of probands contacted are lower than that of probands not contacted.

Great variation in IDDM risk has recently been shown within Italy; a high migration rate from Sardinia to Piedmont Region was evident 20-30 years ago. Preliminary data from the registry showed no difference between the expected and the observed number of incident cases born in Sardinia, based on estimates from the 1981 census. If no difference is found using recent data from the 1991 census, it will be further evidence supporting the hypothesis of environmental determinants of the disease.

REFERENCES

1. A. Green, E.A.M. Gale, C.C. Patterson, for the EURODIAB ACE Study Group, Incidence of childhood-onset insulin-dependent diabetes mellitus: The EURODIAB ACE Study, *Lancet* 339:905 (1992).
2. WHO DIAMOND Project Group, WHO Multinational Project for Childhood Diabetes, *Diabetes Care* 13:1062 (1990).
3. The WHO Multinational Project for Childhood Diabetes, Familial insulin-dependent diabetes mellitus (IDDM) epidemiology: Standardization of data for the DIAMOND Project, *Bull WHO* 69:767 (1991).
4. G. Bruno, F. Merletti, E. Pisu, G. Pastore, et al., Incidence of IDDM during 1984-1986 in population aged <30 yr residents of Turin, Italy, *Diabetes Care* 10:1051 (1990).

THE RISK OF INSULIN-DEPENDENT DIABETES IN OFFSPRING OF PARENTS WITH INSULIN-DEPENDENT DIABETES

Toomas Podar,[1] Jaakko Tuomilehto,[2] Eva Tuomilehto-Wolf,[2] Ronald E. LaPorte,[3] and Janice S. Dorman[3]

[1]Department of Medicine
University of Tartu
202400 Tartu
ESTONIA

[2]Department of Epidemiology and Health Promotion
National Public Health Institute
SF-00510 Helsinki
FINLAND

[3]Department of Epidemiology
Graduate School of Public Health
University of Pittsburgh
Pittsburgh, PA, 15261
USA

INTRODUCTION

Some diseases occur more frequently in relatives of affected individuals than in the general population. They do not always follow a pattern within family pedigrees which would fit adequately any known genetic model.[1] Insulin-dependent diabetes mellitus (IDDM) is one such disorder. In the pathogenesis of IDDM, specific environmental factors appear to operate on a susceptible genetic background which leads to the occurrence of disease. Although the mode of inheritance of IDDM remains unclear, there is no doubt that the genetic factors are major determinants of the risk of IDDM. The environmental factors responsible for the occurrence of IDDM are less well characterized than the genetic determinants of the disease.

Only a minority of offspring who are born to IDDM parents will ever develop diabetes.[2-4] However, the higher risk of IDDM for offspring compared to the general population is well recognized to be of concern to IDDM parents. A difficulty is that it is not possible to predict precisely the probability of IDDM developing in the offspring of a parent

Standardization of Epidemiologic Studies of Host Susceptibility
Edited by J. Dorman, Plenum Press, New York, 1994

with IDDM. At present we can estimate approximately how many children may develop IDDM, but not who will develop it. If the pattern of inheritance of IDDM were precisely understood, then it would be possible to more accurately identify those individuals who will later develop the disease.[5] This information is obviously important for the potential prevention of IDDM.

At the present time, the prediction of IDDM in offspring is limited to empirical data and the presence of various risk factors. The most common risk markers for relatives of IDDM patients include HLA-identity or haploidentity, presence of islet cell, insulin or 64K antibodies and diminished insulin secretory response to an oral or intravenous glucose challenge[6] or their combination. None of the proposed or known risk markers for IDDM can yet provide reliable prediction of IDDM in healthy individuals.[7] With the exception of genetic markers, the others reflect the ongoing diabetogenic process and are, therefore, not independent risk indicators of IDDM.

The present paper is a review of available data on the risk of IDDM for offspring of an IDDM parent. We have summarized studies which have especially investigated this matter, and extracted data from studies that have provided sufficient descriptive data on the study population. From studies conducted prior to 1980, before the clear distinction of IDDM and non-insulin-dependent diabetes (NIDDM), we have attempted to extract data about pathology that is now known as IDDM.

RATIONALE FOR STUDYING IDDM IN PARENTS AND OFFSPRING

Genetic counseling in various forms has been given to potential parents suffering from IDDM as long as insulin therapy has been available. To give reliable advice, empirical data using a standardized approach has to be collected. The present state of knowledge, based on limited empirical risk estimates is, however, still insufficient for appropriate advice in this respect. It is not yet established what extent the risk to offspring varies between populations, or whether it reflects the underlying geographical differences in incidence for the general population. Differences in parent-offspring transmission of IDDM between populations could be one of the reasons for the reported variations in the occurrence of IDDM in many communities. It is possible that in some populations, the risk of IDDM for offspring of IDDM parents is much higher than for others.

The parent-offspring studies of IDDM can potentially be employed to identify the environmental factors responsible for converting genetic susceptibility to overt disease. For instance, comparison of the risk for offspring born prior to and after the onset of IDDM in a parent might provide important clues as to where to look for environmental agents causing the disease. A clue to the etiology of IDDM might also be found by examining the period of pregnancy. Clinical and experimental findings show that the congenitally acquired rubella virus can destroy beta cells and cause IDDM, although it is probably responsible for only a minority of IDDM cases.[8] It is possible that the intrauterine environment in a diabetic pregnancy or congenital infections might contribute to the preferential restriction of β cell functional capacity that would later manifest as diabetes. Alternatively, the exposure of fetuses to a diabetic mother might somewhat protect them from developing IDDM.[9] The in utero effect of environmental factors could be investigated by comparing the risk for dizygotic twins versus siblings. The genetic closeness of both pairs of siblings is identical, but their in utero influences are different.

The prospective studies among offspring of IDDM parents would potentially be more informative than studies among siblings for the identification of the environmental "hit" triggering IDDM, but harder to conduct. Siblings of cases with childhood IDDM usually live in the same household when IDDM develops and, thus, they are also likely to be exposed to the same environmental factors. In contrast, offspring of IDDM parents are born many years after the environmental event which caused IDDM in the parent.

From the genetic viewpoint, the correct estimation of the IDDM risk in offspring of families where one or both parents have IDDM will be useful for establishing the mode of inheritance of the disease.[9] Parent-offspring studies enable one to test which one of the two parental haplotypes are necessary for developing the disease. The diabetic haplotypes carry the putative susceptibility gene for IDDM and are transmitted from a diabetic parent to a diabetic child. These high-risk haplotypes are still being defined. Accumulation of data on parent-offspring transmission of IDDM would form the essential database needed for proposing and testing hypotheses about the mode of inheritance of IDDM.

Linked with the above mentioned needs for research in this area are the organization of appropriate health services and the development of preventive measures. Primary prevention of IDDM can be potentially carried out by implementing preventive measures only in individuals or groups that are at a specific high risk of developing IDDM in the future.[10] There is evidence to suggest that damage to beta cells, eventually leading to IDDM, takes place over a period of months to years.[11] Thus, it allows time to influence the pathogenesis of the disease in the high risk group once such a group is adequately identified.

EPIDEMIOLOGIC STUDIES OF IDDM IN PARENTS AND OFFSPRING

Various study designs and forms of data presentation have been used to assess the parent-offspring transmission of IDDM (Table 1). Empirical data have been obtained through: 1) point-estimations of the prevalence of IDDM in parents at the time of diagnosis of IDDM in offspring,[12-15] or later,[9,16,17] 2) point-estimations of the occurrence of IDDM in offspring of parents with IDDM,[2,9,18,19] 3) prospective follow-up studies of the offspring of IDDM parents,[3,4,20-25] and 4) retrospective estimates of the cumulative risk of IDDM in offspring.[1,2,9,24,26]

Table 1. Approaches used for the epidemiologic study of parent-offspring transmission of IDDM.

I.	Point-estimations of familial aggregation of IDDM
	A. At the time of the diagnosis of IDDM in an offspring, or later in an IDDM registry or clinic-based series
	1) Proportion of IDDM parents among parents of IDDM offspring
	2) Proportion of IDDM offspring having an IDDM parent
	3) Proportion of bigenerational IDDM families of all families
	B. The prevalence of IDDM in offspring of IDDM parents
II.	Prospective follow-up studies of offspring of parents with IDDM
	A. IDDM mothers only
	B. IDDM mothers and fathers
III.	Estimation of the cumulative risk of IDDM in offspring by age
	A. Empirical estimation
	B. Theoretical estimation
IV.	Comparison of risk to offspring of a parent with IDDM with:
	A. Background population
	B. Siblings and other relatives

The point estimation information about parent-offspring distributions of IDDM can be obtained from studies that employ IDDM offspring as probands and evaluate the prevalence of IDDM in parents. Some studies have not investigated the matter in particular, but provided sufficient descriptive data on this subject to allow the estimates to be obtained. Cross-sectional data about parent-offspring transmission of IDDM may be presented in the form of three different proportions: 1) the proportion of IDDM parents among all parents of IDDM offspring, 2) the proportion of IDDM offspring having an IDDM parent, and 3) the proportion of multigenerational IDDM families of all families participating in a certain study. The most common way of data presentation is the first, that is, essentially the prevalence of IDDM among parents of offspring with IDDM. For the purpose of unification and comparability, we have interpreted data from different studies and presented data as the prevalence of IDDM among parents of IDDM probands (Table 2). Although the estimation of IDDM prevalence in parents does not permit obtaining direct risk estimates for offspring of probands, the implication is clear: the higher the prevalence in parents, the more familial is the disease in a given community and the higher the risk for offspring of probands.

Table 2. Prevalence of IDDM (%) among parents of IDDM offspring and incidence* of childhood IDDM.

Reference	Incidence of IDDM	Population	Number of probands	Either Parent	IDDM Fathers	Prevalence Mothers
West et al. 1979[29]	9.8	Canada	518	3.2		
Chern et al. 1982[14]	17.1	USA MN	493	3.0		
Gorsuch et al. 1982[14]	8.7	UK	160	3.5	4.5	1.3
Tajima et al. 1985[13]	1.7	Japan	1572	0.4		
Tillil, Köbberling 1987[9]	7.0	Germany	554	2.5	4.1	1.7
Dahlquist et al. 1989[15]	23.2	Sweden	3503	4.3	6.2	2.3
McCarthy et al. 1991[17]	15.6	USA PA	1774	2.0	3.0	1.0
Tuomilehto et al. 1992[27]	28.8	Finland	801	4.1	5.7	2.6

*The incidence of childhood IDDM was obtained from Reference 41 and may not refer to the period when the parent-offspring was conducted. In Reference 14, the incidence data refer to Leicestershire. Parent-offspring data from Reference 9 refer to the former Federal Republic of Germany and incidence data to the former German Democratic Republic.

The proportion of parents with IDDM among parents of IDDM probands varies from 0.2% to 4.3%.[15,26] The lowest value has been reported from Canada[29] and the highest from Sweden.[15] This variation seems to correspond roughly to the variation in the incidence of childhood IDDM (Table 2). In virtually all studies, the prevalence of parental IDDM has

IDDM among fathers of newly diagnosed children with IDDM ranges from 3.0% to 6.2%,[15,18] and among mothers from 1.0% to 2.6%.[18,27] The most extensive study so far comes from Pittsburgh, PA, involving 1,774 probands with IDDM diagnosed between 1950 and 1981. The prevalence of IDDM among parents was found to be 2%, among fathers 3% and mothers 1%.[17] These estimates were obtained following the diagnosis of IDDM in the proband. We believe that part of this observed variation between countries in the familial clustering of IDDM may be due to the differences in ascertainment of the IDDM status and in temporal trends in IDDM since the incidence of IDDM has been reported to be increasing in a number of populations.

The parent-offspring transmission of IDDM investigated by the proportion of multigenerational IDDM families in Montreal, Canada and Minnesota, US (about 500 families in both surveys) was found to be approximately 6%.[14,28] Approximately 8% of all families participating in the Finnish nationwide insulin-dependent diabetes study were found to be multigenerational.[12]

Another way to estimate the familial risk of IDDM is to determine the proportion of IDDM offspring having an IDDM parent among all offspring with IDDM in a defined population. There are few studies, mostly from high-incidence populations, in which data have been presented in this way. The lowest value (6.4%) has been reported in Wisconsin,[29] the highest value (11%) is based on only 35 families studied.[30] Approximately 9% of the children with IDDM had a diabetic parent in a nationwide study from Sweden.[15] In families with more than one diabetic child, the proportion of children with parental IDDM was higher. Paternal IDDM was found in 6.6% of diabetic boys and 5.7% of diabetic girls. Maternal IDDM among diabetic children of both sexes was found in 2.3% of cases.[15] Male predominance among the IDDM parents of diabetic offspring has been documented in other studies.[12,31,32] In accord with the earlier findings are the data from Pittsburgh which show that the secondary attack rate of IDDM in children of IDDM fathers is much higher than in children of IDDM mothers.[33] Parental IDDM seems to have no effect on sex distribution of IDDM in offspring,[15,26] but a recent study from Pittsburgh, reported higher cumulative risk for male than female offspring of IDDM fathers, but not for mothers.[17] An issue to be considered with this type of investigation is that studies conducted years after the onset of IDDM in offspring usually yield higher prevalence estimates for parents than at the time of diagnosis in offspring.

Fewer studies have investigated the risk of developing IDDM in offspring than in parents (Table 3). The estimates on the occurrence of IDDM obtained for offspring of IDDM fathers are 2.6%[18] and 2.9%,[9] and for mothers 1.0%[16] and 1.2%.[8] In an earlier series, Köbberling and Brüggeboes (1980) analyzed the occurrence of IDDM in 464 offspring of 311 IDDM mothers.[2] Of the 8 diabetic children, 3 did not require insulin treatment; thus only 1.2% of offspring were estimated to have IDDM. Irvine et al. (1977) observed IDDM in 2.5% of offspring to diabetic parents by the mean age of 20.6 years.[19] This type of study provides risk estimates for offspring that are useful for genetic counseling and comparable to prospective studies. However, the parent may not have been diagnosed with IDDM at the time the offspring was born. Estimates of the IDDM risk for offspring born before versus after the parent was diagnosed with IDDM would be relatively simple to conduct and would provide important information regarding the etiology of the disease.

The most accurate and informative method for studying risk of IDDM among offspring of parents with IDDM is prospective follow-up of offspring of IDDM patients in a defined population (Table 3). Insulin only became available for the treatment of IDDM in 1922. Before that, survival and the prognosis of having offspring for IDDM patients was poor. Therefore, it has been possible to carry out offspring studies among IDDM patients for only a relatively limited period. In addition, the period of follow-up must be long, 15-20 years at least, to cover the period when the incidence of IDDM is highest. More data are available on the risk of IDDM for offspring of diabetic mothers than fathers. The highest proportion of offspring of IDDM mothers who developed IDDM, about 3.8%, has been reported from Sweden during a 10-13 year follow-up of 53 offspring.[22] In a larger series from the United Kingdom, the percentage of offspring of IDDM mothers manifesting IDDM was found to be considerably lower, only approximately 0.9%. Nevertheless, the risk of developing diabetes in these offspring was found to be 22 times higher than in the general population.[3] A recent study of 55 children who were in utero when their mothers developed

Table 3. Cross-sectional and prospective studies of the risk of IDDM in offspring of IDDM parents.

Ref. Design Population	Parents' gender	Number of parents	Number of offspring	Number of offspring with IDDM	Prevalence of IDDM in offspring	Mean age offspring
I. Cross-sectional studies						
Germany	Female	424	545	5	0.9	9
	Male	130	104	3	2.9	
	Both	554	649	8	1.2	18
Canada	Female		316	3	1.0	
	Male		228	6	2.6	
	Both		544	9	1.7	19
UK	Both	126	158	4	2.5	20.6
II. Prospective studies						
UK	Female	293	229	2	0.9	3
Multinational	Female	55	55	1	1.8	10.4
Sweden	Female	53	53	2	3.8	11
	Both	187	420	13	3.1	4
Denmark	Female	89	57	1	1.8	24
	Male	98	121	4	3.3	
	Both	187	178	5	2.8	20
USA (PA)	Female	155	249	5	2.0	25
	Male	177	363	11	3.0	16.3
	Both	332	612	16	2.6	13.6

IDDM revealed only 1 diabetic offspring by the mean age of 10 years. The authors suggested that the diabetogenic process in the pregnant mother may not affect the beta cells of the fetus.[20] However, the risk of 2% is approximately the same as that reported from other study designs.

There are only a few studies which have provided prospective data on the risk of IDDM for offspring of both mothers and fathers with IDDM.[4,23,24] While the percentage of IDDM in offspring to IDDM mothers ranges from 0.9%[3] to 3.8%,[22] in IDDM fathers the estimates are considerably higher, from 30.0% to 6.1%.[23,25] In most of the prospective studies, the duration of the follow-up period and the mean age of offspring rarely exceeds 20 years.[3,4,20,22-24] Thus, to study the long-term offspring risk, one approach is to employ theoretical risk calculations based on limited experimental follow-up data. [2,9,24,34] This is useful for genetic counseling.

Degnbol and Green[24] calculated the cumulative risk of developing diabetes for children of probands in whom diabetes was diagnosed before 20 years of age. It was 2.8%

by the age of 25 years. Köbberling and Brüggeboes, using a method of age-correction, estimated the total prevalence of IDDM to be 1.5% in the offspring of IDDM mothers at the age of 25 years.[2] In their more recent analysis, the risk by the same age was estimated to be somewhat higher, 2.5%.[9] The lifetime risk of IDDM for children of an individual with IDDM (assuming that all persons will live to 80 years) was evaluated to be 4.9%.[9] The estimated age-corrected lifetime risk for offspring of all male IDDM probands was approximately 7% and for female probands, approximately 4%, but this difference was not statistically significant. In agreement with these data are those from Denmark, where the estimated lifetime risk of developing IDDM for offspring of IDDM fathers was 6.1%.[34]

Empirical data among children from families where both the father and mother have IDDM are virtually lacking. In a study from Germany, two of the six offspring (33%) of conjugal IDDM parents had developed IDDM, thus the risk was much increased.[9] Without empirical data, Green[34] has estimated the lifetime risk of IDDM for offspring of such parents to be 39%. It would be important world-wide to identify these rare families and to pool the data.

The incidence of IDDM among the offspring of IDDM parents has been estimated to be 4 to 40 times higher than in the background population.[1,2,4,24,26] The highest estimates have been reported by Simpson (1968) who estimated the risk of offspring developing diabetes separately for male and female probands in Canada. Among the living offspring of male probands whose diabetes had been diagnosed under 20 years of age, the increase in IDDM risk was approximately 40-fold, and for female probands approximately 25-fold.[26] On the other hand, assuming that data from Denmark[24] were comparable to the studies in Pittsburgh[35] and Sweden,[32] it has been estimated that the risk for children of an IDDM parent is 4 to 14 times the risk in the general population.[1] Krolewski and Warram[1] and Degnbol and Green[24] estimated the risk for offspring to be lower than that for siblings. According to Tillil and Köbberling,[9] the risk for offspring is slightly higher than for siblings, but not significantly different. Simpson[26] has confirmed higher risks for offspring than siblings. All these estimates have been obtained using different methodologies. None of them have taken into account the genetic background, i.e., HLA haplotype sharing status.

CONCLUSIONS

The main difficulties in performing epidemiological parent-offspring studies of IDDM are the use of a retrospective definition of the type of diabetes in parents, insufficient duration of the follow-up and potential population bias. A clear distinction between IDDM and NIDDM was made only since 1980.[36] Earlier investigators did not discriminate between IDDM and NIDDM. Studies addressing the era before 1980 raise questions about the nosologic entity and homogeneity of the study population. It is probable that the bias in the definition of the type of diabetes has been in favor of including NIDDM patients into the group of IDDM patients. Hence the interpretation of some studies of familial diabetes is difficult.[37] The closer to the 1980's the studies have been performed, a clearer and more reliable distinction has been made between IDDM and NIDDM. Simultaneously, the duration of the follow-up period has decreased. The sufficient length of the follow-up period of offspring is of great importance for the reliable assessment of the risk of IDDM in offspring. It has been shown that the risk to offspring might depend on the age of the mother at the birth of the child,[38,39] which varies in different communities. In the evaluation of results derived from different time periods, some knowledge of eventual secular variations in the incidence of IDDM is necessary.[24] For example, a study from Pittsburgh reported a reduction by half of IDDM risk for the siblings of probands diagnosed during the 1950's compared with probands diagnosed in the 1970's.[40] We do not know whether secular variation also affects the risk for offspring in a similar manner.

Thus, in retrospective interpretation of studies where the authors had not distinguished between the subclasses of diabetes, we have used the age of onset of diabetes and treatment with insulin as criteria for distinguishing between different types of diabetes.[3,4,16,18,19,22,24,26,28] We acknowledge that use of each of these criteria can lead to errors in the ascertainment of the type of diabetes. A number of earlier studies on familial aggregation of diabetes have been omitted from this paper, as the form of data presentation did not permit us to extract information relevant to IDDM with reasonable accuracy. Often the diagnostic criteria applied to parents and offspring have been different. A more detailed analysis of diagnostic criteria has been given by Wagener et al.[16] However, diagnostic problems cannot be avoided in offspring studies of IDDM, particularly in relation to the era of diagnosis of parents.

The majority of the studies completed until now are hospital or clinic based rather than population based, yielding results that cannot be fully generalized. The size of the study population varied to a large extent. Investigators have used various methods of patient recruitment, even including appeals in mass media.[26] Survey of case histories of a single clinic have been the most common method.[4] Some methods have favored the recruitment of preferentially multigenerational IDDM families as the ones more likely to respond to appeals and return questionnaires. Case ascertainment procedures are not always described sufficiently. There is no need to continue the list of these obvious problems with many of the earlier studies.

A factor influencing the proportion of IDDM in offspring is the number of children in the family of an IDDM parent. The denominator in the risk calculation for the offspring of IDDM parent is dependent on the family size of the given population. Family size of an IDDM parent is probably related to sex, psychological and socioeconomic features of the group studied, peculiarities of era and severity of diabetes. Still not long ago, physicians were deliberately discouraging women with diabetes to have children. Due to better care, the earlier fears have been removed to a considerable extent, and the average family size of IDDM patients has increased.

The studies of parent-offspring transmission of IDDM come primarily from high and medium IDDM incidence populations and address the subject mainly in whites. Data from Japan are an interesting exception to this. The available data suggest that the susceptibility to IDDM may more frequently be transmitted to offspring in high childhood IDDM incidence populations like Finland and Sweden. In Japan, where the incidence of childhood IDDM is one of the lowest in the world, the degree of transmission of IDDM susceptibility seems also low.[13,41] The differences reported in parent-offspring transmission of IDDM between populations are much less than those in the overall worldwide IDDM incidence. Though the parent-offspring studies performed have neither been population-based, nor standardized, the results correspond fairly well with the observed differences in the incidence of childhood IDDM. However, reliable conclusions cannot be drawn based on the results of these non-standardized studies.

It is considerably easier for health care and investigators to link a diabetic child to the mother than to the father. Hence, several prospective follow-up studies of IDDM subjects have concentrated on IDDM mothers.[3,20-22] There are only a few prospective studies of the risk of IDDM in offspring of both IDDM fathers and mothers. Nevertheless, one of the most interesting epidemiologic patterns in risk to offspring is that IDDM is transmitted more often through the father. In some studies, this observed differential in the IDDM risk for offspring of IDDM mothers and fathers has been found statistically significant.[23]

No generally accepted theory resolving this disparity has been proposed so far. It has been suggested that the IDDM susceptible fetuses are selectively lost during the pregnancy or perinatal period,[23] but a more recent study by the same group concluded that there was no evidence that selective loss of diabetes-susceptible fetuses in perinatal deaths was a mechanism for the lower incidence of IDDM in the offspring of mothers with IDDM than in those of fathers with IDDM.[42] Also, the protection of fetuses from the later manifestation of IDDM by exposure to a diabetic mother during gestation has been speculated.[9] If this were the case, there would be a difference in the risk for offspring born before and after the mother was diagnosed with IDDM. In a recent study it was found that the risk for the children born before the mother's onset of diabetes was higher than that for those exposed in utero to her diabetes, but the difference did not reach statistical significance.[39] In contrast, the intrauterine environment of NIDDM mothers seems to increase the risk of NIDDM in offspring.[43]

On the other hand, diabetic fathers might be more likely to transmit high-risk haplotypes to their offspring and diabetic mothers low-risk haplotypes that do not show significant association with IDDM at the population level.[12] The possible lower frequency of recombination between HLA loci in men than in women[23] might explain some of the sex differential. There are no data to support the possible role of genomic imprinting.[17]

It is accepted that the best way to study risk factors and putative etiologic agents is by prospective studies.[44] However, prospective studies are usually costly and difficult to carry out. The determination of the risk of IDDM in offspring of IDDM parents prospectively is particularly problematic because of the low incidence of the disease, difficulties in the retrospective IDDM definition and the long follow-up period required. The risk of IDDM in offspring of a parent with IDDM has been assessed only in some non-standardized studies, the results of which are difficult to compare. Therefore, unquestionably, more studies are needed in this field.

One of the goals of the recently initiated WHO DIAMOND Project is to study familial aggregation of IDDM using a well-defined standardized approach.[45] Hopefully, this study will make a substantial contribution to the scarce empirical database of parent-offspring transmission of IDDM.

It is well documented that the incidence of IDDM shows marked differences worldwide. A child in Finland is at a 40 times higher risk of developing IDDM than a child in Japan and at a 2-3 times higher risk than a child in the U.S.[41] It would be interesting to know if the risk to the offspring of IDDM parents varies in the same way and whether there is some similarity between the risk to offspring and secular trends. Moreover, are the large differences observed in the incidence of IDDM between different parts of the world to some extent attributable to the differences in the transmission of susceptibility to the offspring of IDDM parents? For the time being, these questions remain unanswered because of insufficient data and lack of standardization of study methods in this field.

REFERENCES

1. A.S. Krolewski, J.H. Warram, Epidemiology of diabetes mellitus, *in:* "Joslin's Diabetes Mellitus," 12th ed., A. Marble, L.P. Krall, R. Bradley, A.R. Christlieb, J.S. Soeldner, eds., Lea & Febiger, Philadelphia (1985).
2. J. Köbberling, B. Brüggeboes, Prevalence of diabetes among children of insulin-dependent diabetic mothers, *Diabetologia* 18:459 (1980).
3. J.W. Farquhar, Prognosis for babies born to diabetic mothers in Edinburgh, *Arch Dis Child.* 44:36 (1969).
4. M.S. Gottlieb, Diabetes in offspring and siblings of juvenile and maturity-onset-type diabetics, *J Chronic Dis.* 33:331 (1980).
5. D.A. Pyke, Diabetes: The genetic connections, *Diabetologia* 17:333 (1979).
6. R.C. McEvoy, B. Fedun, L.Z. Cooper, et al., Children at high risk of diabetes mellitus: New York studies of families with diabetes and of children with congenital rubella syndrome, *in:* "Prediabetes", R.A. Camerini-Davalos, H.S. Cole, eds., Plenum Press, New York (1988).
7. A.C. Tarn, J.M. Thomas, B.M. Dean, et al., Predicting insulin-dependent diabetes, *Lancet* 1:845 (1988).

8. M.A. Menser, J.M. Forrest, R.D Bransby, Rubella infection and diabetes mellitus, *Lancet* 1:57 (1978).
9. H. Tillil, J. Köbberling, Age-corrected empirical genetic risk estimates for first degree relatives of IDDM patients, *Diabetes* 36:93 (1987).
10. J. Tuomilehto, E. Wolf, Primary prevention of diabetes mellitus, *Diabetes Care* 10:238 (1987).
11. A.N. Gorsuch, K.M. Spencer, J. Lister, et al., Evidence for a long pre-diabetic period in Type I (insulin-dependent) diabetes mellitus, *Lancet* 2:1363 (1981).
12. E. Tuomilehto-Wolf, J. Tuomilehto, Z. Cepaitis, et al., New susceptibility haplotype for Type I diabetes, *Lancet* 2:299 (1989).
13. N. Tajima, R.E. LaPorte, I. Hibi, et al., A comparison of the epidemiology of youth-onset insulin-dependent diabetes mellitus between Japan and the United States (Allegheny County, Pennsylvania), *Diabetes Care* 8:17 (1985).
14. N.M. Chern, V.E. Anderson, J. Barbosa, Empirical risk for insulin-dependent diabetes (IDD) in sibs: Further definition of genetic heterogeneity, *Diabetes* 31:1115 (1982).
15. G. Dahlquist, L. Blom, T. Tuvemo, et al., The Swedish childhood diabetes study - Results from a nine-year case register and a one-year case-referent study indicating that Type I (insulin-dependent) diabetes mellitus is associated with both Type II (non-insulin-dependent) diabetes mellitus and autoimmune disorders, *Diabetologia* 32:2 (1989).
16. D.K. Wagener, J.M. Sacks, R.E. LaPorte, J.M. MacGregor, The Pittsburgh study of insulin-dependent diabetes mellitus. Risk for diabetes among relatives of IDDM, *Diabetes* 31:136 (1982).
17. B.J. McCarthy, J.S. Dorman, C.E. Aston, Investigating genomic imprinting and susceptibility to insulin-dependent diabetes mellitus: An epidemiologic approach, *Genet Epidemiol.* 8:1 (1991).
18. N.E. Simpson, Heritabilities of liability to diabetes when sex and age at onset are considered, *Ann Hum Genet.* 32:283 (1969).
19. W.J. Irvine, A.D. Toft, D.E. Holton, et al., Familial studies of Type-I and Type-II idiopathic diabetes mellitus, *Lancet* 2:325 (1977).
20. K. Buschard, C. Kuhl, L. Molsted-Pedersen, et al., Investigations in children who were in utero at onset of insulin-dependent diabetes in their mothers, *Lancet* 1:811 (1989).
21. B. Persson, Longterm morbidity in infants of diabetic mothers, *Acta Endocrinol.* Suppl 277: 156 (1986).
22. B. Persson, J. Gentz, E. Moller, Follow-up of children of insulin-dependent (Type I) and gestational diabetic mothers, *Acta Paediatr Scand.* 73:778 (1984).
23. J.H. Warram, A.S. Krolewski, M.S. Gottlieb, C.R. Kahn, Differences in risk of insulin-dependent diabetes in offspring of diabetic mothers and diabetic fathers, *N Engl J Med.* 311:149 (1984).
24. B. Degnbol, A. Green, Diabetes mellitus among first- and second-degree relatives of early onset diabetics, *Ann Hum Genet.* 42:25 (1978).
25. B. McCarthy, J. Dorman, C. Aston, S. Gollin, The incidence of insulin-dependent diabetes mellitus (IDDM) among offspring of diabetic parents: A prospective study, Abstract. 14th IDF Congress satellite, Williamsburg, Virginia (1991).
26. N.E. Simpson, Diabetes in the families of diabetics, *Cancer Med Assoc J.* 98:427 (1968).
27. J. Tuomilehto, R. Lounamaa, E. Tuomilehto-Wolf, et al., Epidemiology of childhood diabetes mellitus in Finland - Background of a nationwide study of Type I (insulin-dependent) diabetes mellitus, *Diabetologia* 35:70 (1992).
28. R. West, M. Belmonte, E. Colle, et al., Epidemiologic survey of juvenile-onset diabetics in Montreal, *Diabetes* 28:690 (1979).
29. C. Allen, M. Palta, D.J. D'Alessio, Risk of diabetes in siblings and other relatives of IDDM subjects, *Diabetes* 40:831 (1991).

30. R.B. Tattersall, S.S. Fajans, A difference between the inheritance of classical juvenile-onset and maturity-onset type diabetes of young people, *Diabetes* 24:44 (1975).
31. E. Wolf, J. Tuomilehto, R. Lounamaa, the Study Group of Childhood Diabetes, Finland: Can the high risk of Type I diabetes in Finland be explained by familial aggregation and by HLA haplotype distribution?, *in:* "Prediabetes", R.A. Camerini-Davalos, H.S. Cole, eds., Plenum Press, New York (1988).
32. G. Dahlquist, K.H. Gustavsson, G. Holmgren, et al., The incidence of diabetes mellitus in Swedish children 0-14 years of age. A prospective study 1977-1980, *Acta Paediatr Scand.* 71:7 (1982).
33. J.A. Gavard, J.S. Dorman, R.E. LaPorte, An increased secondary attack rate of insulin-dependent diabetes mellitus for children of diabetic fathers compared with children of diabetic mothers, *Am J Epidemiol.* 130:822 (1989).
34. A. Green, Empirical risk estimation for relatives of insulin-dependent diabetics: Theoretical considerations and practical applications, *in* "The Genetics of Diabetes Mellitus", J. Köbberling, R. Tattersall, eds., Proceedings of the Serono Symposia, Vol 47. Academic Press, London (1982).
35. D. Wagener, L. Kuller, T. Orchard, et al., Pittsburgh diabetes mellitus study II. Secondary attack rates in families with insulin-dependent diabetes mellitus, *Am J Epidemiol.* 115:868 (1982).
36. WHO Expert Committee on Diabetes Mellitus: WHO Tech Rep Ser 646, WHO, Geneva (1980).
37. J.W. Farquhar, Juvenile diabetes mellitus, *Arch of Dis Child.* 54:569 (1979).
38. R.E. LaPorte, K.J. Cruickshanks, Incidence and risk factors for insulin-dependent diabetes, *in:* "Diabetes in America", NIH Publication No. 85-1468, III.
39. J.H. Warram, B.C. Martin, A.S. Krolewski, Risk of IDDM in children of diabetic mothers decreases with increasing maternal age at pregnancy, *Diabetes* 40:1679 (1991).
40. J.A. Gavard, J.S. Dorman, R.E. LaPorte, A reduction of IDDM risk to siblings since 1950: Evaluation of family histories of 1,806 IDDM probands, *in:* Abstract book from Satellite Congress "Epidemiology of Diabetes" of the 13th International Diabetes Federation Congress, Hobart 1988; 18. Tasmania, Nov 26-28.
41. M. Rewers, R.E. LaPorte, H. King, J. Tuomilehto, Trends in the prevalence and incidence of diabetes: Insulin-dependent diabetes mellitus in childhood, *World Health Stat Q.* 41:179 (1988).
42. J.H. Warram, A.S. Krolewski, C.R. Kahn, Determinants of IDDM and perinatal mortality in children of diabetic mothers, *Diabetes* 37:1328 (1988).
43. D.J. Pettitt, K.A. Aleck, H.R. Baird, et al., Congenital susceptibility to NIDDM. Role of intrauterine environment, *Diabetes* 37:622 (1988).
44. T.J. Orchard, A.L. Rosenbloom, The development of insulin-dependent diabetes mellitus among relatives, *Diabetes Care* 8 (suppl. 1): 45 (1985).
45. The WHO Multinational Project for Childhood Diabetes Group: Familial insulin-dependent diabetes mellitus (IDDM) epidemiology: Standardization of data for the DIAMOND Project, *Bull WHO* 69:767 (1991).

DIFFERENTIAL HLA CONTRIBUTION FROM FATHERS VERSUS MOTHERS TO IDDM SUSCEPTIBILITY

L. Leigh Field

Department of Pediatrics
University of Calgary
Calgary, Alberta
T2N 4N1
CANADA

INTRODUCTION

It is well established that people with insulin-dependent diabetes mellitus (IDDM) who are of European ancestry have elevated frequencies of phenotypes positive for HLA-DR3 and DR4. In other words, HLA-DR3 and DR4 are *associated with* increased susceptibility to developing IDDM, either because these genes themselves predispose to disease or genes linked to and in disequilibrium with DR3 and DR4 predispose to disease. For example, in our dataset of 231 diabetics ascertained through the Alberta Children's Hospital Diabetes Clinic in Calgary, frequencies of patients positive for HLA-DR3 and HLA-DR4 were 63% and 70%, respectively,[1] compared with frequencies in the general population of approximately 20% and 29%.[2] Only 7% of our diabetics were negative for both DR3 and DR4, compared with approximately 54% of individuals in the general population. Prior to the discovery of insulin, insulin-dependent diabetes was a lethal disorder usually resulting in death prior to the reproductive years. Thus, genes producing susceptibility to IDDM were selected against. It is probable that other compensating *advantages* of these genes helped to maintain them at relatively high population frequencies.

In 1986, Vadheim et al.[3] reported that in some situations, parents of diabetic children transmit IDDM-associated genes to *non-diabetic* children more frequently than expected by chance, suggesting greater survival of gametes or zygotes possessing IDDM-associated genes compared with those lacking such genes. Specifically, they reported preferential transmission (that is >50%) of HLA-DR3 from heterozygous mothers and fathers to *non-diabetic* children as well as to diabetic children, and preferential transmission of HLA-DR4 from heterozygous *fathers* to both diabetic and non-diabetic children, but not from heterozygous *mothers* to either diabetic or non-diabetic children. These findings were surprising, since one would expect increased "transmission" of DR3 and DR4 from heterozygous parents to *diabetic* children (because of the well known association of IDDM with these DR alleles) but no increased transmission to *non-diabetic* children (assuming no natural selection). One would also expect no differences between mothers and fathers in "transmission" rates. We have used quotation marks when discussing "transmission" to *diabetic* children in order to emphasize that apparent differences in transmission rates could result from phenomena other than true segregation distortion (see below). However, any increased transmission to *non-diabetic* children would constitute true segregation distortion and indicate a selective advantage of those DR alleles.

We have previously suggested that there were methodological problems with the Vadheim et al. study.[4] Furthermore, we have shown that at least one diabetes dataset, the Genetic Analysis Workshop 5 (GAW5) data,[5] does *not* show evidence for non-random segregation of DR3 or DR4 to unaffected children.[6] However, these data *do* show significant differences between mothers and fathers in DR4 transmission to diabetic children.[6] In the present paper, we examine data from three independent sources (including GAW5) to further evaluate evidence for non-random transmission of HLA alleles to non-diabetic children from DR3 or DR4 heterozygous parents, and evidence for transmission differences between fathers versus mothers to either diabetic or non-diabetic children.

MATERIALS AND METHODS

In our analyses, we used combined segregation information from three different IDDM datasets: 1) the Genetic Analysis Workshop 4 dataset (GAW4),[7] consisting of 169 simplex families from Germany; 2) the GAW5 dataset,[5] consisting of 94 multiplex families from 12 centres in Europe and North America, and 3) our Calgary dataset of 87 trios of mother, diabetic child and unaffected child. The GAW4 and GAW5 data reside in the public domain. For each of the three datasets, we scored transmission of DR3 and DR4 from parents who were heterozygous DR3/X or DR4/X (where X does not equal DR3 or DR4). That is, we did not include transmissions from DR3/3, DR4/4, or DR3/4 parents. In addition, we only scored transmissions from parents who were not themselves diabetic. An example of the scoring method is shown in Figure 1. In this example, the DR3/X mother is scored as transmitting DR3 to both the diabetic son and the non-diabetic daughter, while the DR4/X father is scored as transmitting DR4 to the diabetic son and not transmitting DR4 to the non-diabetic daughter. The frequency of positive transmissions for DR3 and DR4 were then calculated across all families and datasets as the number of positive transmissions relative to the total number of positive and negative transmissions, for diabetic and non-diabetic children separately and for mothers and fathers separately. Thus, if there were a total of 100 unaffected children of DR3/X mothers and 45 of these received DR3 while 55 received DRX, then the frequency of DR3 transmission from mothers to unaffected children would equal 45/100 = 45%. The observed transmission frequencies were then compared with a random frequency of 50% using a χ^2 test of significance.

It was hypothesized that: 1) for unaffected children, there would be no difference in transmission frequencies from the 50% expected, but 2) for diabetic children, there would be >50% transmission frequencies due to the association of DR3 and DR4 with IDDM. In other words, distortion of transmission to diabetic children would reflect ascertainment bias rather than true segregation bias. One might think that the transmission of IDDM-associated DR alleles to unaffected children should be significantly <50%, however, due to the low penetrance of susceptibility genes, this is not so. Thomson at the University of California, Berkeley, has estimated that the frequency of transmission to unaffected children should be at least 48% (G. Thomson, personal communication). Nevertheless, transmission to unaffected children should not significantly *exceed* 50%. It was also hypothesized that: 3) there would be no difference between mothers and fathers in the transmission frequencies of either DR3 or DR4 to unaffected or to diabetic children.

An alternative method for examining the possibility that susceptibility genes are transmitted to diabetic children differently from fathers versus mothers is to consider the parental origin of HLA haplotypes in affected sibling pairs who share only one HLA haplotype. This method has the advantage that one can either consider transmission across all HLA haplotypes (ie. disregarding HLA-DR type), or one can look at transmission specific to HLA-DR type. The frequency of origin from fathers versus mothers of the single shared HLA haplotype is compared to the expected proportion of 50:50 using a χ^2 test of significance.

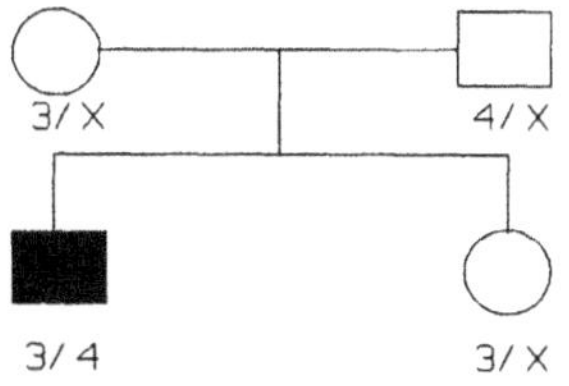

Figure 1. Transmission of DR3 and DR4 (Example).

RESULTS

Transmission of HLA-DR3 or DR4 alleles from heterozygous parents to unaffected children is shown in Table 1. For HLA-DR3, the total frequency of transmission from heterozygous mothers was 45/92 = 49%, and from heterozygous fathers was 48/83 = 58%. Neither of these values is significantly different from the 50% expected if transmission were random. For HLA-DR4, the total frequency of transmission from heterozygous mothers was 64/127 = 50%, and from heterozygous fathers was 70/132 = 53%. Again, neither of these values is significantly different from the 50% expected. In addition, for neither DR3 nor DR4 is the transmission frequency significantly different between mothers and fathers (last column, Table 1).

Table 1. Transmission of DR3 and DR4 to unaffected children.

Data	DR3 from mothers	Vs. 50% χ^2_1 (p)	DR3 from fathers	Vs. 50% χ^2_1 (p)	Mo vs. Fa χ^2_1 (p)
GAW4	18/41 = 44%		26/45 = 58%		1.63 (.20)
GAW5	16/32 = 50%		22/38 = 58%		0.44 (.51)
Calgary	11/19 = 58%		---		
Total	45/92 = 49%	0.04 (.84)	48/83 = 58%	2.04 (.15)	1.39 (.24)
	<u>DR4 from mothers</u>		<u>DR4 from fathers</u>		
GAW4	28/56 = 50%		36/65 = 55%		0.35 (.56)
GAW5	21/44 = 48%		34/67 = 51%		0.10 (.75)
Calgary	15/27 = 56%		---		
Total	64/127 = 50%	0.01 (.93)	70/132 = 53%	0.48 (.49)	0.18 (.67)

Transmission of DR3 or DR4 from heterozygous parents to diabetic children is shown in Table 2. For DR3, the total frequency of transmission from heterozygous mothers was 84/104 = 81%, and for heterozygous fathers 68/78 = 87%. Both of these values are significantly >50%, having p values of <0.0001 (Table 2). The rate of transmission of DR3 to diabetic children does not differ between mothers and fathers (p = 0.25, Table 2). For DR4, the total frequency of transmission from heterozygous mothers was 109/137 = 80%, and from heterozygous fathers was 117/125 = 94%. As with DR3, both values for transmission of DR4 are significantly >50%, having p values of <0.0001. However, unlike DR3, *the frequency of transmission of DR4 was significantly different between mothers and fathers* (χ^2_1 = 10.87, p = 0.001, Table 2).

Table 2. Transmission of DR3 and DR4 to diabetic children.

Data	DR3 from mothers	Vs. 50% χ^2_1 (p)	DR3 from fathers	Vs. 50% χ^2_1 (p)	Mo vs. Fa χ^2_1 (p)
GAW4	29/36 = 81%		30/35 = 86%		0.34(.56)
GAW5	38/49 = 78%		38/43 = 88%		1.87(.17)
Calgary	17/19 = 89%		---		
Total	84/104 = 81%	39.38 (<.0001)	68/78 = 87%	43.13 (<.0001)	1.33(.25)
	DR4 from mothers		DR4 from fathers		
GAW4	37/44 = 84%		52/54 = 90%		4.33(.04)
GAW5	51/66 = 77%		65/71 = 92%		5.37(.02)
Calgary	21/27 = 78%		---		
Total	109/137 = 80%	47.89 (<.0001)	117/125 = 94%	95.05 (<.0001)	10.87 (p=.001)

Table 3 presents the parental origin of the shared HLA haplotype in affected sibling pairs from the GAW5 dataset who share *only one HLA haplotype*. Disregarding the HLA-DR type of the shared haplotype, 76% of 33 affected sib pairs sharing only one HLA haplotype obtained the shared haplotype from their father. This frequency is significantly different from the 50% expected by chance (χ^2_1 = 8.76, p = .003). Considering only HLA-DR4 haplotypes, 81% of shared haplotypes are of paternal origin; considering only HLA-DR3 haplotypes, 71% are of paternal origin. These results demonstrate, using an alternate method to the transmission rate analyses presented above, that the strongest difference between mothers and fathers in "transmission" of IDDM-associated alleles to diabetic children is for HLA-DR4.

Table 3. Parental origin of shared HLA haplotype in affected sib pairs sharing only 1 haplotype.

	% Paternal Origin	
Any HLA Haplotype	25/33 = 76%	(vs. 50%, χ^2_1 = 8.76, p = .003)
HLA-DR4 Haplotypes	17/21 = 81%	(vs. 50%, χ^2_1 = 8.05, p = .005)
HLA-DR3 Haplotypes	5/7 = 71%	
HLA-DRX Haplotypes (where X≠ 3, 4)	3/5 = 60%	

CONCLUSIONS

In summary, using combined data from three independent datasets, we have shown that: 1) neither DR3 nor DR4 is transmitted significantly more often than 50% from heterozygous parents to unaffected siblings of diabetics; 2) both DR3 and DR4 are transmitted significantly more often than 50% to diabetic children from both fathers and mothers; and 3) while both mothers and fathers transmit DR4 significantly more often than 50%, *mothers transmit DR4 significantly less often than fathers* to diabetic children (p = .001).

The lack of preferential transmission of HLA-DR3 and DR4 from heterozygous parents to unaffected children is not unexpected, and suggests that if there are selective advantages of these IDDM-associated alleles, then they are not of the magnitude detectible in samples of the size used in this study. This is consistent with results of other studies which have attempted to demonstrate selective advantages of disease alleles which occur at high frequencies in the general population. For example, families studies have not been able to demonstrate segregation distortion of the cystic fibrosis allele. The finding of increased "transmission" of HLA-DR3 and DR4 from heterozygous parents to diabetic children is also consistent with expectation. These families were ascertained for study on the basis of having at least one IDDM child and it is well established that DR3 and DR4 are associated with the IDDM phenotype. Thus, these families will demonstrate preferential "transmission" of DR3 and DR4 from parents to the diabetic children.

The unexpected finding in this study is the significant difference in the frequency of "transmission" of DR4 from mothers versus fathers to diabetic children. What possible mechanisms might explain this difference? Pre-zygotic mechanisms may be invoked. In other words, there may be a selective advantage of DR4-bearing sperm. However, if this is the case, then one would expect to see the same effect in DR4 transmission to unaffected children. Since the frequency of transmission of DR4 from fathers to unaffected children is not significantly greater than 50%, this would appear to rule out a pre-zygotic mechanism. Possible post-zygotic mechanisms include interaction between a DR4 mother and a DR4 fetus leading to either: 1) protection from IDDM in the fetus ("a protective effect"), or 2) death of the fetus or child ("selective loss effect"). In attempting to find evidence for or against the latter two mechanisms, consideration of the studies by Warram et al. from the Joslin Diabetes Center is instructive. First, these investigators have shown that the frequency of IDDM in children of IDDM mothers is significantly less (1.3%) than the frequency of IDDM in children of IDDM fathers (6.1%).[8] These results from *diabetic* parents are consistent with our findings of "transmission" differences from *healthy* parents. Either genes transmitted through mothers are not as diabetogenic as genes transmitted through fathers, or children with diabetes genes from their mother are more likely to die before or after birth. Secondly, Warram et al. have gone on to study spontaneous abortion rates and perinatal death rates in children of IDDM mothers and have found that there is no correlation between these factors and the frequency of IDDM in the children.[9] These results argue against the "selective loss effect" (although they do not exclude very early selective loss before recognition of pregnancy). At the moment, we, therefore, tend to favor the interactive "protective effect" theory, which postulates that some form of interaction between the DR4 mother and the DR4 fetus protects against subsequent development of IDDM in the child. This interactive protective effect would probably occur in utero, but could also involve the transfer of specific antibodies during breast feeding in infancy.

Another possible mechanism for the differential contribution of mothers versus fathers to IDDM predisposition in their offspring is genomic level imprinting, wherein equivalent genes transmitted through mothers have different zygotic expression than the same genes transmitted through fathers. Extended family studies will be necessary to further assess this possibility for HLA-associated IDDM susceptibility.

If IDDM genes transmitted through mothers are less diabetogenic or more lethal than IDDM genes transmitted through fathers, the genetic contribution to the development of IDDM in living offspring must come more often from the father than the mother. Results of analyzing affected sibling pairs who share only one HLA haplotype support this conclusion; the shared haplotype (which is very likely to contain IDDM-predisposing genes) originates from the father in more than ¾ of the pairs. The largest difference in contribution is for HLA-DR4-associated susceptibility. An interesting consequence of the more frequent origin of DR4 from fathers is that diabetics having the high risk genotype HLA-DR3/4 should derive the DR4 from their father and the DR3 from their mother more often than the reverse. In the GAW5 dataset, 62% of the 77 diabetics who were HLA-DR3/4 received the DR4 from their father rather than their mother (versus 50% expected, $\chi^2_1 = 4.69$, $p = .03$). On the other hand, among 31 normal siblings who were HLA-DR3/4, 49% obtained the DR4 from their father rather than their mother.

In our Calgary dataset, we have an intriguing family in which *both* children of an IDDM mother are themselves affected with IDDM.[10] This would appear to contradict the low risk of diabetes to children of IDDM mothers. However, the mother does not carry either DR3 or DR4, although she has transmitted the same non-DR3, non-DR4 allele to both of her affected children. Perhaps this constitutes a case of *absence* of a maternal-fetal interactive "protective effect".

In summary, this study of transmission of HLA susceptibility genes from parents to diabetic children has demonstrated a clear difference in the frequency of HLA-DR4 "transmission" from fathers versus mothers. We hypothesize that a fetus who inherits HLA-DR4-associated susceptibility genes from its mother (whether or not she is diabetic herself) has a *reduced risk* of developing IDDM, compared with a fetus who inherits these genes from its father, due to a maternal-fetal interactive "protective effect". The protective effect may occur as a result of interaction during gestation or during breast-feeding. We suggest that intensive efforts should be focused on identifying the nature of this protective maternal-fetal interaction which modifies subsequent risk of IDDM.

Acknowledgements

We would like to thank all families and investigators who contributed to the GAW4 and GAW5 diabetes datasets, as well as the diabetics and their families from the Calgary area who contributed to our own local studies. The expert technical assistance of Ms. Rose Tobias and Elzbieta Swiergala are greatly acknowledged. Many thanks also to Ms. Rosemary Hodorek for careful preparation of this manuscript. This research has been supported by grants from the Medical Research Council of Canada and the Alberta Children's Hospital Foundation. Dr. Field is an Alberta Heritage Medical Scientist.

REFERENCES

1. L.L. Field, Non-HLA region genes in insulin-dependent diabetes mellitus (IDDM), *in*: "The Genetics of Diabetes", B.D. Tait, L.C. Harrison, eds., W.B. Saunders Co., *Baillière's Clin Endocrinol Metab.* 5:413 (1991).
2. L.L. Field, R.G. McArthur, The genetics of susceptibility to insulin-dependent diabetes: Possible new markers, *Clin Invest Med.* 10:437 (1987).
3. C.M. Vadheim, J.I. Rotter, N.K. Maclaren, W.J. Riley, C.E. Anderson, Preferential transmission of diabetic alleles within the HLA gene complex, *N Engl J Med.* 315:1314 (1986).
4. C.T. Falk, L.L. Field, S.E. Hodge, Transmission of diabetic alleles within the HLA complex, (Letter-to-Editor) *N Eng J Med.* 317:768 (1987).
5. M.P. Baur, R. Fimmers, C. Fritsche, et al., Genetic analysis of IDDM: The GAW5 multiplex family dataset, *Genet Epidemiol.* 6:15 (1989).
6. L.L. Field, Genes predisposing to IDDM in multiplex families, *Genet Epidemiol.* 6:101 (1989).
7. M.P. Baur, Genetic Analysis Workshop IV: Insulin-dependent diabetes mellitus-summary, *Genet Epidemiol* (suppl.) 1:299 (1986).
8. J.H. Warram, A.S. Krolewski, M.S. Gottlieb, C.R. Kahn, Difference in risk of insulin-dependent diabetes in offspring of diabetic mothers and diabetic fathers, *N Engl J Med.* 311:149 (1984).
9. J.H. Warram, A.S. Krolewski, C.R. Kahn, Determinants of IDDM and perinatal mortality in children of diabetic mothers, *Diabetes* 37:1328 (1988).
10. R.G. McArthur, L.L. Field, J-W. Yoon, N. Boyd, Diminished complications in a nonDR3 DR4 family with IDDM, *Clin Invest Med.* 10:470 (1987).

HLA-DR3 AND DR4 ALLELE SEGREGATION IN IDDM FAMILIES AND IN CONTROLS

Cristina Daielli,[1] Renata Lorini,[2] Miryam Martinetti,[1] Luisella Cortona,[2] Mariaclara Cuccia,[3] Letizia Vitali,[2] Gabriele Devoti,[4] Giuseppe d'Annunzio,[2] and Maria T. Tenconi[4]

[1]Laboratorio HLA
Servizio di Immunoematologia e Trasfusione
IRCCS Policlinico San Matteo
Pavia
ITALY

[2]Dipartimento di Pediatria
[3]Dipartimento di Genetica e Microbiologia
[4]Dipartimento di Medicina Preventiva,
Occupazionale e di Comunita
Sezione di Igiene
Università degli Studi
Pavia
ITALY

INTRODUCTION

Segregation distortion of HLA alleles in IDDM families is still a matter of discussion. Some evidence for preferential transmission of DR3 and DR4 from parents to both diabetic and non-diabetic progenies has been given.[1] However, other studies investigating the same issue have not confirmed this differential mode of inheritance.[2]

To examine in our population whether HLA-DR3 and DR4 are transmitted to offspring with a frequency which deviates from the Mendelian expectation, we analyzed a sample of complete nuclear IDDM and control families selected according to the presence of at least one parent heterozygous for HLA-DR3 and/or DR4. Homozygous parents were not included in the study.

Standardization of Epidemiologic Studies of Host Susceptibility
Edited by J. Dorman, Plenum Press, New York, 1994

MATERIALS AND METHODS

IDDM Families

We analyzed a sample of 121 complete nuclear families with at least one parent heterozygous for DR3 and/or DR4, ascertained through an IDDM child. The allele segregation has been evaluated according to the siblings' sex and disease status (73 IDDM males and 48 IDDM females, 38 healthy brothers and 42 healthy sisters).

Control Families

The data from IDDM families were compared to those obtained from 50 informative pedigrees with at least one parent heterozygous for HLA-DR3 and/or DR4, selected from blood donors. A total of 105 siblings, 51 males and 54 females, were evaluated.

Laboratory Methods

HLA serological typing for class I and II antigens was performed for all family members, by the NIH classical techniques.

Statistical Analysis

Chi square analysis, Fisher's exact test and estimates of the relative risk were employed to evaluate the statistical significance of the data.

RESULTS

HLA-DR3 and DR4 Phenotype Frequency

As illustrated in Table 1, the frequency of HLA-DR3 was generally higher in male than in female IDDM children (63.0% vs. 50.0%), and was higher in fathers than in mothers (42.1% vs. 35.5%) and in healthy brothers than in sisters (52.6% vs. 33.3%). The same trend was observed within the control families (34.0% in fathers and 16.0% in mothers; 29.4% in sons and 22.2% in daughters). When the whole male population (N=333) was compared to the female population (N=315), a significant difference was observed in the frequency of HLA-DR3 (44.7% vs. 32.1%, χ^2=10.99; p<0.001; RR=1.72).

The HLA-DR4 distribution showed a slight, but not significant, increase in females (Table 1).

Table 1. HLA-DR3 and DR4 frequencies according to the sex and diabetes status.

	HLA-DR3			
	Male		Female	
	N	%	N	%
IDDM Probands	46/73	63.0	24/48	50.0
Parents of IDDM	51/121	42.1	43/121	35.5
Healthy Siblings of IDDM	20/38	52.6	14/42	33.3
Control Parents	17/50	34.0	8/50	16.0*
Control Offspring	15/51	29.4	12/54	22.2
Total	149/333	44.7	101/315	32.1**

	HLA-DR4			
	Male		Female	
	N	%	N	%
IDDM Probands	37/73	50.7	24/48	58.3
Parents of IDDM	35/121	28.9	41/121	33.9
Healthy Siblings of IDDM	11/38	28.9	13/42	30.9
Control Parents	15/50	30.0	17/50	34.0
Control Offspring	19/51	37.2	19/54	35.2
Total	117/333	35.1	118/315	37.5

*$p<0.05$
**$p<0.001$

HLA-DR3 and DR4 Segregation

Paternal and maternal HLA-DR3 transmission in IDDM families is shown in Tables 2A and 2B, respectively. A strong distortion was found in the affected sibs who inherited this allele; 88.2% were from the father and 81.4% were from the mother, instead of the expected proportion of 50%, based on Mendelian inheritance. On the contrary, considering the non-affected progeny of IDDM families, we observed a preferential transmission of DR3 from fathers to male children (62.5%, p=n.s.).

Table 2A. HLA-DR3 segregation from heterozygous fathers in IDDM families.

	Male		Female		Total	
	N	%	N	%	N	%
IDDM	32/37	86.5**	13/14	92.9***	45/51	88.2**
No IDDM	10/16	62.5	11/22	50.0	21/38	55.3

Observed percentage versus expected Mendelian (50%): **$p<0.01$, ***$p<0.005$

Table 2B. HLA-DR3 segregation from heterozygous mothers in IDDM families.

	Male		Female		Total	
	N	%	N	%	N	%
IDDM	22/28	78.6**	13/15	86.7**	35/43	81.4**
No IDDM	11/21	52.5	6/14	42.9	17/35	48.6

Observed percentage versus expected Mendelian (50%): $^{**}p<0.01$, $^{*}p<0.05$

When the HLA-DR4 segregation in IDDM families was considered, a distortion was noted in the IDDM patients; 85.7% were of paternal origin (Table 3A), 95.1% were of maternal origin (Table 3B), significantly higher than the expected 50%. An excess of DR4 transmission from both fathers and mothers to healthy male sibs was observed (61.5% and 75%, respectively), but the comparisons with the Mendelian expected value did reach statistical significance.

Table 3A. HLA-DR4 segregation from heterozygous fathers in IDDM families.

	Male		Female		Total	
	N	%	N	%	N	%
IDDM	15/17	88.2**	15/18	83.3**	30/35	85.7**
No IDDM	8/13	61.5	6/11	54.5	14/24	58.3

Observed percentage versus expected Mendelian (50%): $^{**}p<0.01$, $^{*}p<0.05$

Table 3B. HLA-DR4 segregation from heterozygous mothers in IDDM families.

	Male		Female		Total	
	N	%	N	%	N	%
IDDM	24/25	96.0**	15/16	93.7**	39/41	95.1**
No IDDM	3/4	75.0	8/19	42.1	11/23	47.8

Observed percentage versus expected Mendelian (50%): $^{**}p<0.01$, $^{*}p<0.05$

Examination of HLA-DR3 segregation in control families revealed a distortion in the progeny, mainly when the source of DR3 was the father: 69.2% of sons and 62.5% of daughters inherited this paternal allele (Table 4A).

Considering the HLA-DR4 segregation in healthy families, a slight increased frequency was observed for females irrespective of parental origin (55.0% and 56.2%, for paternal and maternal alleles, respectively) (Table 4B).

Table 4A. Offspring of HLA-DR3 heterozygous parents in control families.

	Offspring					
	Male		Female		Total	
Parent	N	%	N	%	N	%
Father	9/13	69.2	10/16	62.5	19/29	65.5
Mother	4/7	57.1	4/8	50.0	8/15	53.3
Total	13/20	65.0	14/24	58.3	27/44	61.4

Table 4B. Offspring of HLA-DR4 heterozygous parents in control families.

	Offspring					
	Male		Female		Total	
Parent	N	%	N	%	N	%
Father	9/18	50.0	11/20	55.0	20/38	52.6
Mother	10/25	40.0	9/16	56.2	19/41	46.3
Total	19/43	44.2	20/36	55.5	39/79	49.4

CONCLUSIONS

It is known that when a parent is heterozygous for HLA alleles, the probability of transmitting one haplotype to the progeny is 50%, according to Mendelian inheritance. In the present study, HLA-DR3 and DR4 segregation was studied in IDDM families according to sex and diabetes status, and in 50 informative non-diabetic control families. Parents homozygous for HLA-DR3 or DR4, or who were affected by diabetes, were not included in the analyses.

Our sample was composed of a higher number of HLA-DR3 positive fathers than mothers in both IDDM and control families. This discrepancy was maintained in the progeny; the proportion of HLA-DR3 positive subjects was higher in healthy sons than in daughters, and in IDDM males compared to IDDM females. On the contrary, HLA-DR4 seemed to be more prevalent in female population in both IDDM and control families.

A segregation distortion was observed in IDDM patients; over 80% of DR3 or DR4 parents transmitted the antigen. This should be an obvious conclusion, assuming that these alleles are in linkage disequilibrium with the diabetogenic genes.

A preferential transmission of HLA-DR3 from fathers to sons was observed in the healthy progeny of both IDDM and control families (Tables 2A and 4A). On the contrary, HLA-DR4 segregation was different between the healthy siblings in IDDM families (higher from fathers to sons) (Tables 3A-3B) and among siblings of controls (higher from mothers to daughters) (Table 4B).

From this study, it appears that: 1) HLA-DR3 alleles are more frequent in males than in females, whereas HLA-DR4 is more common in females, 2) pre- or post-zygotic selection is occurring in mothers and/or fathers to favor both DR3 and DR4, 3) for this reason,

HLA-DR3 or DR4 bearing haplotypes are at an increased frequency compared to other haplotypes, and 4) HLA-DR3 is preferentially transmitted from father to male children (79.2% vs. 66.7% in IDDM families and 69.2% vs. 62.5% in controls). These observations suggest HLA control during the gametogenesis and agree with our previous findings on the possible role of MHC genes in the sex selection.[3]

REFERENCES

1. C.M. Vadheim, J.I. Rotter, N.K. Maclaren, W.J. Riley, C.E. Anderson, Preferential transmission of diabetic alleles within the HLA gene complex, *N Eng J Med.* 315:1314 (1986).
2. J.M. Martin-Villa, J.L. Vicario, J. Martinez-Laso, M. Serrano-Rios, G. Lledo, A. Damiano, F. Hawkins, J.R. Regueiro, A. Arnaiz-Villena, Lack of preferential transmission of diabetic HLA alleles by healthy parents to offspring in Spanish diabetic families, *J Clin Endocrinol Metab.* 70:346 (1990).
3. P. Astolfi, M. Martinetti, F. Gigli-Berzolari, M. Cuccia, The effect of parental and maternal-fetal histocompatibility at MCH on sex ratio in offspring, *Tissue Antigens* 35:172 (1990).

PRELIMINARY RESULTS OF HLA SEROLOGICAL TYPING OF INSULIN-DEPENDENT DIABETES MELLITUS FAMILIES IN PUERTO RICO

Teresa Frazer de Llado,[1] Ramiro Milan,[1] Gladys Veray,[1] Magalys Soto,[1] Arthur Hupka,[1] Peggy Bull,[2] John Gerlach,[2] and Robert Bull[2]

[1]Department of Pharmacology
Ponce School of Medicine
Ponce
PUERTO RICO

[2]Department of Medicine and Pathology of Michigan State University
East Lansing, Michigan 48824
USA

INTRODUCTION

Country of birth is the highest risk factor associated with development of insulin-dependent diabetes mellitus (IDDM).[1] However, the relative contributions of the environment and host susceptibility to this risk are unknown. The Caribbean Basin offers a unique opportunity to study the genetic and epidemiological aspects of IDDM for the following reasons: 1) there is a marked difference in the incidence rates of the islands varying from 2/100,000/yr in Cuba to 12/100,000/yr in Puerto Rico,[2] 2) the islands are geographically separated, and 3) the ancestry of each island population is historically defined for the past 500 years.

Puerto Rico, located at 18 degrees latitude north of the equator, is a land mass of 3,500 square miles with a population of 3.5 million. The population is racially mixed, and it is believed that 60% of the mixture derives from persons from the Iberian Peninsula, 35% from Africa and 5% from other populations. The incidence of IDDM in Puerto Rico appears to be approximately equal to that of the major ethnic population of origin, Spain, with an incidence 11-12 cases per 100,000.[3]

This study was designed as a pilot study to determine if the HLA alleles that have been reported to be associated with IDDM in other populations in the world are also found in the Puerto Rican IDDMs.

POPULATION

The study population includes 36 Puerto Ricans IDDM patients and 43 non-IDDM controls. The criteria for cases to participate in this study is residency in Puerto Rico at the time of diagnosis, the use of insulin at the time of diagnosis of IDDM, diagnosis of IDDM not secondary to any other disease, and at least one Puerto Rican parent. The criteria for the controls are also Puerto Rican residency, no immediate family history of IDDM, and one Puerto Rican parent. For the 21 IDDM families studied, at least one member of the family had to meet the criteria of IDDM. Since this was a pilot project and because no registry existed at its conception, these patients were not randomly chosen, but rather selected from a private pediatric endocrine practice and recruited from the Convencion de Diabetes del Sociedad de Diabeticos del Sur y Oeste de Puerto Rico.[4]

METHODS

Venous blood samples for HLA typing were collected in vacutainers containing heparin or acid-citrate dextrose (ACD). T and B cells were isolated from samples by centrifugation over ficoll gradients. B-cells were separated from T cells by use of B-cell monoclonal antibody bound to ox cells.[5]

Antisera sources were the basic and extended trays from at least two commercial sources, either One Lambda (Terisaki), Gen-Trak or BIOTEST. Each individual typing was characterized for at least 19 HLA-A, 23 HLA-B, and 5-6 HLA-C locus antigens.

The isolated B lymphocytes were serologically DR typed utilizing commercial trays from either One Lambda, Gen-Trak or Biotest. The typing trays identified HLA-DR, DR52 and DR53, DQl, DQ2, and DQ3 epitopes.

All HLA serological typings, whether for HLA-A, B, C or HLA-DR and DQ, was done by the standard microdroplet lymphocytotoxic reaction using rabbit serum as a complement source.

RESULTS

Allele Frequency

Results of DR typing are shown in Table 1, presented as the percentage of subjects in which the antigen is found. DR2, a protective allele, was present in 3% of the IDDM, and in 16% of the control subjects. DR3 was present in 39% of diabetics, but only 9% of controls, whereas DR4 was found in 69% of IDDM and only 33% of controls. The combination of DR3/4 was found in 25% of IDDM, and in no control subjects. Table 2 compares the Puerto Rican data with other populations in the Western Hemisphere. This table is an extraction of data from the noted references for the purpose of comparison of the antigen frequency in the different ethnic populations.[6-9]

DQ allele percentages are shown in Table 3. DQ2 was found in 61% of the diabetics, but only 34% of the controls; DQ3 was found in 86% of diabetics and 62% of controls. Both DQ2/DQ3 were found in 47% of IDDM, but only 16% of controls. Only one diabetic patient did not express either DQ2 or DQ3.

Table 1. HLA specificities DR2, DR3, and DR4 in IDDMs and control subjects of a southern Puerto Rican population.

	DR2	DR3/X	DR4/X	DR3/4	DR3 or DR4	DRX/X
Diabetics (N = 36)	1(3%)	14(39%)	25(69%)	9(25%)	31(86%)	5(14%)
Controls (N = 43)	7(16%)	4(9%)	14(33%)	0(0%)	18(42%)	25(58%)

Table 2. HLA specificities DR2, DR3, and DR4 in Western Hemisphere Caucasians, Hispanics, and Afro-American IDDM populations.[6-9]

Study	Number of Subjects	DR2 %	DR3/X %	DR4/X %	DR3/4 %
Puerto Ricans	IDDM (n=36)	3	39	69	25
	CONT (n=43)	16	9	33	0
Canadians[6] (Solow et al. 1979)	IDDM (n=62)	18	39	44	NA
	CONT (n=54)	38	18	18	NA
N. European Heritage[7] (Rubinstein et al. 1990)	IDDM (n=123)	10	59	64	33
	CONT (n=83)	36	12	27	NA
Afro-American[8] (Rodey et al. 1979)	IDDM (n=40)	0	73	73	50
	CONT (n=67)	37	30	25	6
Mexican Americans[9] (Zeidler et al. 1982)	IDDM (n=85)	0	39	65	18
	CONT (n=209)	15	22	39	5
Hispanics (N.Y.)[7] (Rubinstein et al. 1990)	IDDM (n=49)	4	35	67	16
	CONT (n=39) (sibs)	26	23	31	NA

Table 3. DQ2 and DQ3 antigens in Puerto Rico.

Antigens	DQ2	DQ3	DQ2 & DQ3	DQ2 or DQ3	Neither
Diabetics (n=36)	22(61%)	31(86%)	17(47%)	35(97%)	1(3%)
Controls (n=43)	15(34%)	27(62%)	7(16%)	35(81%)	8(19%)

HLA Haplotypes

Frequency. Twenty-one families of the thirty-six diabetics have been studied. No single haplotype of class I and class II antigens predominated. Three haplotypes, A2, B18, DR3, DQ2; A24, B50, DR6, DQ1 and A23, B44, DR4, DQ3 were noted in two diabetics each. No Puerto Rican IDDM was found to have the haplotype A1, B8, DR3, DQ2 commonly reported in Baltic populations.[10]

Multiple Case Families. In this non-randomized population, four families had multiple cases of IDDM: two had an affected father and son and two had two affected sisters. Only one sister/sister pair was haplo-identical, but in this family, the parents' class II haplotype was also identical.

HLA-DQ Molecular Typing. By use of family data, it was possible to compare serologically-determined DQ alleles to molecular subtyping. Genotypes described by Baisch et al.[11] in Texan diabetics and controls are listed in Table 4 with the serological equivalent. The 24 haplotypes found in Texan IDDMs and controls are equivalent to 18 serological typings, 14 of which can be unequivocally assigned as either protective or non-protective. Only four serological typings are indeterminate, because the serological antigens would describe both a protective and a non-protective molecular genotype.

Table 4. Conversion of molecular genotype reported in Texas population to serological typing.[11]

Molecular Genotype		57th Asp	Serology Genotype	
Protective				
DQ 1.2	DQ 1.2	+,+	DQ 6	DQ 6
DQ 1.2	DQ 2	+,-	DQ 6	DQ 2
DQ 1.2	DQ 1.1	+,-	DQ 6	DQ 5
DQ 1.12	DQ 1.2	+,+	DQ 6	DQ 6
DQ 1.AZH	DQ 2	-,-	DQ 5	DQ 2
DQ 1.1	DQ 1.1	-,-	DQ 5	DQ 5
DQ 7	DQ 7	+,+	DQ 7	DQ 7
DQ 1.2	DQ 7	+,+	DQ 6	DQ 7
DQ 1.2	DQ 8	+,-	DQ 6	DQ 3(N7)
DQ 1.2	DQ 1.1	+,-	DQ 6	DQ 5
DQ 1.2	DQ 7	+,+	DQ 6	DQ 7
Neutral				
DQ 7	DQ 1.1	+,-	DQ 7	DQ 5
DQ 7	DQ 2	+,-	DQ 7	DQ 2
DQ 1.1	DQ 2	-,-	DQ 5	DQ 2
DQ 2	DQ 2	-,-	DQ 2	DQ 2
DQ 1.12	DQ 2	+,-	DQ 6	DQ 2
Significant risk				
DQ 7	DQ 8	+,-	DQ 7	DQ 3(N7)
DQ 1.1	DQ 1.AZH	-,-	DQ 5	DQ 5
DQ 1.AZH	DQ 8	-,-	DQ 5	DQ 3(N7)
DQ 4	DQ 8	+,-	DQ 4	DQ 3(N7)
DQ 4	DQ 2	+,-	DQ 4	DQ 2
DQ 1.1	DQ 8	-,-	DQ 5	DQ 3(N7)
DQ 8	DQ 8	-,-	DQ 3(N7)	DQ 3(N7)
DQ 2	DQ 8	-,-	DQ 2	DQ 3(N7)

Table 5 shows the assignment HLA-DQ genotypes for the Puerto Rican IDDMs to the specific serological typings. Table 6 compares the Puerto Rican data and the Texan data. In contrast to the Texan population where only 3% of the IDDM's genotypes were classified as protective, we found that 10% of the Puerto Ricans have DQ alleles that would be considered protective. Eighty percent of the Puerto Rican IDDMs DQ alleles are classifiable into non-protective haplotypes, while 10% are indeterminant. It is important to note that these loci are indeterminant because of the DQ1, not the DQ3.

Table 5. Classification of Puerto Rican IDDM by DQ serology.

Protective		N*	Nonprotective		N*
			Neutral		
DQ 6	DQ 6	0	DQ 7	DQ 5	0
DQ 6	DQ 5	0	DQ 7	DQ 2	1
DQ 6	DQ 7	1	DQ 5	DQ 2	0
DQ 6	DQ 3(N7)	1	DQ 2	DQ 2	0
DQ 7	DQ	0			
			Susceptible		
Indeterminant			DQ 7	DQ 3(N7)	3
DQ 6	DQ 2	2	DQ 5	DQ 3(N7)	2
DQ 5	DQ 2	0	DQ 4	DQ 2	0
DQ 5	DQ 5	0	DQ 3(N7)	DQ 3(N7)	2
DQ 5	DQ 7	0	DQ 2	DQ 3(N7)	8

*N=Number of Puerto Rican IDDM with haplotypes.

Table 6. Frequencies of HLA-DQ haplotypes in Puerto Rican and Texan IDDMs.

	PR(%)	Texan(%)
Protective	10	3
Non-Protective	80	97
Indeterminant	10	

CONCLUSIONS

These preliminary data indicate that IDDM in the Puerto Rican population is associated with some of the same HLA-DR/DQ markers identified in other populations. This implies that DQ haplotypes with substitutions of the aspartic acid at the 57th position of the HLA-DQ beta chain may have a role in the Puerto Rican population's susceptibility to IDDM. Because no data are available for the other Caribbean populations, it is not possible to conclude if it is the differences in the frequencies of the DQ beta chain alleles which cause the differences in the incidence of IDDM in the Caribbean, or whether this is due to other factors.

Familial serological data is insufficient to characterize the DQ6 alleles that appear to be more frequently associated with IDDM in our population than in Northern European and American populations. Two possibilities exist: IDDM populations could be expressing unusual or new DQ6 antigens or DQ antigens that are rarely expressed (<3%) in IDDM's in more Northern climates might be commonly found in IDDM's in Puerto Rico (20%). Molecular subtyping of our patients will be necessary to answer these questions. The high percentage of unusual alleles in the Puerto Rican IDDM population makes it an interesting ethnic group to study the role of susceptibility and protection of the HLA alleles and associated gene products.

Acknowledgements

This work supported by MBRS 5 SO6-GM08239-07 & RCMI 5 G12-RR03050-07.

REFERENCES

1. R. LaPorte, "Incidence of IDDM", Cambridge Symposia of Epidemiology of Diabetes, Cambridge University (1990).

2. WHO Multinational Project for Childhood Diabetes, "Epidemiology of IDDM in Iberian Heritage Populations", Newsletter, Jan. 1992.
3. A. Goday, C. Castell, R. Tresserras, G. LLoaveras and the Catalan Epidemiology Diabetes Study Group, Incidence of insulin-dependent diabetes in Catalonia, Spain, *Diabetes* 40:317a (1991).
4. A. Cartegena, Presidente de Sociedad de Diabeticos del Sur y Oeste de Puerto Rico, personal communication.
5. J. Cicciarelli, G. Ayoub, P. Terasaki, R. Billing, Improved HLA-DR typing of dialysis patients using monoclonal antibodies, *Transplantation* 33:558 (1982).
6. H. Solow, R. Hidalgo, D. Singal, Juvenile-onset diabetes: HLA-A, -B, -C and -DR allogantigens, *Diabetes* 28:1 (1979).
7. P. Rubinstein, M. Walker, N. Mollen, et al., No excess of DR*3/4 in Ashkenazi Jewish or Hispanic IDDM patients, *Diabetes* 39:1138 (1990).
8. G. Rodey, N. White, T. Frazer, et al., HLA-Dr specificities in American Black, juvenile-onset diabetics, *N Engl J Med.* 301:810 (1979).
9. A. Zeidler, D. Fraser, D. Kumar, et al., Histocompatibility antigens and immunoglobulin G insulin antibodies in Mexican-American insulin-dependent diabetic patients, *J Clin Endocrinol Metab.* 54:569 (1982).
10. E. Tuomilehto-Wolf, "HLA Family Studies", *in:* NATO Advanced Research Workshop, Standardization of Epidemiologic Studies of Host Susceptibility, Pittsburgh, 1992.
11. J. Baisch, T. Weeks, R. Giles, et al., Analysis of HLA-DQ genotypes and susceptibility in IDDM, *N Engl J Med.* 322:1837 (1990).

CASE-CONTROL MOLECULAR EPIDEMIOLOGY STUDIES: STANDARDS FOR THE WHO DIAMOND PROJECT

Janice S. Dorman,[1] Mirjana Kocova,[2] Leslie A. O'Leary,[1] Mary Kaye Kramer,[1] and Massimo Trucco[2]

[1]Department of Epidemiology
Graduate School of Public Health
University of Pittsburgh
Pittsburgh, PA 15261
USA

[2]Department of Pediatrics
Division of Immunogenetics
Children's Hospital of Pittsburgh
Pittsburgh, PA 15213
USA

INTRODUCTION

The development of standardized incidence registries has led to one of the most striking epidemiologic findings regarding IDDM. Unlike most other chronic childhood diseases, there is more than a 30-fold difference in incidence across populations, with the rates being very high in Scandinavian countries, such as Finland, but very low in the Oriental populations, such as Korea and Japan.[1] Among Caucasians, there is also marked variability in the annual incidence of IDDM. Rates are much lower in the central and southern parts of Europe, such as France and Italy, than in northern Europe.[2,3] In addition, major racial differences in incidence within populations have been reported, where Caucasians consistently have higher rates of IDDM than African Americans from the same geographically-defined area.[3]

With the development of IDDM registries through the WHO Multinational Project for Childhood Diabetes, known as the DIAMOND Project,[4] population-based resources are available for more in-depth studies of the etiology of disease. Research is now focusing on the identification of factors responsible for the global patterns of IDDM.[3-5] Geographic variation in incidence may be due to differences in the environment, such as viruses,[6,7] nutritional factors,[8-10] or other lifestyle habits.[11,12] However, major genetic variation also exists across racial groups and countries.[13]

Recent studies of the HLA-DQ beta chain have revealed that the presence of DNA sequences coding for an amino acid other than aspartate in the 57th position of the molecule (non-Asp-57) were highly associated with IDDM susceptibility.[14-16] In contrast, an aspartic acid allelic sequence in position 57 (Asp-57) appeared to be protective against the development of the disease. The presence of arginine in position 52 of the DQα chain is also significantly related to increased IDDM risk, particularly in the presence of non-Asp-57.[17-19] The heterodimers formed by the combination of DQB1*non-Asp-57 and DQA1*Arg-52 alleles appear to have the strongest quantitative diabetogenic effect. Although the presence of HLA-DQα and ß high risk alleles are unlikely to be the only genes associated with IDDM susceptibility,[20-22] they are currently the genetic markers which are most strongly related to the disease.

The identification of these striking HLA class II molecular associations to IDDM, and the well-documented geographic variability in the frequency of HLA antigens[13] provided the rationale for the DIAMOND Case-Control Molecular Epidemiology Sub-Project.[23] This multinational study will test the hypothesis that the worldwide patterns of IDDM incidence are due to population variability in the prevalence of HLA class II genes. The specific aims of the project are:

1) To determine the degree to which the distribution of HLA class II genes vary across racial groups and countries,
2) To evaluate the contribution of differences in the frequency of HLA class II genes to the geographic and racial variations in IDDM incidence.

These specific aims will be accomplished by studying IDDM cases and non-diabetic controls from high, moderate and low risk populations and in different ethnic groups.

OBJECTIVES OF CASE-CONTROL STUDIES

Case-control studies, which are also called case-referent or retrospective studies, are among the most frequently used epidemiologic approaches for investigating associations between potential risk factors and disease. In case-control studies, selected individuals who have the disease of interest (i.e, IDDM cases) and a comparable group of unaffected persons (i.e., non-diabetic controls) are evaluated to assess differences in exposure to potential risk factors. In this manner, new and previously evaluated etiologic hypotheses, including an examination of the interaction between potential risk factors, are tested. Case-control studies can "provide a valid, and reasonably precise estimate of the strength of at least one hypothesized cause-effect relationship".[24] They also permit the assessment and control of factors which could confound the evaluation of potential etiologic associations. The advantages of case-control studies are listed in Table 1.

Table 1. Advantages of case-control molecular studies.

1.	Permit evaluation of etiologic hypothesis, including interaction between potential risk factors
2.	Allow for the assessment and control of confounding variables
3.	Efficient design in terms of time and cost
4.	Ideal for rare diseases
5.	No recall bias when genetic markers are evaluated
6.	Provide risk factor-specific incidence rates when population-based

To achieve these objectives, case-control studies must address two major issues: *1) validity*, and *2) generalizability*. Valid case-control studies are not necessarily generalizable, nor are generalizable studies always valid. In fact, it is often difficult to achieve both validity and generalizability.

Validity refers to the selection of a group of well-defined cases and the identification of comparable controls for accurate and unbiased estimates of the strength of potential disease associations.[24] To achieve validity, one must restrict the case series to a homogeneous group from an etiologic perspective. This can be accomplished by identifying a precise set of inclusion criteria to define the disease under investigation. Similarly, criteria for the unaffected controls must be carefully identified to avoid possible selection biases that may influence exposure to the specific risk factors of interest. Validity, therefore, ensures that true differences between the group of cases and controls, should they exist, can be properly identified.

Generalizability refers to the selection of cases, using population-based methods, such that they reflect the total group of affected individuals.[24] Similarly, the controls must be chosen to be representative of the general population at-risk. For disorders that are heterogenous, generalizability will permit extrapolation to a larger group of cases with a similar etiology. When representative case groups are compared to population-based controls, one may assess the absolute, as well as relative differences in risk between exposed and non-exposed individuals. Thus, one may describe the risk of disease for various sub-groups of the population, characterized according to the presence or absence of specific risk factors. This is possible *only* when generalizable population-based case-control studies are conducted.

For the DIAMOND Case-Control Molecular Epidemiology Sub-Project, generalizability will permit the genetic characterization of a well-defined, randomly selected sample of IDDM cases identified from an incidence registry. Moreover, the selection of population-based controls will provide essential data regarding the frequencies of HLA class II alleles in the general population from the same geographically-defined area. This will permit estimates of genotype-specific incidence rates, as well as an evaluation of the contribution of host susceptibility to the incidence of IDDM within and between populations. Generalizability is, therefore, essential for testing the hypotheses proposed by DIAMOND.

To achieve validity and generalizability, the DIAMOND Case-Control Molecular Epidemiology Sub-Project requires a standardized research model which can be applied internationally. Standards are needed for: 1) the selection of IDDM cases, 2) the selection of non-diabetic controls, 3) HLA molecular typing, and 4) methods for data handling. Each of these standards will be described in detail.

Standards for the Selection of IDDM Cases

All IDDM cases will be identified from standardized population-based incidence registries from DIAMOND Participating Centers for the years 1979 to the present time. A uniform definition of IDDM is being employed for the DIAMOND Incidence Study and consists of the following: 1) diagnosed by a physician as diabetic, 2) placed on daily insulin injections before the 15th birthday, and 3) a resident in the area of registration at the time of first insulin administration. These criteria accurately identify an etiologically homogenous group of children with IDDM and have permitted standardization of the disease definition across countries. Case validation, which is also being performed as part of DIAMOND, will confirm that the identified cases actually have IDDM. Primary and secondary sources of case ascertainment are needed for determining the incidence of IDDM for ages 0-14 years. It is only through the accurate evaluation of the risk of disease that it will be possible to assess the contribution of the molecular markers to the global patterns of incidence.

Each racial or ethnic group will be evaluated separately. For each racial or ethnic group, a simple random sample of 150 IDDM cases (age 0-14 years at onset and diagnosed between 1979 to the present time) will be selected from the registry case listings using a computer program. Registry listings should be sorted by year of diagnosis and must contain the case's identification numbers, date of birth, date of diagnosis, sex and race/ethnic group. The use of a computer program to select a simple random sample will guarantee standardization of the sampling procedures and identify a representative sample of all registered IDDM cases in the area.

From the list of 150 selected IDDM cases, the first 100 cases who consent will be recruited for the study. The identification of 50 potential cases beyond those to be included in the investigation will compensate for the lowest acceptable participation rate of 67%. A sample size of 100 cases is required for precise estimates of the frequency of HLA class II susceptibility alleles. In areas where less than 100 cases are registered, a minimum of 50 IDDM cases will be recruited.

Standards for Selection of Non-Diabetic Controls

The best control group for a population-based case series is a random sample of the population at-risk. However, logistical problems were encountered when trying to obtain blood samples from non-diabetic children. Many parents were unwilling to allow their healthy young child to donate a blood sample for genetic marker studies. Therefore, by requiring that eligible controls be at least 15 years of age or older, we will minimize refusals and enhance the feasibility of the study design. This will also ensure that the identified controls have lived through the high-risk period (0-15 years), during which time IDDM is most likely to develop.

The demographic characteristics (current age and sex) of 100 non-diabetic controls will be determined using a computer program which simulates a simple random sample from the age and sex distribution of the population at-risk for the time period covered by each registry. Only individuals who were older than 15 years of age in 1992 will be considered eligible. Potential controls with the identified characteristics will be recruited from consecutive hospital admissions for injuries or from local populations. They will be considered ineligible if: 1) they did not reside in the area geographically defined by the registry during 1979 to the present time, or 2) they have any type of diabetes. Controls selected in this manner will be representative of the population at-risk for each country, rather than the group of individuals hospitalized for injuries. As it is unlikely that HLA class II genes are associated with injuries, the identification of controls from hospitalized injury cases would not be biased with regard to the risk factor under consideration.

Although the sampling procedure for the controls is different from that employed for the case selection, the control groups will be comparable across countries and will reflect the general population in each area, which is a primary consideration for the study. The difference in the age distribution of the cases and controls will also not confound the analyses because age is not related to HLA genotype. Moreover, these control populations will be far more representative and standardized than those typically employed by genetic marker studies.

Standards For HLA Molecular Typing: Technology Transfer And Quality Control

After obtaining informed consent, two-7 ml blood samples will be collected in a purple top EDTA tube for the genetic marker analyses and a 10 ml blood sample will be collected for storage of sera. Blood samples will be processed according to the standardized procedures described in the Methods of Operation Manual for the Project.[25] Mononuclear cells will be isolated from the purple top blood samples and at least one-half of the cells will be cryopreserved and stored at -70°C as a back-up cell source. Serum samples (~5ml) from the red top blood sample will be stored at -20°C. The remaining cell sample will be employed for DNA extraction and molecular analyses of HLA class II genes, as described in a separate chapter of the Proceedings from this NATO Workshop.

Standards for HLA class II molecular typing are essential for international collaborative studies and will be achieved by having all DNA analyses performed by a single laboratory for the centers participating in NIH grant (RO1-DK42316). However, these arrangements are not feasible for the DIAMOND Project in general. Thus, alternative approaches to standardization are needed through technology transfer.

At the present time, a number of researchers from various countries are being trained at the Coordinating Center Laboratory in Pittsburgh. They will be transferring the molecular technology utilized in Pittsburgh to their respective countries. As a result, several centralized, standardized core laboratories, directed by an individual who has received training at the Coordinating Center Laboratory, will be developing across the world. These laboratories will employ standard protocols and will be available to other DIAMOND Participating Centers for training and collaboration. Arrangements can be made with the directors of the laboratories for sample transfer and analyses. These core laboratories will utilize the identical methodology that is being developed by the Coordinating Center Laboratory and they will participate in quality control procedures, such as exchanging samples, assuring standardization across countries.

Standardization will also be achieved by having many laboratories employ identical, standardized procedures. Workshops similar to those conducted for the HLA serological typing will be held to ensure that all laboratories are correctly applying these techniques for the molecular analyses of class II genes. The organization of formal training courses in which core laboratories provide instruction and standardized materials to local facilities in various countries is currently being undertaken. The availability of the new HLA molecular testing kits, as described in the Proceedings of this NATO Workshop, will greatly facilitate the standardization of molecular analyses for both research and clinical practice. This will permit the transfer of technology to multiple centers across the world. Moreover, these efforts will provide a model for the development and standardization of molecular epidemiologic studies of other chronic diseases.

To further promote quality control for the HLA class II molecular analyses, all Participating Centers will first conduct a 'pilot' study. A batch of six specimens will be obtained from local staff or volunteers at the Participating Centers and processed *exactly* as outlined in the Methods of Operational Manual.[25] They will be analyzed at the Coordinating Center Laboratory or a standardized core laboratory according to the specified protocol. This procedure will permit pilot testing of all phases of specimen handling, including their collection, processing, shipping and analyses. Potential problems will likely be identified by this procedure and can be corrected prior to the collection and shipment of specimens from IDDM cases and non-diabetic controls. After the 'pilot' study is complete, quality control will be maintained by analyzing duplicate specimens and conducting sample exchange. In addition, analyses of specimens will be performed 'blind' to assure that accurate data will be obtained.

Standards for HLA Molecular Typing: Definition of Host Susceptibility

For any case-control study, the potential risk factors must be evaluated in a precise manner, permitting the distinction between individuals on the basis of their genetic susceptibility. This will also provide an approach for the evaluation of a possible dose-response. HLA antigens have been assessed in many previous investigations of IDDM, where the definition of susceptibility was based on the presence or absence of particular alleles that were more prevalent among affected cases than among healthy non-diabetic controls. With advances in molecular biology, studies of the association between HLA and IDDM are being conducted at the DNA level, using RFLP analyses and the polymerase chain reaction (PCR). However, these sophisticated methods have made it more difficult to define susceptibility to the disease, and, these definitions have not been standardized across studies.

At the present time, susceptibility to IDDM in case-control studies is generally characterized by the presence or absence of particular HLA-DQA1 and DQB1 alleles, alone or in combination. These alleles, which have been previously shown to be significantly associated with IDDM, are listed in Table 2, and are defined by the presence of arginine in position 52 of the DQα chain and the presence of an amino acid other than aspartate in position 57 of ß chain.

Table 2. HLA-DQA1 and DQB1 alleles associated with IDDM susceptibility.

DQA1 Alleles•	DQB1 Alleles•
DQA1 * 0301	DQB1 * 0201
DQA1 * 0401	DQB1 * 0302
DQA1 * 0501	DQB1 * 0501
DQA1 * 0601	DQB1 * 0502
	DQB1 * 0604

•DQA1 * Arg-52 Alleles
•DAB1 * non-Asp-57 Alleles

These two loci should be evaluated independently, as the frequencies of the alleles vary across populations. When the genes at these loci are considered in combination, one can determine the number of possible "diabetogenic" heterodimers that can be formed given an individual's genotype. For example, homozygotes for DQA1*Arg-52 and DQB1*non-Asp-57 alleles can make four 'diabetogenic' heterodimers, whereas individuals who are heterozygous at one of these loci can make only two heterodimers, double heterozygotes can produce one, and DQA1*non-Arg-52/DQB1*Asp-57 homozygotes can produce no such heterodimers. This definition of susceptibility can be employed to evaluate a potential dose-response in terms of IDDM susceptibility. As the DIAMOND Project progresses, other allelic combinations and/or haplotypes may provide better definitions of susceptibility to IDDM than the DQA1 and DQB1 alleles listed in Table 2. These new markers will also be employed to define IDDM susceptibility for the DIAMOND Sub-Project.

Standards for Methods of Data Handling

From each IDDM case and non-diabetic control, standard demographic information will also be obtained. This will include name, place of residence, date of birth, date of IDDM diagnosis, sex, race, socio-economic status, family history of diabetes, ethnic group of the study subjects, their parents, etc. In addition, a brief questionnaire pertaining to other autoimmune diseases will be collected. This information will be obtained by direct interview with the co-investigator and/or a member of his/her staff.

Programs for data entry and data analyses in dBASEIV are being developed at the DIAMOND Coordinating Center in Pittsburgh. They will be distributed to each Participating Center for their own data entry, statistical analyses and publication of the results. This will assure standardization of data analyses within and across populations. Comparative analyses will be conducted at the Coordinating Center, and manuscripts will be written according to the publication policy for the DIAMOND Project.

CONCLUSIONS

Case-control studies are efficient and cost-effective study designs; they are ideally suited for investigating the etiology of rare diseases. However, one potential problem with case-control studies is that of recall bias, as information about exposure to possible risk factors is typically collected after the onset of disease. This limitation does not exist with the evaluation of genetic markers as possible risk factors for disease. Molecular analyses of HLA class II alleles permits one to define IDDM susceptibility by evaluating genes which were operative prior to disease onset. Thus, one is able to establish true cause-effect genetic relationships from such case-control studies. Moreover, the standard definition of IDDM and the selection of population-based cases and controls has led to the development of valid and generalizable research model, which can be standardized and applied within and across populations. This will permit the evaluation of the relative risk associated with particular genotypes, as well as genotype-specific IDDM incidence rates.

To address concerns regarding feasibility, which is a key issue with multinational studies, a preliminary study of HLA-DQß alleles in five populations was conducted. Data from China,[26] Caucasians and African-Americans from Allegheny County, Norway,[27] and Sardinia revealed that HLA-DQß genotype distributions varied significantly among IDDM cases ($p<0.0001$) and non-diabetic controls ($p<0.001$).[28] In each of the five populations tested, odds ratios for non-Asp-57 homozygotes relative to Asp-57 homozygotes ranged from 14 to 111. Heterozygotes in Allegheny County and China also had a significantly increased IDDM risk relative to Asp-57 homozygotes. Genotype-specific incidence rates for Allegheny County, PA were estimated and suggestive of a dose-response in IDDM susceptibility. The absolute risk was highest for non-Asp-57 homozygotes (47.6/100,000 per year), intermediate for heterozygotes individuals (13.0/100,000 per year) and lowest for Asp-57 homozygotes (0.45/1000,000 per year). These genotype-specific rates were then used to predict the overall incidence rates in the remaining populations, which fell within the 95% confidence intervals of the actual rates established by incidence registries (Figure 1). The results of this study were consistent with the hypothesis that population variation in the distribution of non-Asp-57 alleles may explain much of the geographic variation in IDDM incidence and demonstrated the feasibility of the international research approach.[28]

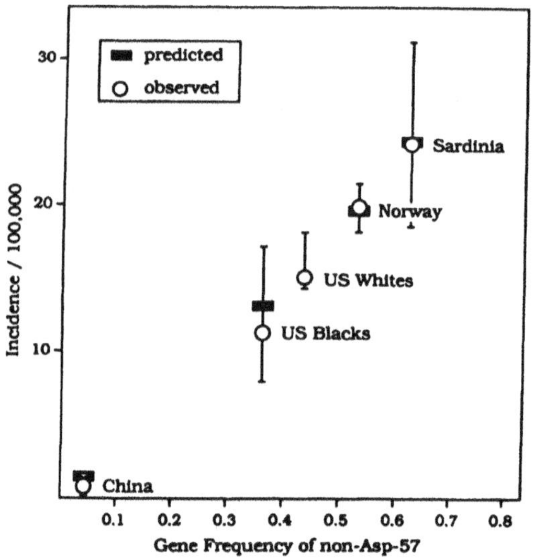

Figure 1. Association between actual and predicted age-adjusted annual IDDM incidence rates.

Additional published case-control studies of the associations between HLA-DQß polymorphisms and IDDM in African[29,30] and Japanese[31-33] populations have yielded less consistent results. These differences may be due to true population variation in host susceptibility, the influence of DQA1 and other genetic markers for the disease, or methodologic differences between studies. Both DQB1*non-Asp-57 and DQA1*Arg-52 alleles have been found to be independent markers of IDDM, with significantly higher frequencies among the IDDM cases compared to non-diabetic controls from the general population of Madrid.[18] However, when DQB1*non-Asp-57 and DQA1*Arg-52 were considered in combination, the absolute risk of developing IDDM was markedly increased (101/100,00 per year), supporting a synergistic interaction between these two markers. The attributable risk was estimated as 62%. Thus, approximately two-thirds of the incidence of IDDM in Madrid could be explained by these two high-risk alleles. This study provided evidence for a quantitative effect on IDDM susceptibility and emphasizes the importance of the complete molecule rather than a single chain alone in the etiology of the disease.

The foundation of the DIAMOND Case-Control Molecular Epidemiology Sub-Project is a funded international NIH grant (R01-DK42316). The study will evaluate the relationship between HLA class II molecular polymorphisms and IDDM in a standardized manner from population-based cohorts in high, intermediate and low-risk areas and in different racial groups. The Coordinating Center for the Project is in Pittsburgh, PA, USA. Countries representing areas of high disease incidence (Finland) were selected, as were those with lower IDDM incidence rates (USA, Spain, New Zealand, Croatia). This will permit comparisons among Caucasian populations which vary in IDDM incidence. Within the USA, Allegheny County, PA, and Jefferson County, AL, were selected because two different racial groups (Caucasians and African Americans) from the same geographically-defined area are represented in their registries. The risk of IDDM among African Americans from Alabama is considerably less than that for the African American population of Allegheny County. These USA populations will be contrasted with the African population from the Sudan. Finally, Mexico City and three Oriental populations (Japan, Korea and China) were chosen to represent the areas in the world where the incidence of IDDM is low.

Additional countries, not included in the NIH grant, will collaborate through the DIAMOND Project by employing the standardized methods outlined in this manuscript and the Methods of Operation Manual for the Project.[25] These include Iberian-heritage countries,[34] Asian-heritage population,[35] the Arab-heritage group[36] and numerous other groups in Africa, Europe and across the world. Thus, the populations to be included in the Project will vary widely with respect to ethnicity, race, geographic location and reflect the global and racial patterns of IDDM incidence. International comparative research based upon the epidemiologic and laboratory standards developed during this NATO Workshop will permit an accurate evaluation of the contribution of the genetic factors to the global patterns of IDDM incidence. Moreover, the transfer of DNA technology to other countries for purposes of standardizing the HLA molecular analyses will be one of the most important contributions of this DIAMOND Sub-Project, as molecular technology can be employed for diagnosis and research for many other infectious and chronic diseases.[37]

The research model for the proposed investigation, including the selection of the populations, the identification of the cases and controls, and the methods of analyses is molecular epidemiology.[23] With this research approach, it will be possible to determine the contribution of a specific risk factor, identified at the molecular level, to the overall incidence of IDDM within and across populations. A new interface between molecular biology and traditional epidemiologic methods is being developed to study the etiology of IDDM and to test the specific hypotheses of the proposed investigation for the DIAMOND Project.

REFERENCES

1. R.E. LaPorte, N. Tajima, H.K. Akerblom, N. Berlin, J. Brosseau, M. Christy, A.L. Drash, H. Fishbein, A. Green, R. Hamman, M. Harris, H. King, Z. Laron, A. Neil, Geographic differences in the risk of insulin-dependent diabetes mellitus: The importance of registries, *Diabetes Care* 8 (suppl. 1):101 (1985).
2. Diabetes Epidemiology Research International Group, Geographic patterns of childhood insulin-dependent diabetes mellitus, *Diabetes* 37:1113 (1988).
3. A. Green, E.A.M. Gale, C.C. Patterson, Incidence of childhood onset insulin-dependent diabetes mellitus: The European ACE study, *Lancet* 339:905 (1992).
4. WHO DIAMOND Project Group, WHO multinational project for childhood diabetes, *Diabetes Care* 13:1062 (1990).
5. J.S. Dorman, R.E. LaPorte, IDDM epidemiology - Next generation of research, *Diabetes Care* 13:184 (1990).
6. J.W. Yoon, U.R. Ray, Perspectives on the role of viruses in insulin-dependent diabetes, *Diabetes Care* 8 (suppl. 1):39 (1985).
7. J.E. Banatvala, G. Schernthaner, E. Schober, L.M. DeSilva, J. Bryant, M. Borkenstein, D. Brown, M.A. Menser, M. Silink, Coxsackie B, mumps, rubella, and cytomegalovirus specific IgM responses in patients with juvenile-onset insulin-dependent diabetes mellitus in Britain, Austria and Australia, *Lancet* i:1409 (1985).
8. T. Helgason, M.R. Jonasson, Evidence for a food additive as a cause of ketosis-prone diabetes, *Lancet* ii:716 (1981).

9. K. Borch-Johnsen, T. Mandrup-Poulsen, B. Zachau-Christiansen, G. Joner, M. Christy, K. Kastrup, J. Nerup, Relation between breast-feeding and incidence rates of insulin-dependent diabetes mellitus, *Lancet* ii:1083 (1984).
10. J.N. Kostraba, J.S. Dorman, R.E. LaPorte, F.W. Scott, A.R. Steenkiste, M. Gloninger, A.L. Drash, Early infant diet and risk of IDDM in Blacks and Whites, *Diabetes Care* 15:626 (1992).
11. N. Robinson, J.H. Fuller, Role of life events and difficulties in the onset of diabetes mellitus, *J Psychosom Res.* 29(6):583 (1985).
12. R.E. LaPorte, T.J. Orchard, L.H. Kuller, D.K. Wagener, A.L. Drash, B.B. Schneider, H.A. Fishbein, The Pittsburgh insulin-dependent diabetes mellitus registry: The relationship of insulin-dependent diabetes mellitus incidence to social class, *Am J Epidemiol.* 114:379 (1981).
13. J.L. Tiwari, P.I. Teraski. "HLA and Disease Associations," Springer-Verlag, New York (1985).
14. J.A. Todd, J.I. Belle, H.O. McDevitt, HLA-DQ gene contributes to susceptibility and resistance to insulin-dependent diabetes mellitus, *Nature* 329:599 (1988).
15. P.A. Morel, J.S. Dorman, J.A. Todd, H.O. McDevitt, M. Trucco, Aspartic acid at position 57 of the DQ chain protects against Type I diabetes: A family study, *Proc Natl Sci. USA* 85:1 (1988).
16. M. Trucco, J.S. Dorman, Immunogenetics of insulin-dependent diabetes mellitus in humans, *Crit Rev Immunol.*, 9:201 (1989).
17. I. Khalil, I. Deschamps, V. Lepage, R. Al-Daccak, L. Degos, J. Hors, Dose effect of Cis and trans-encoded HLA-DQα6 heterodimers in IDDM susceptibility, *Diabetes* 41:378 (1992).
18. M.D. Gutierrez-Lopez, S. Bertera, M.T. Chantres, C. Vavassori, J.S. Dorman, M. Trucco, M. Serrano-Rios, Susceptibility to Type I diabetes in Spanish patients correlated qualitatively with expression of HLA-DQA Arg-52 and HLA-DQB non-asp-57 alleles, *Diabetologia* 35:583 (1992).
19. M. Trucco, To be, or not to be Asp 57, that is the question, *Diabetes Care* 15:705 (1992).
20. G. Thomson, W.P. Robinson, M.K. Kuhner, S. Joe, M.J. MacDonald, J.L. Gottschall, J. Barbosa, S.S. Rich, J. Bertrams, M.P. Baur, J. Partanen, B.D. Tait, E. Schober, W.R. Mayr, J. Ludvigsson, B. Lindblom, N.R. Farid, C. Thompson, I. Deschamps, Genetic heterogeneity, modes of inheritance, and risk estimates for a joint study of caucasians with insulin-dependent diabetes mellitus, *Am J Hum Genet.* 43:799 (1988).
21. H.A. Erlich, T.L. Bugawan, S. Scharf, G.T. Nepom, B. Tait, R.L. Griffith, HLA-DQß sequence polymorphism and genetic susceptibility to IDDM, *Diabetes* 39:96 (1990).
22. L.L. Field, Non-HLA region genes in insulin-dependent mellitus, *in*: "Genetics of Diabetes," Part II, L.C. Harrison, B.D. Tait, eds., Bailliere Tindall, London, (1991).
23. J.S. Dorman, Genetic epidemiology of insulin-dependent diabetes mellitus: International comparisons using molecular genetics, *Ann Med.* 24:393 (1992).
24. N.E. Breslow, N.E. Day. "Statistical Cancer Research," Volume II, Oxford University Press, Oxford (1987).
25. J.S. Dorman, WHO DIAMOND Project, Genetic etiologic investigations - Case-control study, Methods of Operation, Part VIIIA, Version 3, June 1992.
26. M.Z. Bao, J.X. Wang, J.S. Dorman, M. Trucco, HLA-DQ non-asp-57 allele and incidence of diabetes in China and the USA, *Lancet* ii:497 (1989).

27. L.S. Ronningen, T.S. Halstensen, A. Spurkland, E. Thorsby, The amino acid at position 57 of the HLA-DQ beta chain and susceptibility to develop insulin-dependent diabetes mellitus, *Hum Immunol.* 26:215 (1989).
28. J.S. Dorman, R.E. LaPorte, R.A. Stone, M. Trucco, Worldwide differences in the incidence of Type I diabetes are associated with amino acid variation as position 57 of the HLA-DQβ chain, *Proc Natl Acad Sci. USA* 87:7370 (1990).
29. J.A. Todd, C. Mijovic, J. Fletcher, D. Jenkins, A.R. Bradwell, A.H. Barnett, Identification of susceptibility loci for insulin-dependent diabetes mellitus by trans-racial gene mapping, *Nature* 338:587 (1989).
30. C.H. Mijovic, D. Jenkins, K.H. Jacobs, M.A. Penny, J.A. Fletcher, A.H. Barnett, HLA-DQA1 and DQB1 alleles associated with genetic susceptibility to IDDM in a Black population, *Diabetes* 40:748 (1991).
31. T. Awata, T. Kuzuya, A. Matsuda, Y. Iwamoto, Y. Kanazawa, M. Okuyama, T. Juji, High frequency of aspartic acid at position 57 of HLA-DQ β-chain in Japanese IDDM patients and nondiabetic subjects, *Diabetes* 39:266 (1990).
32. J.A. Todd, Y. Fukui, T. Kitagawa, T. Sasazuki, The A3 allele of the HLA-DQA1 locus is associated with susceptibility to Type I diabetes in Japanese, *Proc Natl Acad Sci. USA* 87:1094 (1990).
33. K.S. Ronningen, H.A. Gjertsen, T. Iwe, A. Spurkland, T. Hansen, E. Thorsby, Particular HLA-DQα6 heterodimer associated with IDDM susceptibility in both DR4-DQ1w4 Japanese and DR4-DQw8/DRw8-DQw4 whites, *Diabetes* 40:759(1991).
34. Diabetes Epidemiology Research International Group, Evaluation of epidemiology and immunogenetics of IDDM in Spanish and Portuguese-heritage registries, A key to understanding the etiology of IDDM?, *Diabetes Care* 12:487 (1989).
35. N. Tajima, R.E. LaPorte, Incidence of IDDM outside Europe, *in:* "Pediatric Adolescent Endocrinology", Z. Laron, P. Tikva, eds., S. Karger, Basel, Switzerland (1992).
36. Diabetes Epidemiology Research International Study Group, Childhood diabetes in Arab countries, *Bull WHO* 68:231 (1990).
37. J.S. Dorman, C. Gorodezky, Molecular epidemiology, DNA technology transfer and disease prevention in developing countries (submitted).

GENETIC DETERMINANTS OF IDDM INCIDENCE: AN INTERNATIONAL STUDY OF HLA HAPLOTYPES IN MULTIPLEX PARENT - CHILD FAMILIES

Eva Tuomilehto-Wolf and Jaakko Tuomilehto

Department of Epidemiology and Health Promotion
National Public Health Institute
SF-00510 Helsinki
FINLAND

INTRODUCTION

The incidence of insulin-dependent diabetes mellitus (IDDM) varies enormously among populations.[1,2] The geographic variation in IDDM incidence is higher than that for coronary heart disease and for cancer. A 30-fold difference in IDDM risk has been demonstrated between the country with the highest incidence of IDDM in the world, Finland, compared with a country having one of the lowest reported incidence rates, Korea.[3]

Genetic factors are playing an important role in determining the differences in IDDM incidence, although environmental factors are without doubt necessary to convert the underlying genetic susceptibility into overt disease. Whereas the environmental factors causing diabetes have so far remained elusive, it is well established that the major genetic susceptibility to IDDM is conferred by a gene or by genes in the HLA region which is located on the short arm of chromosome 6 (band 6p21.3).

In recent years, there has been considerable interest in evaluating the relationship between HLA genetic factors and the incidence of IDDM. An early ecologic analysis of the relationship between the frequency of specific HLA class II antigens (DR4, DR3 and DR2) in the general population of 13 countries and IDDM incidence revealed no significant association.[4] In contrast, a recent case-control study using molecular methods to determine the frequency of class II alleles with an aspartic acid in position 57 of the HLA-DQ beta chain (Asp-57) versus an amino acid other than aspartic acid in this position (non-Asp-57) in five populations found that variation in the distribution of these alleles was a primary determinant of the global pattern of IDDM incidence.[5] However, recent studies from Madrid, Spain,[6,7] and Finland[8,9] have reported a higher frequency of non-Asp-57 alleles than would be predicted based on the incidence of the disease. An international comparative case-control molecular epidemiology study of IDDM is currently being conducted as part of the WHO Multinational Project for Childhood

Diabetes, known as DIAbetes MONDial or the DIAMOND Project, in order to determine the contribution of population differences in the prevalence of HLA class II alleles to the worldwide patterns of IDDM incidence.[10] This investigation is described in detail elsewhere in the Proceedings.

The focus of the current paper is an international comparative study of HLA haplotypes found in multiplex parent-child families and defined by conventional serology, which is also being carried out as part of the DIAMOND Project. This study will be the first to examine the variation in "IDDM susceptibility" HLA haplotype distribution and its association with IDDM incidence. As the HLA system is the most polymorphic system known in man and a very good anthropologic marker, we hope that this international family study will help to elucidate how much of the geographic variation of IDDM is due to major genetic differences among populations.

INTERNATIONAL STUDY OF MULTIPLEX PARENT-CHILD IDDM FAMILIES

The main objective of this international study is to evaluate the transmission of HLA haplotypes in multiplex families where a parent and a child have IDDM. Several populations with different ethnic backgrounds and varying IDDM incidence are participating. The aim is to determine whether the difference in frequencies between "diabetic" haplotypes and "non-diabetic" haplotypes across populations reflects the varying incidence of IDDM in these populations. "Diabetic" haplotypes are those which are found in IDDM patients. "Non-diabetic" haplotypes are defined as parental haplotypes which are not found in IDDM patients, nor in their parents or siblings with IDDM.[11]

By tissue typing all members of a family, HLA genotypes can be determined. HLA genotypes give the correct proportions at a given locus because homozygosity can be defined in related people. Each genotype is made up of two haplotypes, one inherited from the father and one from the mother. In other words, a haplotype is the genetic contribution of one parent coded by genes on a single chromosome.

Due to the dominant en bloc inheritance of HLA antigens, parent and child are by definition haplo-identical. The haplotype which the affected child has inherited from the affected parent - the "transmitted" haplotype - must carry the disease susceptibility gene (or genes) for IDDM except in the rare event of recombination.

The DIAMOND parent-child family study will use conventional serology to determine the alleles of the HLA-A, C, B, DR, DQ loci on haplotypes which were transmitted from a parent with IDDM to the diabetic child ("transmitted" haplotypes). It will be determined how much variation exists with regard to "transmitted" haplotypes, and how their frequencies and haplotype-specific risks vary between different populations. This international genetic HLA study of multiplex parent-child families will be carried out on a strict epidemiologic basis in order to avoid selection bias. A prerequisite for participation is a standardized population-based registry[3] and the use of a standard family history survey developed for the DIAMOND Project.[12] The above mentioned objectives are based on the following hypotheses.

Hypothesis 1: The entire HLA region from the A to the DR/DQ loci is important in determining susceptibility to IDDM.

Whole HLA haplotypes will be used as markers for the genetic susceptibility to IDDM under the hypothesis that the entire HLA region is important in IDDM, including all HLA loci from the A to the DR/DQ loci. (The DP locus is not taken into account, as there is a recombination hot spot between DQ and DP and no association has been found between IDDM and DPA1 or DPB1 alleles). This hypothesis is based on the assumption that no single allele coded by any single HLA locus is specific for IDDM, but that certain combinations of alleles coded by the various HLA loci (detectable as haplotypes) are specific for IDDM. It is also based on the fact that siblings who were defined as being HLA identical, i.e.,having inherited the same two parental haplotypes, defined by conventional HLA serology, are also identical on the molecular level. This implies that determination of restriction fragment length polymorphisms using HLA probes, oligonucleotide typing and amino acid sequencing of the HLA genes themselves are just more refined and detailed ways of defining specific small parts of the HLA haplotype.

Our hypothesis that whole HLA haplotypes are important for conferring susceptibility to IDDM stands in contrast to the theory put forward by Thorsby and his group.[13] They believe that the primary association is with the DQ molecules themselves and not with closely linked genes in the HLA region. Certain DQA1 and DQB1 genes in cis positions (same chromosome) or in trans positions (different chromosomes) explains the HLA association with IDDM according to their theory. The existence of trans-associated heterodimers was originally proposed by a French group.[14]

We want to test the hypothesis that from the class II alleles, not only the DQ but also the DRB1 plays a role in IDDM.[15] In addition, certain class I alleles also contribute to IDDM susceptibility. "IDDM susceptibility" haplotypes which only differ at the A locus, but which are identical for the C, B, DR, DQ loci have different haplotype-specific risks.[8]

The merits of conventional HLA serology versus molecular methods in defining HLA alleles have been discussed in detail elsewhere.[16] Serology remains the solid foundation of all molecular genetic studies of IDDM. Increasing the resolution of markers on HLA haplotypes through molecular methods may increase specificity, but may not increase the predictive power for determining IDDM risk. So far, all siblings who were defined by conventional serology as being HLA identical have also been identical on the DNA level. The crucial question to be answered is whether HLA identical siblings or monozygotic twins always form the same DQ trans-associated heterodimers.

Hypothesis 2: Only a limited number of "IDDM susceptibility" haplotypes exist.

Another hypothesis behind this international genetic study is that only some of the "diabetic" haplotypes (i.e.,both haplotypes found in IDDM patients) confer susceptibility to IDDM. Only a limited number of so called "IDDM susceptibility" haplotypes exist which occur in all populations. The specific antigen combinations which make up the "IDDM susceptibility" haplotypes have evolved due to environmental pressures in the past and are maintained together as certain fixed highly selected combinations. If this hypothesis is correct, then the genetic distance of "IDDM susceptibility" haplotypes between populations of varying genetic backgrounds will be smaller than the genetic distance of "non-diabetic" haplotypes between the same populations. Genetic distance will be calculated between different populations using the formula developed by Cavalli-Sforza.[17]

It has already been shown that linkage disequilibrium, which is a very important feature in the HLA region, is stronger on "IDDM susceptibility" haplotypes than on "non-diabetic" haplotypes.[18,19] This means that certain HLA alleles occur even more often together on haplotypes found in IDDM patients and their first degree family members than among unrelated people in the general population. Two-locus and three-locus delta values, which are a measure of linkage disequilibrium, will be calculated.[20]

Hypothesis 3: "Non-diabetic" haplotypes can be used as control haplotypes.

"Non-diabetic" haplotypes, as defined above, will be used as control haplotypes as they in principle reflect the genetic situation found in the background population.[11,18] They display a much greater variety of alleles and are much more diverse than "IDDM susceptibility" haplotypes. No non-diabetic "healthy" unrelated control subjects or control families from the same ethnic background are needed using this approach. This is a real advantage compared with any case-control study as selecting controls for HLA studies is especially difficult in genetically mixed and heterogeneous populations.

Hypothesis 4: The best way of assessing the relative importance of different "diabetic" haplotypes is the frequency of transmission seen in unselected multiplex parent-child families.

The problem in studying the underlying genetic susceptibility to IDDM is that each diabetic patient possesses two HLA haplotypes - two "diabetic" haplotypes. Whether they both play a role, or whether one of the two haplotypes is more important for conferring IDDM susceptibility needs to be assessed for the most frequent haplotype combinations. Most studies of multiplex families have concentrated on families with two or more affected siblings. This has led to many interesting observations. For instance, linkage between the HLA system and the putative susceptibility gene (or genes) for IDDM has clearly been demonstrated using this affected sibling-pair method.[21,22] As the majority of affected siblings were HLA identical in nearly all studies, the genetic impact of both haplotypes for IDDM has not been assessed. Studying multiplex parent-child families might be a good way to determine the relative importance of different "diabetic" HLA haplotypes. Therefore, one of the objectives of this international study is to ascertain whether the frequency of transmission of a particular "diabetic" haplotype from an IDDM parent to an IDDM child in randomly selected families reflects the relative importance of "diabetic" haplotypes.

The "transmitted" haplotype has to carry the putative disease susceptibility gene (or genes) for IDDM except in the rare event of recombination. The susceptibility gene (or genes) themselves cannot yet be determined directly not even by molecular methods. The HLA alleles are taken as markers for them as it is very likely that the IDDM susceptibility genes are in linkage disequilibrium with certain HLA alleles.

Hypothesis 5: Several haplotypes found in IDDM patients are "bystander"haplotypes.

From this international comparative family study we can also determine whether certain "diabetic" HLA haplotypes are "bystander" haplotypes, which are just found in diabetic patients as they are very common in the background population, but which do not play a role in conferring IDDM susceptibility. It will be important to identify these "bystander" haplotypes because they create "false positives". "Bystander" haplotypes are not taken into account in most molecular studies, and are included in all genetic models based on trans-complementation of DQ alleles.[13,14]

Evidence for the existence of "bystander" haplotypes comes from the first population-based HLA study of IDDM families carried out in Finland recently.[23,24] This genetic study of 757 unselected IDDM families indicated that in most cases one "IDDM susceptibility" haplotype was enough to determine the genetic predisposition for IDDM. Those diabetic patients who possessed two "IDDM susceptibility" haplotypes either had first degree relatives with IDDM, i.e.,belonged by definition to multiplex families, or had second or third degree relatives with IDDM, i.e. had a positive family history of IDDM (unpublished data).

Hypothesis 6: IDDM is a disease of Caucasian origin.

An additional aim is to test the hypothesis that IDDM is mainly a Caucasian disease and that the susceptibility to IDDM is basically inherited through a limited number of Caucasian haplotypes only. The hypothesis is that typical African haplotypes (for example A30, Cwx,B42, DR3-18, DQ4) or Mongoloid haplotypes (for example A2, Cw11, B46, DR9, DQ9) will not be found as "transmitted" haplotypes. IDDM will segregate together with typical Caucasian haplotypes which appeared in African American and Mongoloid populations through genetic admixture.

There is clearly epidemiologic evidence for this theory even though the data is scarce and not yet standardized. IDDM is rare in Africans.[25] In countries on the East coast of Africa where the prevalence of IDDM is slightly higher, for instance in Tanzania[26], Arab admixture very likely introduced Caucasian genes. In the West Indies, the ethnic admixture is well documented.[27] The best evidence comes from the African Americans where it is known from blood groups studies that a 20-30% admixture of Caucasian genes evolved over the 400 years in which Caucasians and originally West Africans lived on the same continent.[28,29] MacDonald found that the differences in IDDM frequency between African Americans and American Caucasians were of the same order as the estimated admixture of Caucasian genes.[30,31] Roseman and his group very elegantly showed that the estimated Caucasian admixture for African Americans with IDDM was greater than for African American controls.[32]

IDDM is also rare in Mongoloid populations.[1] There is tantalizing evidence that in the North of Japan in Hokkaido where some admixture with Portuguese travelers had occurred in the past the incidence of IDDM is higher.[33] It is also higher in that province of China where the Silk Road started (N. Tajima, personal communication).

Hypothesis 7: The "IDDM susceptibility" haplotype A2,Cw1,B56,DR4,DQ8 is only found in the Finnish population.

Another objective of this international study is to determine whether the new susceptibility haplotype consisting of A2,Cw1,B56,DR4,DQ8, which was recently found in Finnish diabetic children is also present in other populations.[11] This haplotype was the most common "transmitted" haplotype (19%) in multiplex parent-child families in a population-based nationwide study done in Finland (DiMe study).[24] The high frequency of the new A2,Cw1,B56,DR4,DQ8 haplotype as a "transmitted" haplotype indicates that it plays a major role in IDDM in Finland. This haplotype, which is the third most common haplotype found in Finnish diabetic children, provides partly a genetic explanation why Finland has the highest incidence of IDDM in the world, which during the DiMe study period was 35 per 100,000 per year.

This haplotype has not as yet been found in any other population. It was not seen in IDDM patients in the North of Sweden (50 families),[34] nor in Norway (94 unrelated patients),[35] where the IDDM incidence is also high, nor in IDDM patients from Estonia (34 families),[36] nor from Hungary (19 families) (personal communication). Estonians and Hungarians are linguistically and genetically related to the Finns. The incidence of IDDM in Estonians is approximately one third[37] and in Hungary approximately one fourth[38] of that in Finland.

Hypothesis 8: Is there a genetic reason why offspring of IDDM fathers are more likely to become diabetic than offspring of IDDM mothers?

The majority (77%) of all high-risk "IDDM susceptibility" haplotypes were transmitted from IDDM fathers in the Finnish DiMe study.[39] High-risk haplotypes were defined as being either DR3 or DR4 positive and consisting of B locus antigens which showed strong associations with IDDM in population studies, such as B8, B62, B56 and B60. Low-risk "IDDM susceptibility" haplotypes were defined as haplotypes which were either DR3 or DR4 positive and carried B locus antigens which did not show significant associations with IDDM in population studies, or were haplotypes which carried a DR antigen other than DR3 or DR4. The finding that IDDM fathers transmitted high-risk HLA haplotypes more frequently than IDDM mothers in the DiMe study might provide a genetic explanation for the finding of earlier epidemiologic studies that offspring of IDDM fathers were more likely to become diabetic than offspring of IDDM mothers.[40] The aim of the planned international study of multiplex parent-child families is to see whether IDDM fathers are more likely to transmit high-risk "IDDM susceptibility" haplotypes to their diabetic children in other populations, especially in those where the incidence of IDDM is low. The transmission frequency of high-risk haplotypes will be compared with that of low-risk haplotypes, assuming enough families can be collected where the "transmitted" haplotype is a low-risk one. In addition to testing these hypotheses, we might also be able to address the frequency of recombination in these IDDM families and the mode of inheritance of IDDM.

Recombination. Collecting HLA genetic data from multiplex parent-child families will also enable us to collect data on informative meiosis in order to determine the recombination rate in females and males in IDDM in different populations. This study might make it possible to assess potential recombination hotspots between the A and C loci, as postulated. DNA sequence data indicate that the physical distance between these loci is less than the observed recombination data suggest. A recombination hotspot between the A and C loci might have lead to the A locus diversity of certain "IDDM susceptibility" haplotypes (Cw3,B62,DR4,DQ8 and Cw7,B8,DR3,DQ2) whereas other "IDDM susceptibility" haplotypes (for instance the A2,Cw1,B56,DR4,DQ8 haplotype in the Finnish IDDM patients) remained preserved. It will be interesting to see whether the female to male recombination rate is the same in these multiplex parent-child families as in non-diabetic families. In non-diabetic families, the female to male recombination ratio in the A to DQ segment of chromosome 6 is approximately 1.3:1. In contrast, the recombination between the B locus and the locus coding for enzyme glyoxalase (GLO) is 3 times more frequent in females than males.

The distance between the A and the DP locus is about 2.5 - 3 centimorgans which is the distance giving 1% of recombination in the offspring. A recombination hotspot is found in the HLA region between the DQ and the DP locus. This means that recombination occurs more often between these loci than their physical distance would indicate. Therefore, no linkage between the DP locus and IDDM was observed in several studies and this is the reason why the DP locus will not be studied in this international study of multiplex parent-child families, as DP does not contribute to the determination of susceptibility to IDDM.

Mode of inheritance. This international study of multiplex parent-child families might also enable us to test the mode of inheritance of IDDM to a certain extent. The hypothesis based on the results of the Finnish DiMe study is that DR3 positive "IDDM susceptibility" haplotypes were transmitted from IDDM parents to the IDDM child with a frequency similar to the haplotype frequency found for DR3 in the diabetic probands. This would support a recessive inheritance of DR3 positive "IDDM susceptibility" haplotypes. In contrast, DR4 positive "IDDM susceptibility" haplotypes were transmitted from IDDM parents with a frequency greater than the haplotype frequency found for DR4 in IDDM probands (unpublished data). This finding would support a dominant inheritance of DR4 positive "IDDM susceptibility" haplotypes as suggested by MacDonald.[31,41]

PATIENTS AND THEIR FAMILIES

A prerequisite for including a family where a parent and a child (or children) have IDDM in this international HLA genetic family study is that one of the family members with IDDM had been diagnosed under the age of 15 years (either IDDM parent or IDDM child). This limited age at diagnosis is given through the design of the IDDM registries, and will be imposed so that the results from this international family study can be directly compared to the only truly population-based family study done so far, the Finnish DiMe Study.[24] It will be important to estimate from the standardized DIAMOND registries how many multiplex parent-child families were diagnosed during the year the proband was diagnosed. This way we can calculate how many of the parent-child families registered during a particular year have been HLA genotyped and how many were not typed.

METHOD OF SEROLOGIC HLA TYPING AND QUALITY CONTROL

Some centers participating in this international study have never done HLA genotyping before. For those centers which do not have tissue typing facilities, blood samples from all participating family members will be sent to the Tissue Typing Laboratory of the National Public Health Institute in Helsinki, Finland. From each family member, 10ml of blood in ACD (acid citrate-dextrose) is taken in a sterile way and the tubes sent as quickly as possible to Helsinki. This way many of the HLA typings of the multiplex parent child families will be done in Helsinki, Finland, using the same HLA antisera and the same test. Therefore, quality control will be centralized and easy to carry out using hidden controls like duplicates of the same serum and duplicates of the same blood sample using a different name.

For some centers, a small team of HLA serologists will travel to the participating center with microscope and all necessary equipment for tissue typing the parent-child families locally. Quality control will be the same as above, although additional logistical problems are expected.

For centers which do have tissue typing facilities and which will carry out the HLA genotyping of the multiplex parent-child families themselves, the following requirements are needed to assure comparability of the HLA typing results: 1) definition of the various HLA antigens according to the 10th International Histocompatibility Workshop held in New York, USA, in 1987, and 2) copies of the raw serologic results of the genotyped family sent to Helsinki. HLA typing results of all family members are in themselves the best way to check reproducibility and quality control. All serologic reactions present in the children have to be also present in one of the parents, as HLA antigens are inherited in a dominant way. Sera containing non-HLA cytotoxic antibodies are extremely rare and can be spotted easily as they do not segregate according to Mendelian rules.

Some of the participating centers have existing HLA genotype data of multiplex parent-child families. Copies of the HLA genotype results of the families will be sent to Helsinki.

DATA MANAGEMENT AND ANALYSIS

All HLA genetic data will be collected and stored at the Department of Epidemiology in Helsinki, Finland, where the incidence data for the DIAMOND study will be analyzed. The first priority will be to calculate what proportion of the registered multiplex parent-child families have actually been HLA genotyped. This is important as selecting only certain families for tissue typing inevitably leads to biased results.

Special computer programs have been developed to analyze HLA family data and to select "non-diabetic" parental haplotypes which will be used as control haplotypes as outlined above (Hypothesis 3).

Genetic distance will be calculated for "IDDM susceptibility" and "non-diabetic" haplotypes in the different populations.[17] Two-locus and three-locus delta values will be estimated in each population.[20]

REFERENCES

1. M. Rewers, R.E. LaPorte, H.O.M. King, J. Tuomilehto, for the Diabetes Epidemiology Researchers Study Group (DERI), Insulin-dependent diabetes mellitus in childhood: International patterns and trends, *World Health Stat Q.* 41:179 (1988).
2. Diabetes Epidemiology Research International Group, Geographic patterns of childhood insulin-dependent diabetes mellitus, *Diabetes* 37:1113 (1988).
3. The WHO Multinational Project for Childhood Diabetes Group, Familial insulin-dependent diabetes mellitus (IDDM) epidemiology: Standardization of data for the DIAMOND project, *Bull WHO.* 69:767 (1991).
4. R.E. LaPorte, N. Tajima, H. Åkerblom, et al., Geographic differences in the risk of insulin-dependent diabetes mellitus: The importance of registries, *Diabetes Care* 8:101 (1985).
5. J.S. Dorman, R.E. LaPorte, R.A. Stone, M. Trucco, Worldwide differences in the incidence of Type I diabetes are associated with amino acid variation at position 57 of the HLA-DQ beta chain, *Proc Natl Acad Sci. USA* 87:7370 (1990).
6. J.L. Vicario, J. Martinez-Laso, A. Corell, J.M. Martin-Villa, P. Morales, G. Lledo, O.G. Segurado, D. de Juan, A. Arnaiz-Villena, Comparison between HLA-DRB and DQ DNA sequences and classical serological markers as Type I (insulin-dependent) diabetes mellitus predictive risk markers in the Spanish population, *Diabetologia* 35:475 (1992).
7. M.D. Gutierrez-Lopez, S. Bertera, M.T. Chantres, C. Vavassori, J.S. Dorman, M. Trucco, M. Serrano-Rios, Susceptibility to Type I (insulin-dependent) diabetes mellitus in Spanish patients correlate quantitatively with expression of HLA-DQalpha Arg 52 and HLA-DQbeta non-Asp 57 alleles, *Diabetologia* 35:583 (1992).
8. P.J. Tienari, E. Tuomilehto-Wolf, J. Tuomilehto, L. Peltonen, and the DIME Study Group, HLA haplotypes in Type I (insulin-dependent) diabetes mellitus: Molecular analysis of the HLA-DQ locus, *Diabetologia* 35:254 (1992).
9. H. Reijonen, J. Ilonen, M. Knip, H.K. Åkerblom, HLA-DQB1 alleles and absence of Asp 57 as susceptibility factors of IDDM in Finland, *Diabetes* 40:1640 (1991).
10. J.S. Dorman, Genetic epidemiology of insulin-dependent diabetes mellitus: International comparisons using molecular genetics, *Ann Med.* 24:393 (1992).
11. E. Tuomilehto-Wolf, J. Tuomilehto, Z. Cepaitis, R. Lounamaa, and the DIME Study Group, A new susceptibility haplotype in Type I diabetes, *Lancet* 2:299 (1989).
12. WHO Multinational Project for Childhood Diabetes, "Familial IDDM Epidemiology: Collection and Analyses of Family History Information. Methods of Operation", Version I, June 1992.
13. E. Thorsby, K.S. Ronningen, Role of HLA genes in predisposition to develop insulin-dependent diabetes mellitus, *Ann Med.* 24:523 (1992).

14. I. Khalil, L. d'Auriol, M. Gobet, et al., A combination of HLA-DQbeta Asp57-negative and HLA-DQalpha Arg52 confers susceptibility to insulin-dependent diabetes mellitus, *J Clin Invest.* 85: 1315 (1990).
15. M.J. Sheehy, S.J. Scharf, J.R. Rowe, M.H. Nemw de Gimenez, L.M. Meske, H.A. Ehrlich, B.S. Nepom, A disease-susceptibility HLA haplotype is best defined by a combination of HLA-DR and -DQ alleles, *J Clin Invest.* 83:830 (1989).
16. E. Tuomilehto-Wolf, J. Tuomilehto, HLA antigens in insulin-dependent diabetes mellitus, *Ann Med.* 23:481 (1991).
17. L.L. Cavalli-Sforza, Human diversity, *Proc Int Congr Genet.* 3:405 (1969).
18. E. Wolf, K.M. Spencer, A.G. Cudworth, The genetic susceptibility to Type I (insulin-dependent) diabetes: Analysis of the HLA-DR association, *Diabetologia* 24:224 (1983).
19. E. Wolf, Different linkage disequilibrium in Type I diabetic and "healthy" HLA haplotypes, *in*: "Prognosis of Diabetes in Children", Z. Laron, M. Karp, eds., Pediatr Adolesc Endocrinol., Basel, Karger 18:270 (1988).
20. M.P. Baur, M. Neugebauer, E.D. Albert, Reference tables of two-locus haplotype frequencies for all MHC marker loci, *in*: "Histocompatibility Testing 1984," E.D. Albert, M.P. Baur, W.R. Mayr, eds., Springer-Verlag, Berlin 677 (1984).
21. A. Walker, A.G. Cudworth, Type I (insulin-dependent) diabetic multiplex families: Mode of genetic transmission, *Diabetes* 29:1036 (1980).
22. A. Svejgaard, P. Platz, L.P. Ryder, Insulin-dependent diabetes mellitus, *in*: "Histocompatibility Testing 1980," P.I. Terasaki, ed., Los Angeles: UCLA, Tissue Typing Laboratory 638 (1980).
23. E. Tuomilehto-Wolf, J. Tuomilehto, J. Lappivaara, A. Cepaitis, and the DiMe Study Group, Estimation of haplotype specific absolute risk in Type I diabetes in Finland, *Diabetologia* 33:44A (1990).
24. J. Tuomilehto, R. Lounamaa, E. Tuomilehto-Wolf, A. Reunanen, E. Virtala, E.A. Kaprio, H.K. Åkerblom, and the Childhood Diabetes in Finland (DiMe) Study Group, Epidemiology of childhood diabetes mellitus in Finland - Background of a nationwide study of Type I (insulin-dependent) diabetes mellitus, *Diabetologia* 35:70 (1992).
25. S.A. Ajgaonkar, Epidemiology of diabetes in the tropics - An overview, *in*: "Diabetes 1979," W.K. Waldhäusl, ed., Excerpta Medica, Amsterdam, 819 (1980).
26. D.G. McLarty, A.B.M. Swai, H.M. Kitange, G. Masuki, B.L. Mtinangi, P.M. Kilima, W.J. Makene, L.M. Chuwa, K.G.M.M. Alberti, Prevalence of diabetes and impaired glucose tolerance in rural Tanzania, *Lancet* 871 (1989).
27. P. Hugh-Jones, Diabetes in Jamaica, *Lancet* 2:891 (1955).
28. B. Glass, C.C. Li, The dynamics of racial intermixture: An analysis based on the American Negro, *Am J Hum Genet.* 5:1 (1953).
29. R. Chakraborty, M.I. Kamboh, M. Nwankwo, R. Ferrell, Caucasian genes in American Blacks: New data, *Am J Hum Genet.* 50:145 (1992).
30. M.J. MacDonald, Lower frequency of diabetes among hospitalized Negro than white children: Theoretical implications, *Acta Genet Med Gemellol.* 24:119 (1975).
31. M.J. MacDonald, Hypothesis: The frequencies of juvenile diabetes in American Blacks and Caucasians are consistent with dominant inheritance, *Diabetes* 29:110 (1980).

32. P.J. Reitnauer, R.C.P. Go, R.T. Acton, C.C. Murphy, B. Budowle, O. Barger, J.M. Roseman, Evidence for genetic admixture as a determinant in the occurrence of insulin-dependent diabetes mellitus in U.S. Blacks, *Diabetes* 31:532 (1982).
33. N. Matsuura, Hokkaido IDDM registry, Japan Diabetes Foundation Publication Series 1:50 (1991).
34. B. Hägglöf, G. Holmgren, G. Holmlund, B. Lindblom, B. Olaisen, P. Teisberg, Studies of HLA, Factor B(Bf), complement C2 and C4 haplotypes in Type I diabetic and control families from Northern Sweden, *Hum Hered.* 36:201 (1986).
35. K.S. Ronningen, G. Markussen, T. Iwe, E. Thorsby, An increased risk of insulin-dependent diabetes mellitus (IDDM) among HLA-DR4, DQw8/DRw8,DQw4 heterozygotes, *Hum Immunol.* 24:165 (1989).
36. E. Tuomilehto-Wolf, T. Podar, B. Adojaan, I. Kalits, J. Tuomilehto, Can the difference in incidence of Type I diabetes mellitus between Estonia and Finland be explained by genetic reasons? *Diabetologia* 34:A65 (1991).
37. J. Tuomilehto, T. Podar, A. Reunanen, I. Kalits, R. Lounamaa, E. Tuomilehto-Wolf, B. Adojaan, B. Neff, R. LaPorte, Comparison of the incidence of insulin-dependent diabetes mellitus in childhood between Finland and Estonia, 1980-1988, *Diabetes Care* 14:982 (1991).
38. G. Soltesz, L. Madacsy, D. Bekefi, I. Danko, The incidence of Type I (insulin-dependent) diabetes in Hungarian children 0-14 years of age, 1976-1985, *Diabetologia* 31:545A (1988).
39. E. Tuomilehto-Wolf, J. Tuomilehto, R. Lounamaa, and the DiMe Study Group, HLA haplotypes transmitted in Finnish multiplex families from parents with Type I diabetes to children with Type I diabetes, *Diabetologia* 32:550A (1989).
40. J.H. Warram, A.S. Krolewski, M.S. Gottlieb, C.R. Kahn, Differences in risk of insulin-dependent diabetes in offspring of diabetic mothers and diabetic fathers, *N Engl J Med.* 311:149 (1984).
41. M.J. MacDonald, Speculation on the evolution of insulin-dependent diabetes genes, *Metabolism* 37:1182 (1988).

CHALLENGES IN THE GENETIC EPIDEMIOLOGY OF INSULIN-DEPENDENT DIABETES MELLITUS

Anders Green

Genetic Epidemiology Research Unit
Odense University
Winsløwparken 15
DK-5000 Odense C
DENMARK

INTRODUCTION

The advances in human genetics and molecular biology during the last decades have provided new opportunities for a better understanding of the genetic and molecular susceptibility to many diseases of complex aetiology, including insulin-dependent diabetes mellitus (IDDM). This is illustrated by the almost exploding number of publications dealing with new genetic and molecular markers of IDDM. As a consequence of this development, a new front, the molecular epidemiology of IDDM, has emerged in the efforts to unravel the aetiology of IDDM.

This trend is probably going to be of essential importance in IDDM research since it will, with increasing accuracy, allow for the characterization of individual susceptibility to IDDM, thus making it theoretically possible to identify individuals at high risk of disease and therefore candidates for intervention.[1]

There is, however, another - and so far rather unrecognized - side of this otherwise glamorous story. It appears that currently available methods for analysis of information on genetic markers may not be sufficiently developed to deal with the data. Furthermore, basic principles and laws of human genetics are often ignored by researchers when designing, analysing and reporting research on molecular markers of IDDM.

In this paper, I will consider some aspects related to the design, analysis and interpretation of such research.

WHAT IS A GENETIC MARKER?

A marker of a given disease may be defined in general terms as some factor, the presence of which in an individual modifies the risk of developing the disease, as compared to the risk of an individual lacking the marker. A *genetic marker* is a phenotype (or genotype, or even haplotype) characterizing a specific locus (or, in the case of haplotype, a range of linked loci) that confers a particular risk of developing the disease.

The application of genetic markers in molecular epidemiological research involves three different issues:

1) the identification of a genetic marker;
2) the assessment of the effect upon disease risk conferred by the marker;
3) the exploration of the biological mechanisms by which the marker exerts its effect.

With this in mind, it is evident that research involving genetic markers must rely upon methodology from traditional epidemiology, population genetics, as well as molecular biology and related disciplines like immunology. Whereas aspects relating to the biological mechanisms of genetic markers are outside the scope of this review, the remaining parts will be commented upon below.

GENETIC MARKERS: IDENTIFICATION AND ASSESSMENT OF EFFECT

Case-Control Design

Genetic markers are often identified from the finding of an increased prevalence of a specific phenotype/genotype/haplotype among unrelated patients with a disease, as compared with the prevalence of this trait among unrelated controls. In analogy with the case-control design in traditional epidemiology, the effect of the marker is measured from the strength of the association between disease and marker, expressed in terms of the *odds ratio*. The odds ratio measures, under certain assumptions, the *relative risk*, i.e., the increase in disease susceptibility for a marker-positive individual as compared with a marker-negative individual.[2] Also, if the disease risk in the general population is known or assumed, *absolute* disease risks may be estimated for the various marker categories.[3]

Another concept from traditional epidemiology, the *aetiologic fraction* (attributable risk), has been introduced as a measure of the proportion of cases of the disease that have developed due to the effect of the genetic marker.[4]

Comments. The above-mentioned measures of association between a genetic marker and a disease are all empirical, i.e., they are free from assumptions regarding any underlying model of genetic susceptibility. The analogy with the design and analysis in traditional epidemiological case-control studies is obvious, and the interpretation of the measures of effect and association may seem straightforward at first glance. However, there are some particular aspects to consider, due to the fact that the disease determinant under study is a genetic factor, and not an environmental factor.

Consider first the issue of how to select the control individuals. In epidemiology, controls in case-control studies are often recruited from the population of disease-free individuals. In studies of associations between genetic markers and diseases, this is illustrated by the traditional routine of using healthy blood donors as controls. This practice may be questioned. The purpose of the control series is to provide information on the distribution of the disease determinant in the population base from which the cases develop. When the disease determinant is a genetic factor, it may be argued that the control series should represent the *total* population, i.e., including individuals with the disease under study, since, otherwise the estimation of gene frequencies (which is the essential role of the controls) may be biased. Another point is that blood donors are selected because they are healthy and will, therefore, most likely carry disease-associated genetic markers with lower frequencies than apply to the total population. The impact of this potential bias will depend upon the prevalence of the disease under study and may be of less significance for IDDM as opposed to, for example, non-insulin-dependent diabetes mellitus (NIDDM).

The distribution of the genetic marker in the population base, and thus the control series, should be in accordance with population genetics laws. This means that the proportion of individuals who are homozygous for the marker allele, homozygous for the alternative allele(s), and heterozygous should follow the Hardy-Weinberg equation:

$$q^2 + 2q(1-q) + (1-q)^2 = 1$$

where q^2 and $(1-q)^2$ denote the proportions of two homozygous classes and $2q(1-q)$ that of the heterozygous class, with q denoting the marker gene frequency. If there is departure from this equation, the marker system is not in population genetic equilibrium, or the control series may be flawed from biases due to selection or classification errors. Whatever the explanation, it makes interpretation of the results difficult if not impossible. In studies where it is possible to perform genotyping at the marker locus, a test of accepting the Hardy-Weinberg equation should always be performed, but this is virtually never done. For example, a recent study on the association between IDDM and the lack of aspartate at position 57 on the HLA-DQB locus[5] shows a significant departure from the Hardy-Weinberg equation; a close look reveals that there is a relative lack of heterozygotes among the controls. This may indicate a possible classification bias of the control individuals, e.g., that heterozygous may be mis-classified to either of the two homozygous classes when reading the laboratory results.

Finally, the validity of employing the concept of aetiologic fraction (attributable risk) directly to studies of genetic markers may be questioned. The point here is that in traditional epidemiology the definition and estimation procedures of the aetiologic fraction are developed on the basis of an underlying model of disease aetiology that invokes a (large) number of risk factors that in terms of occurrence and effect may be considered independent of each other.[6] This may not necessarily be the case for studies on genetic markers. For example, it is well known that among the HLA-DR specificities both DR3 and DR4 (and other markers) are positively associated with IDDM. Since both DR3 and DR4 belong to the same locus, an individual carrying, say, DR3 will not have the same chance of carrying DR4 as a random individual from the population because the DR3-positive individual has as a maximum only one "seat" left for a possible DR4 allele, whereas a random individual may (or may not) have two. Accordingly, DR3 and DR4 are not distributed independently of each other in the general population and the aetiological fraction for, say DR3, cannot immediately be interpreted without somehow considering the effect of DR4.

The comments above clearly indicate that a critical appraisal of the application of traditional epidemiological case-control technology in molecular epidemiology is greatly needed.

Family Studies

The classical approach to the identification of possible genetic models of disease susceptibility is represented by *segregation analysis* by which the occurrence of the disease trait among the relatives of probands is analysed. Segregation analysis requires carefully ascertained families, information on the descriptive epidemiology of the disease in the population base, and rather sophisticated methods of analysis. In the case of IDDM, attempts of segregation analysis, even with the employment of advanced methods of analysis, have been disappointingly inconclusive although a dominant transmission of disease susceptibility seems to be ruled out.[7]

The identification of genetic markers of IDDM has provided new possibilities since genetic markers allow for *linkage analysis*. The underlying principle is to study the co-segregation of disease and marker allele within families. In addition to the requirements of performing segregation analysis, the inclusion of genetic markers requires knowledge about the frequency of the marker gene(s) in the population base. Family studies on IDDM with the inclusion of genetic markers from the HLA-DR locus have been fairly conclusive, particularly when segregation and linkage parameters have been estimated *simultaneously*.[8,9,10] It has, for example, been possible to derive a simplified model of the genetic susceptibility to IDDM, involving a HLA-linked locus of recessive nature, another - and HLA-unrelated - familial component, as well as residual environmental determination.[9]

Also, from family studies with extended HLA-haplotypes as genetic markers, it has been possible to identify and rank specific marker haplotypes according to their effect on individual disease risk.[11] Nevertheless, a definite model of the genetic susceptibility to IDDM has not been established yet.

Comments. The family study approach is attractive in several respects. It makes it possible to obtain a tentative model of the genetic susceptibility to a disease, with possible estimation of the frequency of disease susceptibility gene(s), of the degree of dominance of the disease gene(s), and of the relative importance of genetic factors as opposed to environmental factors in the total disease aetiology. Furthermore, the inclusion of genetic markers improves the efficiency of such estimations in addition to providing information on the tentative genomic location of the disease susceptibility gene(s). Thus, information from family studies may supplement the information from the case-control studies of genetic markers, from which the impact of the individual marker(s) is assessed in terms of population and individual risks. However, several methodological shortcomings reduce the informative value of the family studies. Linkage and segregation analyses are performed on the basis of many assumptions. For example, in one of the most advanced computer-based programs of combined segregation and linkage analysis, the COMBIN model,[12] it is assumed that the effects of the marker alleles are independent of the effect of other genetic determinants, as well as of the effects of non-genetic (environmental) determinants. With our current knowledge about IDDM aetiology and pathogenesis, this may very well represent an over-simplification of the biological reality. Furthermore, such computer-based programs do not, in general, make it possible to account for interaction between genetic markers from different loci (e.g. cis- or trans-position of markers from extended haplotypes) which with respect to the molecular susceptibility to IDDM may appear to be an important feature.

Therefore, there is a need to improve the methodological basis of the analysis of genetic marker information from family studies. Also, to avoid potential selection biases, families for genetic studies must be ascertained from epidemiologically defined populations rather than from diabetes clinics.

CONCLUSIONS

There is no doubt that the identification of genetic markers of diseases with complex aetiology like IDDM represents an important tool in our attempts to unravel the causation of the disease. It is, however, evident from considerations like those presented here that there is an imbalance between, on one hand, the increasing amount of molecular biological information on genetic markers, and, on the other hand, the methodology available to handle this information in order to provide inference on disease causation. As new markers are identified with increasing speed, this imbalance is probably going to be further enhanced in the future.

Accordingly, one of the major future challenges in the genetic epidemiological research of IDDM will be to solve the methodological problems indicated here. This can only be achieved by close collaboration of epidemiologists, population geneticists and molecular biologists. International research networks like DIAMOND and Eurodiab ACE provide unique opportunities to meet these challenges.

REFERENCES

1. A.C. Tarn, J.M. Thomas, B.M. Dean, et al., Predicting insulin-dependent diabetes, *Lancet* i:845 (1988).
2. A. Green, The epidemiologic approach to studies of association between HLA and disease. I. The basic measures, concepts and estimation procedures, *Tissue Antigens* 19:245 (1982).
3. A. Green, The epidemiologic approach to studies of association between HLA and disease. II. Estimation of absolute risks, etiologic and preventive fraction, *Tissue Antigens* 19:259 (1982).
4. B.O. Bengtsson, G. Thomson, Measuring the strength of associations between HLA antigens and diseases, *Tissue Antigens* 18:356 (1981).
5. I. Vallet-Colom, C. Lévy-Marchal, D. Zarrouk, et al., HLA-DQß1 codon 57 and genetic susceptibility to Type I (insulin-dependent) diabetes mellitus in French children, *Diabetologia* 33:174 (1990).
6. O. Miettinen, Estimability and estimation in case-referent studies, *Am J Epidemiol.* 103:226 (1976).
7. A. Green, N.E. Morton, L. Iselius, et al., Genetic studies of insulin-dependent diabetes mellitus: Segregation and linkage analyses, *Tissue Antigens* 19:213 (1982).
8. N. Risch, Segregation analysis incorporating linkage markers. I. Single-locus models with an application to Type I diabetes, Amer *J Hum Genet.* 36:363 (1984).
9. A. Green, A. Svejgaard, P. Platz, et al., The genetic susceptibility to insulin-dependent diabetes mellitus (IDDM): Combined segregation and linkage analysis, *Genet Epidemiol.* 2:1 (1985).
10. L. Pascoe, S. Sherman, D. Wu, M. Becker, C. Falk, Combined segregation and linkage analysis for IDDM and HLA-DR under several ascertainment assumptions, *Genet Epidemiol.* 6:131 (1989).
11. E. Tuomilehto-Wolf, J. Tuomilehto, Z. Cepaitis, R. Lounamaa, and DIME Study Group, New susceptibility haplotype for Type I diabetes, *Lancet* ii:299 (1989).
12. C.J. MacLean, N.E. Morton, S. Yee, Combined analysis of genetic segregation and linkage under an oligogenic model, *Comp Biomed Res.* 17:471 (1984).

MOLECULAR STUDIES OF GENETIC DISTANCE AND ADMIXTURE

Robert E. Ferrell

Department of Human Genetics
Graduate School of Public Health
University of Pittsburgh
Pittsburgh, PA 15261
USA

INTRODUCTION

The striking variation in incidence or prevalence of common diseases in world populations immediately suggests the possibility of a biological link between ethnicity and disease. However, these populations are separated by language, geography and culture, as well as genes, and most of the common diseases of interest are known to result from the interaction of genetic susceptibility, often determined by genes at more than one genetic locus, and internal (age, gender, etc.) and external (diet, pathogen exposure, etc.) environments. The methods of genetic epidemiology are appropriate for the study of such diseases, which aggregate in families, but do not segregate in a simple mendelian manner. The quantitative assessment of genetic distance and population admixture is a tool to help elucidate the relationship between ethnicity and disease.

GENETIC DISTANCE

Genetic distance is the degree of gene differentiation between populations or species, as measured by some numerical quantity. At the molecular level, this would be the number of nucleotide substitutions in homologous genes between two populations or species. In practice, genetic distance usually refers to gene differences as measured by differences in allele frequencies at polymorphic genetic loci. The idea of measuring the biological distance between groups by some metric which could then be evaluated by rigorous statistical tests is an old one. One of these early measures, Czekanowski's distance[1] was originally proposed for the analysis of quantitative traits, but it can be applied to gene frequency data as well. Czekanowski's distance[1] is given by:

$$C_z = \frac{1}{m} \sum_{i=1}^{m} |x_i - y_i|$$

Standardization of Epidemiologic Studies of Host Susceptibility
Edited by J. Dorman, Plenum Press, New York, 1994

where m is the number of alleles and x_i and y_i are the frequencies of allele i in populations X and Y, respectively.

Numerous variations in methods for estimating genetic distance have been proposed, but they all express the simple idea summarized in Equation 1, that the biological difference between populations can be expressed as a function of quantitative traits or gene frequencies which characterize the populations. These various methods are summarized and discussed in Nei.[2]

GENETIC ADMIXTURE

If we can estimate the genetic distance between two populations using quantitative traits or gene frequencies, then we should be able to estimate the relative contribution of two populations to a third (hybrid) population using these same traits. Measures of gene admixture use this idea to estimate the proportion of genes that one population has received from another by gene flow.

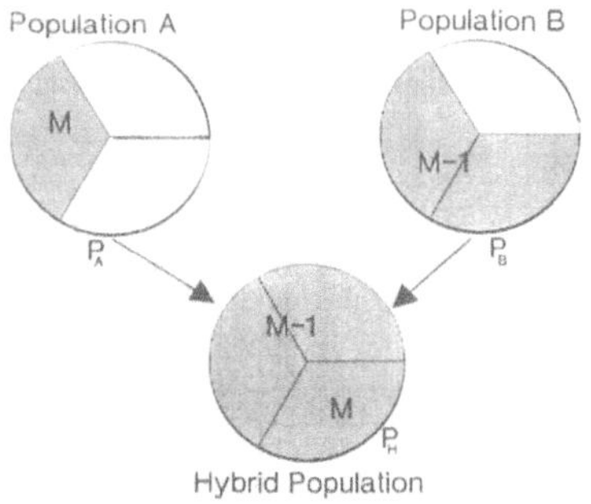

$$P_H = MP_A + (1-M)P_B = P_B + (MP_A - P_B)$$

$$M = \frac{P_H - P_B}{P_A - P_B}$$

Figure 1. Estimating the frequency of an allele (P_H) in a hybrid population and the proportion of admixture (M) in a dihybrid population.

In Figure 1, we have ancestral populations A and B which contribute some fraction M and (1-M), respectively, to a present-day hybrid population. Let P_A and P_B be the frequency of a gene in the parental populations A and B, and P_H be the frequency of the same gene in a present-day hybrid population. The fraction of genes M in the hybrid population that originated in donor population A can be estimated from Equation 2. While the principle expressed in Equation 2 is simple, the accurate estimation of admixture depends on a number of conditions which are seldom met in human population studies. The conditions that must be satisfied to accurately estimate M are:[3]

1. The exact ethnic composition of the parental populations are known and the frequencies of genes used to estimate M are known in the parental populations.

2. There has been no systematic change in gene frequencies in the parental populations since the formation of the hybrid population.

When these two assumptions can be satisfied, there are a number of ways of computing admixture estimates. These have been reviewed extensively by Chakraborty.[4] In addition, adequate random samples for the unbiased estimation of allele frequencies in the parental populations should be available.

In human populations, these two conditions can seldom be strictly satisfied. For example, in the mixture of European Caucasian and aboriginal American populations which gave rise to the Hispanic populations in the Americas, the Caucasian component was itself heterogeneous reflecting centuries of migration and conquest in Europe, and the American Indian contribution varied depending on the nature of the local population with whom the Europeans established sustained contact. In many instances, the native populations who may have contributed to the hybrid population did not survive as populations due to disease and exploitation that followed European contact. In many areas, exploitation of the local population was followed by the forced immigration of Blacks of mixed African origin who contributed to the gene pool of present day hybrid populations. While the mathematics of estimating admixture can be easily extended to trihybrid hybrids, the problem of estimating parental gene frequencies remains unsolved.

The most frequently cited source of a systematic change in gene frequency is selection. For example, the high frequency of the sickle cell hemoglobin gene (HbS) in West African populations is due to selection for resistance to falciparum malaria which is not endemic to North American. Among African populations in North America, the frequency of the HbS gene should decline due to negative selection acting on heterozygotes. An estimate of admixture based on the frequency of the HbS would overestimate the degree of Caucasian admixture in U.S. African Americans if parental allele frequencies of HbS were taken from a West African population under intense malarial selection. While these are only a few such documented cases of selection, population admixture events frequently involve the exposure of one or both parental populations to new pathogens, occasionally with disastrous results for the previously unexposed population. In cases where susceptibility to disease has a genetic component, such epidemics can significantly alter gene frequencies for genes involved in disease susceptibility, as well as for genes which happen to be linked to those that determine susceptibility. This may be particularly relevant to IDDM where genes in the HLA complex are known to be directly involved in determining susceptibility to a variety of diseases.

While the two primary assumptions for estimating admixture can seldom be precisely satisfied, there are conditions where they can be approximated and relative estimates of admixture obtained which are useful in understanding the relationship between population composition and disease prevalence. If it can be shown that modern, relatively unmixed populations living in the same geographical areas as the parental population do not differ from each other greatly in gene frequencies, then using average gene frequencies from these contemporary populations may be satisfactory for practical purposes. Where one of the parental populations was itself heterogeneous, but plausible representative populations can be sampled and their proportional contribution to the parental population estimated from historical data, an average frequency weighted by the proportional contribution of each contributing population can be estimated. This procedure has been used in several attempts to estimate European admixture in African Americans where some rough estimate of the geographical origin of African slaves is available from historical documents.[5] An alternative approach, which has been used extensively to estimate European admixture in American Indian tribes, is to consider only those alleles that are unique to European populations and assume that their presence in Amerindians is due solely to admixture. The problem with this approach is that these diagnostic alleles are often rare in the parental population of interest and large samples from the hybrid population are required to accurately estimate their frequency. Where such population specific alleles occur and their frequencies in the parental population in which they occur can be accurately estimated, they are generally more useful for estimating admixture than alleles that occur at moderate frequencies in both parental populations.

The assumption that there has been no systematic change in gene frequencies due to selection is difficult to evaluate. Despite numerous attempts to detect selection at specific genetic loci in man, convincing evidence of the action of selection at polymorphic genetic loci has seldom be presented. The outstanding exception has been the case of selection favoring the HbS allele in African populations due to malaria, and selection due to malaria for certain other hemoglobin alleles (HbE and α-thalassemia) in Asian and Pacific populations. Despite this, estimates of admixture based on the frequency of HbS in African Americans seem to be only moderately biased compared to estimates of admixture based on other loci in the same populations.[5] This suggests that estimates of admixture based on a reasonable number of randomly selected loci can provide estimates of admixture that are not greatly biased by the action of selection. This probably reflects the fact that selection generally operates over many generations while most of the admixture events of interest are relatively recent events.

ROLE OF MOLECULAR METHODS

The wealth of polymorphic genetic variation revealed by molecular techniques may offer a solution to the problems that have limited prior admixture studies. First, the number of polymorphic markers available for the characterization of populations and the estimation of admixture and genetic distance is for practical purposes infinite. While decades of effort revealed only a few hundred serological and biochemical polymorphisms, molecular techniques have discovered thousands of polymorphisms spanning the entire human genome. While the bulk of the reported allele frequency data for the DNA polymorphisms is from mixed Caucasian populations, an increasing number of population specific alleles are being identified. Perhaps the most informative of these are alleles which involve the specific insertions or deletions of a small number of nucleotides into the coding sequence of a gene without disrupting the expression of a functional gene product. Because such insertions must be precise in order to maintain the proper reading frame for translation of the gene product, such polymorphisms are almost certain to be rare and unique in origin. In cases where these alleles arose after the divergence of the major human racial groups, they are extremely useful for admixture analysis. Examples of such polymorphisms are the unique apolipoprotein B signal peptide insertion only reported in Amerindians and Mexican Americans[6] and the apolipoprotein A-IV-5 allele found only in populations of African origin.[7] Recent studies of the Alu family of repeated sequences suggest that recently inserted Alu family members may serve as diagnostic markers for specific human populations.[8]

Information on the number of DNA polymorphisms, their degree of informativeness in the population, and their population specificity is rapidly expanding. However, the limiting factor in application of molecular markers in worldwide genetic epidemiology studies is the paucity of information on the population distribution of this variation. The collection of epidemiologically well defined populations of diverse geographic and ethnic origin, as proposed by the WHO DIAMOND Project, provides a unique opportunity to expand our knowledge of the worldwide distribution of molecular variation. The banking of DNA from a representative sample of each of the participating populations could provide a baseline measure of genetic variation that would be useful in genetic epidemiology studies of any common disease. Selection of a representative sample from the general population to serve as controls in the proposal case-control study of IDDM will assure the usefulness of this sample for assessing the distribution of molecular variation in human populations and relating genetic variation to ethnic and disease variation.

APPLICATION: NIDDM AND AMERINDIAN ANCESTRY

The Pima Indians of the Southwest U.S. have among the highest recorded incidence and prevalence rates for NIDDM among world populations. A similarly high prevalence of NIDDM is reported for a number of other North American Indian tribes. Hispanic populations of the Southwest U.S., a hybrid population of mixed European and Amerindian heritage have an elevated incidence of NIDDM compared to non-Hispanic Caucasian populations residing in the same geographic area. This distribution of NIDDM suggests that the increased susceptibility to NIDDM observed in Hispanics in the Southwest U.S. is partially determined by genes of American Indian origin. A number of studies have demonstrated a significant correlation between the proportion of Amerindian ancestry, assayed by skin reflectance or by genetic marker typing, and the prevalence of NIDDM. In the Hispanic population of San Antonio, Texas, we have observed a clear correlation between proportion of Amerindian ancestry and NIDDM prevalence among subsets of the population defined by degree of adoption of non-Hispanic cultural practices.[9] These studies illustrate the usefulness of genetic admixture analysis in studies of the pattern of occurrence of common diseases in populations and suggest that hybrid populations may be a unique source of segregating families for the mapping of genes that are involved in determining susceptibility to diseases which are at high frequency in one of the parental populations.

RECOMMENDATIONS FOR THE WHO DIAMOND PROJECT

The precise population history of many of the populations of interest to the DIAMOND investigators is not known, and the parental populations which contributed to each are unknown or unavailable for sampling. Additionally, the admixed populations were not formed by a discrete admixture event but by continuous genetic exchange between populations. This precludes the direct estimation of admixture in terms of exact estimates of the proportion of gene in the contemporary sample contributed by specific parental populations. However, a genetic distance approach to estimating the degree of relationship between these populations may yield insight into the origin of the gene(s) which predispose to IDDM. The HLA region itself is a powerful genetic system for assessing genetic distance because of the large number of alleles at these loci. Because HLA is nonrandomly associated with IDDM, the estimation of genetic distance between study populations must rely on the frequency of HLA alleles in the controls drawn from each population. This requires that the controls be selected to be representative of the population from which the IDDM patients are drawn. Any geographic, cultural, ethnic or other bias in selection of cases must also be reflected in selection of controls. This requirement for comparability between cases and controls applies not only to between population comparisons (genetic distance), but also to within population comparisons (association studies).

If admixture plays a role in determining IDDM susceptibility in hybrid populations derived from the mixing of high incidence and low incidence parental populations, the genetic distance matrix for these populations should be similar to a matrix based on differences in IDDM prevalence between these populations. A significant difference between these two matrices would indicate that admixture has played little role in determining the present day distribution of IDDM, and would suggest the involvement of genes outside HLA complex or different environmental exposures as the major determinants of IDDM prevalence.

The precision of estimates of genetic distance depend on the number of loci tested and their degree of heterozygosity. While the HLA region is highly informative for one particular chromosomal region, the possibility that local selection has systematically altered HLA allele frequencies in different environments must be considered in between population comparisons using HLA region alleles. An independent estimate of genetic distance, based on loci not thought to be subject to local selection, is desirable. Thus, the banking of viable lymphocytes or DNA from cases and controls is recommended. There are a large number of highly variable loci which can be easily assayed by DNA amplification using the polymerase chain reaction which can be used to independently confirm the genetic distance relationships observed using the HLA loci. The power of using number or variable number of tandem repeat (VNTR) loci for estimating genetic distance is supported by the population studies of Deka et al.[10]

REFERENCES

1. J. Czekanowski, Zur differentialdiagnose der Neandertalgruppe, *Korrespondenzblatt Deutsch Ges Anthropol Ethnol Urgesch.* 40:44 (1909).
2. M. Nei, "Molecular Evolutionary Genetics," Columbia University Press: New York, (1987).
3. T.E. Reed, Caucasian genes in American Negroes, *Science* 165:762 (1969).
4. R. Chakraborty, Gene admixture in human populations: Models and prediction, *Yearbook Phys Anthropol.* 29:1 (1986).
5. J. Adams, R.H. Ward, Admixture studies and the detection of selection, *Science* 180:1147 (1973).
6. E. Boerwinkle, S.-H. Chen, S. Visvikis, et al., Signal peptide length variation in human apolipoprotein B gene, *Diabetes* 40:1539 (1992).
7. M.I. Kamboh, E.R. Williams, E. Law, et al., Molecular basis of an unique African variant of human apolipoprotein A-IV and its significance in lipid metabolism, *Genet Epidemiol.* 9:379 (1992).
8. M.A. Batzer, V.A. Gudi, J.C. Mena, et al., Amplification dynamics of human specific Alu family members, *Genomics* 19:3619 (1991).
9. R. Chakraborty, R.E. Ferrell, M.P. Stern, et al., Relationship of prevalence of non-insulin-dependent diabetes mellitus to Amerindian admixture in the Mexican-Americans of San Antonio, Texas, *Genet Epidemiol.* 3:435 (1986).
10. R. Deka, R. Chakraborty, R.E. Ferrell, A population genetic study of six VNTR loci in three ethnically defined populations, *Genomics* 11:83 (1991).

ADDITIONAL READINGS

1. J. Adams, R.H. Ward, Admixture studies and the detection of selection, *Science* 180:1137 (1973).
   Discussion of the impact of selection on admixture estimates and the potential use of single locus admixture estimates to detect the action of selection.
2. L.L. Cavalli-Sfarza, W.F. Bodmer, "The Genetics of Human Populations," Freeman, San Francisco, (1971).
   Discussion of genetic distance and population admixture and the assumptions made in estimating them.
3. R. Chakraborty, M.I. Kamboh, R.E. Ferrell, Unique alleles in admixed populations: A strategy for determining "hereditary" population differences of disease frequencies, *Ethnicity Disease* 1:245 (1991).
   R. Chakraborty, et al. discuss the use of population specific alleles to estimate admixture and their usefulness in the study of common disease.
4. R. Chakraborty, M.I. Kamboh, M. Nwankwo, R.E. Ferrell, Caucasian genes in American Blacks: New data, *Am J Hum Genet.* 50:145 (1992).

5. P.J. Byard, Quantitative genetics of human skin color, *Yearbook Phys. Anthropol.* 24:123 (1981).
Skin color as a quantitative trait for the study of hybrid population.
6. R. Chakraborty, Gene admixture in human populations: Models and predictions, *Yearbook Phys Anthropol.* 29:1 (1986).
Reviews the various methods of estimating admixture and discusses the advantages and disadvantages of various estimation methods. Describes some applications to specific problems.
7. B. Glass, C.C. Li, The dynamics of racial intermixing - An analysis based on the American Negro, *Am J Hum Genet.* 5:1 (1953).
First use of blood group data to estimate the proportion of European genes in American Blacks.
8. C.L. Hanis, R. Chakraborty, R.E. Ferrell, W.J. Schull, Individual admixture estimates: Disease associations and individual risk of diabetes and gallbladder disease among Mexican-Americans in Starr County, Texas, *Am J Phys Anthropol.* 70:433 (1986).
9. C.L. Hanis, D. Hewett-Emmett, T.K. Bertin, W.J. Schull, Origins of U.S. Hispanics. Implications for diabetes, *Diabetes Care* 14 (suppl. 3):618 (1991).
Hanis, et. al., discuss the genetic origins of Mexican-Americans and the usefulness of admixture as a tool in the study of common disease in Hispanic hybrid populations.
10. J. Hutchinson, Relationship between African admixture and blood pressure variation in the Caribbean, *Hum Hered.* 36:12 (1986).
11. J.E. Keil, H.A. Tyroler, S.H. Sandifer, E. Boyle, Jr., Hypertension: Effects of social class and racial admixture, *Am J Publ Hlth.* 67:634 (1977).
12. K.A. Korey, A critical appraisal of methods of measuring admixture, *Hum Biol.* 50:343 (1978).
Critical review of various methods of estimating admixture.
13. F.C. Lees, J.H. Relethford, Admixture estimation using skin reflectance data, *Am J Phys Anthropol.* 49:505 (1978).
Comparison of various methods of estimating admixture based on skin reflectance data.
14. C.J. MacLean, P.L. Workman, Genetic studies on hybrid populations. II. Estimation of the distribution of ancestry, *Ann Hum Genet.* 36:459 (1978).
15. C.J. MacLean, P.L. Workman, Genetic studies on hybrid populations. I. Individual estimates of ancestry and their relation to quantitative traits, *Ann Hum Genet.* 36:341 (1973).
MacLean and Workman present the theoretical basis of the estimation of admixture in an individual.
16. M. Nei, "Molecular Evolutionary Genetics," Columbia University Press, New York (1987).
Discussion of the concept of genetic distance and the various methods of estimating genetic distance.
17. T.E. Reed, Caucasian genes in American Negroes, *Science* 165:762 (1969).
18. T.E. Reed, Critical tests of the hypotheses for race mixture using Gm data on American Caucasians and Negroes, *Am J Hum Genet.* 21:71 (1969).
Reed discusses the problem of reliably estimating admixture and the biological reasons for studying admixture using American Blacks as an example.
19. T.E. Reed, Number of gene loci required for accurate estimation of ancestral population proportions is individual human hybrids, *Nature* 344:575 (1973).
Discusses the practical problems of estimating individual admixture.

20. P.J. Reitnauer, R.C.P. Go, R.C. Acton, C.C. Murphy, B. Budowle, B.O. Barger, J.M. Roseman, Evidence of genetic admixture as a determinant in the occurrence of insulin-dependent diabetes mellitus in U.S. Blacks, *Diabetes* 31:531 (1982).
21. J.H. Relethford, F.C. Lees, Admixture and skin color in the transplanted Tlaxcaltecan population of Saltillo, Mexico, *Am J Phys. Anthropol.* 56:259 (1981).
22. J.H. Relethford, M.P. Stern, S.P. Gaskill, H.P. Hazuda, Social class, admixture, and skin color variation in Mexican-Americans and Anglo-Americans living in San Antonio, Texas, *Am J Phys Anthropol.* 61:97 (1983).
    Relethford, et al.,discuss the use of skin color measures to estimate the degree of admixture in European-Amerindian hybrid populations.
23. E. Szathmary, T.E. Reed, Caucasian admixture in two Ojibwa Indian communities in Ontario, *Hum Hered.* 44:655 (1972).
24. E. Szathmary T.E. Reed, Calculation of the maximum amount of gene admixture in a hybrid population, *Am J Phys Anthropol.* 48:29 (1978).
    Szathmary and Reed discuss the use of unique alleles to estimate the proportion of European admixture in American Indians.

IMMUNE MARKERS IN POPULATION SURVEYS AND FAMILY STUDIES

Bernhard O. Boehm[1] and Werner A. Scherbaum[2]

[1]University of Ulm
Internal Medicine
Robert-Koch-Strasse 8
89070 Ulm/Donau
FEDERAL REPUBLIC OF GERMANY

[2]University of Leipzig
Internal Medicine
Johannisallee 32
04103 Leipzig
FEDERAL REPUBLIC OF GERMANY

INTRODUCTION

Insulin-dependent diabetes mellitus (IDDM) is now perceived as an immune-mediated disease which leads to a progressive loss of the insulin-producing ß-cells.[1] The observation that IDDM may be a feature of polyendocrine autoimmune disease gave the first hint of an autoimmune pathogenesis of IDDM. This suggestion was further substantiated by the observation of Gepts (1965) that the pancreatic islets in children with newly diagnosed IDDM were infiltrated with lymphocytes and plasma cells.[2]

IDDM appears to fulfill the criteria of Witebsky and Milgrom for an autoimmune disease.[3] The most important criteria: 1) the detection of antibodies or lymphocytes specifically reacting with a defined autologous antigen(s), 2) immunization of animals with an auto-antigen should result in an autoimmune disease corresponding to its original tissue distribution, and 3) experimental autoimmune disease should be inducible in healthy recipients by passive transfer of autoantibodies or auto-reactive lymphocytes.

Until 1974, antibodies reacting with the islets were the missing link in the autoimmune hypothesis of IDDM. Bottazzo and colleagues were the first to describe the presence of such antibodies in patients with polyendocrine autoimmunity including insulin-dependent diabetes mellitus.[4] These islet cell antibodies (ICA) have the capacity to bind to cytoplasmic components of normal human islets.[5] Presence of islet cell antibodies alone does not diagnose IDDM. ICA are important markers of the disease providing an important aid to diagnosis, and they may be used to identify individuals at-risk for future disease, especially those with a familial trait.[1,6,7]

When using a humoral marker in population surveys or in family studies, it is of considerable importance to identify antibodies that are related to the disease process.[8] In addition, it is also important to distinguish them from antibodies that have developed as a secondary phenomenon, e.g., as a result of ß-cell destruction, loss of insulin or the resulting chronic hyperglycemia. Markers reflecting an ongoing islet cell or ß-cell destruction should, therefore: 1) be islet cell specific, 2) be of high affinity for their corresponding antigen(s), 3) reveal a high prevalence at onset of IDDM, 4) have a high prevalence in the so-called prediabetic phase of insulitis and become positive in the early phases of autoimmunity.[9] These criteria are fulfilled with cytoplasmic ICA, insulin-autoantibodies (IAA) and antibodies to the 64K islet cell auto-antigen, recently identified as glutamic acid decarboxylase (GAD). Using the conventional assay system (indirect immunofluorescence), it could be demonstrated that ICA: 1) are most prevalent among newly diagnosed IDDM patients, 2) decrease in prevalence and titer with the duration of the disease, and 3) may be present several years before the clinical onset of IDDM.[9]

PREVALENCE OF ICA AMONG PATIENTS WITH IDDM

Table 1 summarizes studies which analyzed the prevalence of islet cell antibodies among patients with IDDM.[4,10-30] Prevalence rates, as described by the classical ICA assay, as well as markers of humoral islet cell immunity such as islet cell surface antibodies (ICSA) or antibody dependent cytotoxicity (ADCC), are also presented. What clearly emerges from these data is the heterogeneity in the observed prevalence of ICA, with proportions ranging from 50% to 80% in recent onset cases. The observed differences might be largely due to the fact that a standardized technology for the determination of ICA was not used in the studies. This lack of standardized methodology has made the interlaboratory comparison of ICA difficult and has complicated the assessment of the significance and the clinical value of ICA testing.[31,32]

The prevalence of ICA in IDDM patients has not been reported for all different ethnic groups. Most reports are from England, Scandinavia, several other European countries and the United States of America. In the populations and ethnic groups studied, ICA was a humoral marker which fulfills the above mentioned criteria for a "disease marker", i.e., specificity and high prevalence at onset of the disease.

PREVALENCE OF ICA AMONG NONDIABETIC HEALTHY CONTROLS

The prevalence rates of ICA in population surveys of healthy individuals are summarized in Table 2.[12-14,33-44] The difficulties concerning comparability due to different assay systems for the determination of ICA also holds for the population-based analysis.[31,32] The detection threshold for ICA varied considerably among different laboratories. Based on the scoring system for ICA in Juvenile Diabetes Federation Units (JDF-U), the old designations "ICA-positive" and "ICA-negative" are unsatisfactory and must be interpreted with caution when studies on the prevalence of ICA among nondiabetic healthy controls are compared.

Starting initially with rather small samples, representative data have been collected from different ethnic groups including population data from high-risk (i.e., Finland), medium-risk (i.e., Germany) and low-risk (i.e., Japan) populations (Table 2).

Table 1. Islet cell antibody prevalence among patients with IDDM.

Reference	Type	Prevalence	Study Population
Bottazzo et al, 1974[4]	ICA	14.5%	IDDM + polyendocrine
MacCuish et al, 1974[10]	ICA	4.8%	IDDM + endocrine dis.
Bottazzo et al, 1976[11]	ICA	38.0%	IDDM + endocrine dis.
		7.8%	endocrine dis.
Lendrum et al, 1975[12]	ICA	48.6%	IDDM-children, at onset
Lendrum et al, 1976[13]	ICA	22.8%	diabetic identical twins
Irvine et al, 1977[14]	ICA	22.0%	IDDM
		26.0%	IDDM + endocrine dis.
		7.0%	IDDM without endocrin.
Del Prete et al, 1977[15]	ICA	45.4%	IDDM recent onset
Kolb et al, 1980[16]	ICA	80.0%	IDDM, at onset
Doberson et al, 1980[17]	ICA	55.6%	IDDM
Madsbad et al, 1980[18]	ICA	36.3%	IDDM, young onset
	ICA	31.1%	IDDM, old onset
Bruining et al, 1984[19]	ICA	80.0%	IDDM, at onset
	ICA	39.4%	IDDM
Betterle et al, 1984[20]	ICA	59.0%	IDDM
Assa et al, 1985[21]	ICA	60.0%	IDDM, at onset
Maclaren et al, 1975[22]	ICSA	87.1%	IDDM
Lernmark et al, 1978[23]	ICSA	67.0%	IDDM, at onset
Doberson et al, 1980[17]	ICSA	63.9%	IDDM, at onset
Eisenbarth et al, 1981[24]	ICSA	56.0%	IDDM, at onset
Van DeVinkel et al 1982[25]	ICSA	93.8%	IDDM, at onset
Toguchi et al, 1985[26]	ICSA	35.0%	IDDM
Maruyama et al, 1984[27]	ADCC	28.6%	IDDM, ICSA positive
Lernmark et al, 1984[28]	ADCC	30.4%	IDDM
Michaelis et al, 1985[29]	ADCC	28.0%	IDDM
Smets et al, 1986[30]	ADCC	12.0%	IDDM

ICA: cytoplasmic islet cell antibodies; ICSA: islet cell surface antibodies; ADCC: antibody dependent cytotoxity.

Table 2. Islet cell antibody prevalence among nondiabetic, healthy individuals.

Reference	Type	Prevalence	Study Population
Lendrum et al, 1975[14]	ICA	1.0%	72 controls from UK
Irvine et al, 1977[16]	ICA	0.5%	434 controls from UK
Rodger et al, 1980[33]	ICA	0.4%	537 Australian children
	ICA	0.2%	3,229 adults
Notsu et al, 1983[34]	ICA	0.5%	1,125 rural Japanese
Betterle et al, 1984[22]	ICA	0.5%	611 controls from Italy
Riley/Maclaren, 1984[35]	ICA	0.4%	2,500 controls from US
Toguci et al, 1985[36]	ICSA	2.0%	110 controls from US
Maclaren et al 1985[37]	ICA	0.84%	5,003 U.S. children
Bergua et al, 1987[38]	ICA	0.35%	2,291 Spanish teenagers
Bruining et al, 1988[39]	ICA	0.24%	2,805 Dutch children
Landin-Olsen et al,[39a]	ICA (≤ 3 JDF-U)	3.0%	321 children from Sweden
Karjalainen, 1990[40]	ICA (≥3 JDF-U)	4.1%	1,212 children/young adults from Finland
Boehm et al, 1991[41]	ICA (≥5 JDF-U)	1.05%	4,287 children from Germany
Boehm et al, 1992[42]	ICA (≤ 5 JDF-U)	0.2%	2,000 adults from Germany
Levy-Marchal et al,[42a]	ICA (≤4.5 JDF-U)	1.8%	8,363 children from France

The upper part summarizes studies that have used the old designations "ICA-positive" versus "ICA-negative". The lower part of the table lists studies that have followed the standard protocol for ICA detection. The detection threshold is given in Juvenile Diabetes Federation-Units (JDF-U).

Four population studies have used the new standardization protocol for the determination of islet cell antibodies[39a,40-42,42a]. These studies came from Finland and Sweden, high-risk countries, as well as from France and Germany, both of which are medium risk countries. Based on the efforts of the Juvenile Diabetes Federation Workshops, standardized protocols for assays and titer measurements were established. By the use of standard-strength patient sera as a reference, major progress in standardization has been achieved.[32] Therefore, the studies from Sweden, Finland, France and Germany are comparable. The data from these studies suggest that differences in IDDM incidences are also reflected by the differences in ICA prevalences in healthy control subjects from the same regions. Differences in the prevalence of ICA between children and adults were observed in Germany with a five- to six-fold higher rate of ICA-positivity in children compared to the adults. Follow-up studies have shown that high-titer ICA are a predictive marker for future IDDM in randomly selected populations.[1,6,7]

THE ULM-FRANKFURT POPULATION STUDY

In a series of prospective family studies and studies of identical twins and triplets, which were discordant for the disease, the concept was established that the presence of ICA may reflect prediabetic insulitis.[1,6,7] In prospective family studies, it looked as if high titer ICA reflected a failure to maintain tolerance towards the ß-cells. The question remained whether the same holds true for randomly selected populations. To address these questions, the Ulm-Frankfurt population study was started in 1988.[41] A large population-based survey in the Alb-Dlonau county with a cross-sectional determination of ICA was performed in more than 4,000 school children. An ascertainment of approximately 50% was achieved in the region studied. HLA-risk markers, as well as metabolic tests (intravenous glucose tolerance test, IVGTT), supplemented the study protocol. The prevalence of ICA in this study was 1.05% (95% CI: 0.8 - 1.4). The highest prevalence was found in the 13 to 15 year age group, with 1.23% ICA positivity (95% CI: 0.7 - 1.9). The analysis of HLA-DRß and -DQß alleles revealed that the specificities found to be increased in IDDM patients were also associated with ICA positivity. HLA-DR3 and HLA-DR4 phenotypes and absence of aspartic acid at position 57 of the HLA-DQ ß-chain (non-Asp-57) were significantly increased in ICA positive compared with control subjects. The HLA risk markers also showed an association with the level of ICA and their persistence in follow-up studies. It can thus be concluded from this study that the combination of DQß determination and ICA testing provided a good predictor for Type I diabetes susceptibility.[41,42]

RELATIONSHIP BETWEEN ICA POSITIVITY AND DIABETES PREVALENCE

The incidence of IDDM varies dramatically across racial groups, with age-adjusted incidence rates of approximately 40/100,000 per year in Finland, but only 0.5/100,000 per year in China.[43] Prevalences of HLA risk markers have been suggested to explain, at least in part, some of the observed differences in incidence rates.[44] Since the results of the Ulm-Frankfurt population study suggested a strikingly similar immunogenetic basis for islet cell autoimmunity (i.e., presence of ICA) as well as the autoimmune disease (i.e., presence of IDDM), we have compared the incidence of the disease with the prevalence of ICA as reported from various populations (Figure 1). Although different assays were used with different sensitivities for ICA, diabetes incidence and ICA prevalence seemed to run in parallel. This result can be interpreted as indirect evidence for the hypothesis that islet cell autoimmunity and the autoimmune disease IDDM share a common genetic basis which is strongly related to the prevalence of HLA risk markers in each population studied.

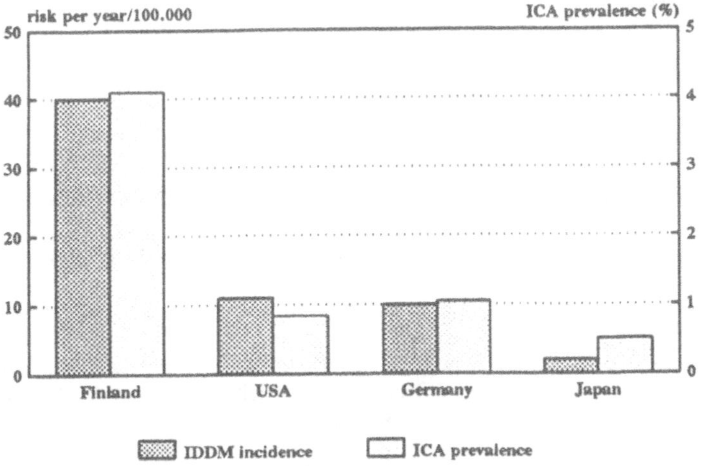

Figure 1. Diabetes incidence and prevalence of islet cell antibodies in selected populations.

NEW TESTS FOR THE DETECTION OF HUMORAL SIGNS OF ß-CELL AUTOIMMUNITY: A PERSPECTIVE

The characterization of islet cell antigens, which are the target of the autoimmune attack in IDDM, may provide new test systems for prediction. While the detection of ICA on pancreatic cryostat sections reflects a response to a plethora of antigens, the use of well defined targets may provide the advantage of more reproducible and easy to handle test systems, with more reliable results. Insulin and proinsulin, glutamate decarboxylase (GAD,64-k), glycolipids, carboxypeptidase H, and islet cell proteins of 37, 38, 52 and 150 k have been described as target antigens in IDDM.[45-52] Test systems which reveal binding to these defined antigens are at present in their infancy. Thus far, they have only been evaluated in selected populations. Most of these systems involve time consuming procedures, such as radioimmunoassay for the detection of insulin autoantibodies or an immunoprecipitation assay for 64k antibodies.[45,51,52]

Over the last few years, it became clear that the target antigens in organ specific autoimmune disease are not simply innocent bystanders. These antigens are molecules, related to critical functions of the cell, that have become the target of an immune-mediated attack. In thyroid autoimmunity, the main target antigens are thyroid peroxidase, an enzyme that directs the organification of iodine, and the TSH receptor.[53] In gastric autoimmunity, one target is the proton-pump of the gastric mucosa, and in autoimmune adrenalitis, steroid 17 alpha-hydroxylase has been described as a major antigen.[54]

A population-based approach will help in the near future to identify which (auto)-antigens and which epitopes from these antigens are relevant at the very beginning of an autoimmune attack against the ß-cells. The plethora of auto-antigens, bound by circulating immunoglobulins in sera from recent-onset diabetics, points to the urgent need to determine which are primary immunologic targets and which antigen(s) become involved secondarily.[55] Based on simple and easy to handle test systems, populations with high incidence rates for IDDM would be well suited for such screening programs. The identification of a common target antigen(s), which is (are) the primary immunologic target(s), could also provide the basis for new strategies to induce tolerance against the primary ß-cell antigen(s).

Acknowledgements

This work was supported by grants from the Deutsche Forschungsgemeinschaft: BO 829/6-1.

REFERENCES

1. G.S. Eisenbarth, Type I diabetes mellitus: A chronic autoimmune disease, *N Engl J Med.* 314:1360 (1986).
2. W. Gepts, Pathological anatomy of the pancreas in juvenile diabetes mellitus, *Diabetes* 14:619 (1965).
3. H.R. Smith, A.D. Steinberg, Autoimmunity - A perspective, *Ann Rev Immunol.* 1:175 (1983).
4. G.F. Bottazzo, A. Florin Christensen, D. Doniach, Islet cell antibodies in diabetes mellitus with autoimmune polyendocrine deficiencies, *Lancet* ii:1279 (1974).
5. W.A. Scherbaum, R. Mirakain, R. Pujol-Borell, B.M. Dean, G.F. Bottazzo, Immunocytochemistry in the study and the diagnosis of organ-specific autoimmune disease, *in*: "Immunocytochemistry," J.M. Polak, S. VanNoorden, eds., Wright, Bristol (1986).

6. A.C. Tarn, J.M. Thomas, B.M. Dean, D. Ingram, G. Schwarz, G. F. Bottazzo, E.A.M. Gale, Predicting insulin-dependent diabetes, *Lancet* i:845 (1988).
7. W.J. Riley, N.K. MacLaren, J. Krischer, R.P. Spillar, J.H. Silverstein, D.A. Schatz, S. Schwartz, J. Malone, S. Shah, C. Vadheim, J.I. Rotter, A prospective study of the development of diabetes in relatives of patients with insulin-dependent diabetes, *N Engl J Med.* 323:1167 (1990).
8. A. Lernmark, M. Markholst, S. Baekkeskov, Circulating signs of autoimmune islet disease, *in*: "The Pathology of the Endocrine Pancreas in Diabetes," P.J. Lefebvre, D.G. Piperleers, eds., Springer-Verlag, Berlin (1988).
9. E. Sigurdsson, S. Baekkeskov, The 64-kDa Beta-cell membrane autoantigen and other target molecules of humoral autoimmunity in insulin-dependent diabetes mellitus, *in*: "Human Diabetes: Genetic, Environmental and Autoimmune Etiology," S. Baekkeskov, B. Hansen, eds., Springer-Verlag, Berlin (1990).
10. A.C. MacCusih, E.W. Barnes, W.J. Irvine, L.J.P. Duncan, Antibodies to pancreatic islet cells in insulin-dependent diabetics with coexisting autoimmune disease, *Lancet* ii: 7896 (1974).
11. G.F. Bottazzo, R. Lendrum, Separate autoantibodies to human pancreatic glucagon and somatostatin cells, *Lancet* ii:873 (1976).
12. R. Lendrum, G. Walker, D.R. Gamble, Islet-cell antibodies in juvenile diabetes mellitus of recent onset, *Lancet* i:880 (1975).
13. R. Lendrum, R.G. Nelson, D.A. Pyke, Islet-cell, thyroid, and gastric autoantibodies in diabetic identical twins, *Br Med J.* 1:553 (1976).
14. W. J. Irvine, C.J. McCallum, R.S. Gray, Pancreatic islet cell antibodies in diabetes mellitus correlated with duration and type of diabetes, coexisting autoimmune disease and HLA type, *Diabetes* 26:431 (1984).
15. G.F. Del Prete, C. Betterle, D. Padovan, G. Erle, A. Toffolo, G. Bersani, Incidence and significance of islet-cell autoantibodies in different types of diabetes mellitus, *Diabetes* 26:909 (1977).
16. H. Kolb, M. Stroheker, J. Biener, "Analysis of the persistence of islet cell antibodies and islet cell type specific antibodies in Type I diabetic children," A. Pinchera, D. Doniach, G.F. Fenzi, eds., Academic Press, London (1980).
17. M.J. Doberson, J.E. Scharf, F. Ginsberg-Fellner, A.L. Notkins, Cytotoxic autoantibodies to beta cells in the serum of patients with insulin-dependent diabetes mellitus, *N Engl J Med.* 303:1493 (1980).
18. S. Madsbad, G.F. Bottazzo, A. Cudworth, B. Dean, O. Faber, C. Binder, Islet cell antibodies and ß-cell function in insulin-dependent diabetics, *Diabetologia* 18:45 (1980).
19. G.J. Bruining, J. Molenaar, C.W. Cuk, Clinical time course and characteristics of islet cell cytoplasmic antibodies in childhood diabetes, *Diabetologia* 26:24 (1984).
20. C. Betterle, F. Zanette, B. Pedini, F. Presotto, L.B. Rapp, C.M. Monciotti, F. Rigon, Clinical and subclinical organ-specific autoimmune manifestations in Type I (insulin-dependent) diabetic patients and their first-degree relatives, *Diabetologia* 26:431 (1984).
21. S. Assa, M. Karp, B. Erster, Cytoplasmic islet cell antibodies in Type I diabetics in Israel and their first-degree relatives, *Isr J Med Sci.* 21:727 (1985).
22. N.K. Maclaren, S.W. Huang, J. Fogh, Antibody to cultured human insulinoma cells in insulin-dependent diabetes, *Lancet* i:997 (1975).
23. A. Lernmark, Z.R. Freedman, C. Hofmann, A.H. Rubenstein, D.F. Steiner, R.L. Jackson, R.J. Winter, H.S. Traisman, Islet cell-surface antibodies in juvenile diabetes mellitus, *N Engl J Med.* 299:375 (1978).
24. G.S. Eisenbarth, M.A. Morris, R.M. Scearce, Cytotoxic antibodies to cloned rat islet cells in serum of patients with diabetes mellitus, *J Clin Invest.* 67:403 (1981).

25. M. Van DeVinkel, G. Smets, W. Gepts, D. Piperleers, Islet cell surface antibodies from insulin-dependent diabetics bind specifically to pancreatic ß-cells, *J Clin Invest.* 70:41 (1982).
26. Y. Toguchi, F. Ginsberg-Fellner, P. Rubinstein, Cytotoxic islet cell surface antibodies (ICSA) in patients with Type I diabetes and their first-degree relatives, *Diabetes* 34:855 (1985).
27. T. Maruyama, I. Takei, I. Matsuba, A. Tsuruoka, M. Taniyama, Y. Ikeda, K. Kataoka, A. Abe, S. Matsuki, Cell-mediated cytotoxic islet cell surface antibodies to human pancreatic beta cells, *Diabetologia* 26:30 (1984).
28. A. Lernmark, Cell mediated immunity in (insulin-dependent) diabetes: Update 84, *in*: "Immunology of Diabetes," D. Andreani, U Di Mario, K.F. Federlin, L.G. Heding, eds., Kimpton Medical Publications, London (1984).
29. D. Michaelis, I. Rjasanowski, B. Hehmke, K.D. Kohnert, W. Hildmann, H. Keilacker, K.V. Richter, Lack of strong association between residual human C-peptide secretion and islet cell antibodies, complement-dependent antibody-mediated cytotoxicity, and HLA antigens in newly diagnosed Type I diabetics, *Biomed Biochim Acta.* 44:99 (1985).
30. G. Smets, W. Gepts, Selective complement fixation by pancreatic ß-cells after binding to islet cell surface antibodies, *J Endocrinol Invest.* 9:43 (1986).
31. B. Marner, A. Lernmark, J. Nerup, J.L. Molenaar, C.W. Tuk, G.J. Bruining, Analysis of islet cell antibodies on frozen sections of human pancreas, *Diabetologia* 25:93 (1983).
32. H. Gleichmann, G.F. Bottazzo, Progress towards standardization of cytoplasmic islet cell antibody assay, *Diabetes* 36:578 (1987).
33. B. Rodger, S. Whittingham, F.I.R. Martin, B.R. Hawkins, R.L. Dawkins, T.A. Welborn, A population survey of pancreatic islet cell antibodies, *Clin Exp Immunol.* 39:125 (1980).
34. K. Notsu, Y. Goto, T. Sakurami, A population study of pancreatic islet cell antibodies and antithyroid antibodies, *Tohoku J Exp Med.* 141:261 (1983).
35. W. Riley, N. Maclaren, Islet cell antibodies are seldom transient (Letter), *Lancet* i:1352 (1984).
36. Y. Toguci, Ginsberg-Fellner, P. Rubinstein, Cytotoxic islet cell surface antibodies (ICSA) in patients with Type I diabetes and their first-degree relatives, *Diabetes* 34:855 (1985).
37. N.K. MacLaren, G. Horn, R.P. Spillar, H. Barbour, D. Harrison, J. Duncan, Islet cell antibodies (ICA) in U.S. school children (Abstract), *Diabetes* 34 (suppl. 1): 84A (1985).
38. M. Bergua, J. Sole, G. Marion, M.C. Perez, A. Recasens, J. Fernandez, R. Casamitjana, R. Gomis, Prevalence of islet cell antibodies, insulin antibodies and hyperglycemia in 2291 schoolchildren, *Diabetologia* 30:724 (1987).
39. G.J. Bruining, D.E. Grobbee, G.J. Scheffer, A.M. de Bruyn, J.L. Molenaar, A. Hofman, H.A. Bruining, H.A. Valkenburg, Ten-year follow-up study of islet cell antibodies and childhood diabetes mellitus, *Lancet* i:1100 (1989).
39a. M. Landin-Olsson, A. Karlsson, G. Dahlquist, Islet cell and other organ-specific autoantibodies in all children developing Type I diabetes mellitus in Sweden during one year and in matched control children, *Diabetologia* 32:387 (1989).
40. J.K. Karjalainen, Islet cell antibodies as predictive markers for IDDM in children with high background incidence of disease, *Diabetes* 39:1144 (1990).
41. B.O. Boehm, B. Manfras, J. Seibler, K. Schöffling, M. Glück, G. Holzberger, S. Seidl, P. Kühnl, M. Trucco, W.A. Scherbaum, Epidemiology and immunogenetic background of islet cell antibody-positive nondiabetic schoolchildren: Ulm-Frankfurt population study, *Diabetes* 40:1435 (1991).

42. B.O. Boehm, J. Sessler, M. Glück, B.J. Manfras, H. Thomas, K. Schmidt, W.A. Rudert, K.H. Usadel, M. Trucco, W.A. Scherbaum, The level and the persistence of islet cell antibodies in healthy schoolchildren are associated with polymorphic residues of the HLA-DQß chain, *Disease Markers* 9:273 (1991).
42a. C. Levy-Marchal, J. Tichet, I. Fajardy, F. Dubois, P. Czernichow, Prevalence of islet cell antibodies in French children from background population, *in:* "Epidemiology and Etiology of Insulin-dependent Diabetes in the Young," C. Levy-Marchal, P. Czernichow, eds., Karger Basel, Munich (1992).
43. Diabetes Epidemiology Research International Group, Geographic patterns of childhood insulin-dependent diabetes mellitus, *Diabetes* 37:113 (1988).
44. J.S. Dorman, R.E. LaPorte, R.A. Stone, M. Trucco, Worldwide differences in the incidence of Type I diabetes are associated with amino acid variation 57 of the HLA-DQß chain, *Proc Natl Acad Sci. USA* 87:7370 (1991).
45. S. Baekkeskov, M. Landin, J.K. Kristensen, S. Srikanta, G.J. Briuning, T. Mandrup-Poulsen, C. deBeaufort, J.S. Soeldner, G. S Eisenbarth, F. Lindgren, G. Sundquist, A. Lernmark, Antibodies to a 64,000 MW human islet cell antigen precede the clinical onset of insulin-dependent diabetes, *J Clin Invest.* 79:926 (1987).
46. D.L. Kaufmann, H.G. Erlander, M. Clare-Salzer, M.A. Atkinson, N.K. Maclaren, A.J. Tobin, Autoimmunity to two forms of glutamate decarboxylase in insulin-dependent diabetes mellitus, *J Clin Invest.* 89:283 (1992).
47. D.U. Rabin, S.M. Pleasic, R. Palmer-Crocker, J.A. Shapiro, Cloning and expression of IDDM-specific human autoantigens, *Diabetes* 41:183 (1992).
48. R.C. Nayak, M.A.K. Omar, A. Rabizadeh, S. Srikanta, G.S. Eisenbarth, "Cytoplasmic" islet cell antibodies: Evidence that the target antigen is a sialoglycocojugate, *Diabetes* 34:617 (1985).
49. B.O. Boehm, C. Rosak, U. Sasse, W.A Scherbaum, K. Schöffling, Western blot analysis of islet cell antibodies in Type I diabetes (Abstract), *Diabetes* 35 (suppl. 1):520 (1986).
50. W. Richter, J. Endl, T.H. Eierman, M. Brandt, R. Kientsch-Engel, C. Thivolet, H. Jungfer, W. A. Scherbaum, Human monoclonal islet cell antibodies from a patient with insulin-dependent diabetes mellitus reveal glutamate decarboxylase as the target antigen, *Proc Natl Acad Sci. USA* 89:8467 (1992).
51. J. Seibler, R. Hering, W. Richter, M. Glück, N. Yassin, R.G. Bretzel, B.O. Boehm, K. Federlin, W.A. Scherbaum, Antibodies to the Mr 64,000 (64K) protein in islet cell antibody positive non-diabetic individuals indicate high risk for impaired Beta-cell function, *Diabetologia* 35:550 (1992).
52. P. Vardi, S.A. Dib, M. Tuttleman, J.E. Conelly, M. Grinberg, A. Radizabeh, W.J. Riely, N.K. MacLaren, G.S. Eisenbarth, J.S. Soeldner, Competitive insulin autoantibody assay: Perspective evaluation of subjects at high risk for development of Type I diabetes mellitus, *Diabetes* 36:1286 (1987).
52a. J. Tuomilehto, P. Zimmet, I.R. Mackay, P. Koskela, G. Vidgren, L. Toivanen, E. Tuomilehto-Wolf, K. Kohtamäki, J. Stengard, M.J. Rowwley, Antibodies to glutamic acid decarboxylase as predictors of insulin-dependent diabetes mellitus before clinical onset of disease, *Lancet* 343:1383 (1994).
53. B. Czarnocka, J. Ruf, M. Ferrand, S. Lissitzky, P. Carayon, Interaction of highly purified thyroid peroxidase with anti-microsomal antibodies in autoimmune thyroid disease, *J Endocrinol Invest.* 9:135 (1986).
54. K. Krohn, R. Uibo, E. Aavik, P. Peterson, K. Savilahti, Identification by molecular cloning of an autoantigen associated with Addison's disease as steroid 17a-hydroxylase, *Lancet* 339:770 (1992).

GEOGRAPHIC DIFFERENCES IN ANTIBODIES TO GLUTAMIC ACID DECARBOXYLASE IN INSULIN-DEPENDENT DIABETES MELLITUS

Paul Zimmet,[1] Ian R. Mackay,[2] Merrill Rowley,[2] William Knowles,[3] Susan W. Serjeantson[4]

[1]International Diabetes Institute
260 Kooyong Road
Caulfield South
Victoria 3162
AUSTRALIA

[2]Centre for Molecular Biology and Medicine
Monash University
Clayton
Victoria 3168
AUSTRALIA

[3]Molecular Diagnostics Inc
West Haven, Connecticut 06516
USA

[4]John Curtin School of Medicine
Canberra
AUSTRALIA

INTRODUCTION

Major ethnic and geographic differences exist for insulin-dependent diabetes mellitus (IDDM) around the world.[1] The disease is most common in people of European descent (Europids) with the highest reported incidence in Scandinavian countries,[1,2] particularly in Finland where the incidence exceeds 30 cases/annum per 100,000 population.[3] The lowest incidence rates are in Asia, notably Korea at 0.5 cases/annum[1] with low rates in Japan[3] and China as well.[4] These geographic and ethnic differences are possibly due to variations in environmental risk determinants, although genetic differences may also be operative.[5]

The geographic and ethnic variations in the incidence of IDDM provide a unique

Standardization of Epidemiologic Studies of Host Susceptibility
Edited by J. Dorman, Plenum Press, New York, 1994

opportunity to obtain a better understanding of the genetic and environmental determinants of IDDM. This knowledge is an essential step to providing a scientific basis for intervention towards the primary prevention of IDDM. Just as important is a sensitive and specific screening test for presymptomatic diagnosis for subjects at high risk of developing IDDM.

Autoantibodies to various islet cell antigens are demonstrable in individuals with IDDM including islet cell cytoplasm,[6] insulin,[7] and the 64k islet cell antigen[8] recently identified as the enzyme glutamic acid decarboxylase (GAD),[9] as well as at least nine other candidate autoantibodies.[10] Are the ethnic differences in IDDM reflected in different prevalence of these autoantibodies in IDDM patients? If so, what does this say about the pathogenesis and natural history of IDDM? The islet cell cytoplasmic autoantibodies (ICA) have been the most extensively studied epidemiologically.[11]

GEOGRAPHIC DIFFERENCES IN ICA AND IAA

There seems to be a wide variation between the high frequency in IDDM in Finns[12] and the low frequency in IDDM in African Blacks in Tanzania.[13] However, these differences may be due to the individuals selected and whether diabetes was classified precisely, as well as variations in the duration of diabetes. Thus, in the Tanzanian Blacks, there was a low frequency of ICA positivity, but the authors' classification of the diabetics[13] raises some doubts as to whether these subjects actually had IDDM, or at least the form of IDDM seen in Europids. This is supported by the fact that in African Blacks from Sudan, a reported prevalence of ICA was 63% in a group of 46 children with a mean age 11.6 years (range 7-16) and a mean duration of diagnosed diabetes of 10 days (range 0-25).[14] Another major problem in comparing studies of ICA prevalence from around the world has been that the ICA assay is both technically demanding and hard to standardize.[15] It is difficult to study large numbers of subjects. Whereas ICA may be found in 60-80% of newly-diagnosed IDDM patients at diagnosis, the rate falls to about 30% after 5 years.[16] Hence, for comparisons to be valid, they must be made between groups of IDDM subjects representative of the population they are taken from, with similar duration of diabetes, and using a similar and well-standardized assay.

A summary of studies in Europid, Japanese and Black populations is provided in Table 1. A more detailed review of the epidemiology of ICA can be found elsewhere.[11]

IAAs have been detected in newly diagnosed IDDM subjects before insulin therapy was commenced.[17] Data on prevalence suffers from limitations similar to those described above for ICA. In several studies of Europid IDDM subjects, the prevalence of IAAs at diagnosis ranged between 37-53%.[17]

ANTIBODIES TO GAD

As a result of the current interest in GAD, particularly as the possible major autoantigen for IDDM and the suggestion that assays for antibodies to GAD hold the greatest promise as a predictive marker for the later development of IDDM,[15] this report will review our current work in this area.

Until recently, the major barrier to carrying out population studies on rates of antibodies to GAD has been that the assay for the antibodies has, like that for ICA, been technically difficult and has required a source of pancreatic islet tissue.[18,19] We have recently developed a sensitive radioimmunoprecipitation assay for antibodies to GAD.[20] The experimental format of the assay is shown in Figure 1.

Table 1. ICA prevalence in IDDM in different ethnic groups with duration of diabetes $\leq$ 3 months.*

Ethnic Group	No. studied	Prevalence (%)	Duration (months)	First author
Europids	33	67	0	Lernmark 1981
Europids	25	80	0-3	Bruining 1984
Europids	172	60	0-5	Assa 1985
Europids	60	75	0	Karjailanen 1986
Europids	74	66	0	Ludvigsson 1988
Japanese	96	50	< 12	Shinhyo 1983
Japanese	24	83	0-1	Kobayashi 1984
Blacks (USA)	100	36	**	Neufeld 1980
Blacks (USA)	132	70	0	Cavender 1990
Blacks (Sudan)	46	63	0-1	Elamin 1992

*Differences in methods for ICA, including cut-off for positivity, and differences in selection of cases in some of these studies render the results not directly comparable.
**Duration 0.03-25 years.

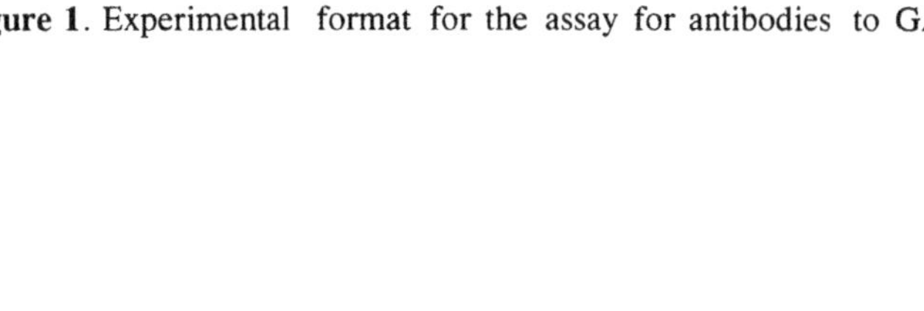

Figure 1. Experimental format for the assay for antibodies to GAD.

Using this assay,[20] we described a 69% frequency of positive tests for antibodies to GAD in Europid (Australian) cases of short duration (0-12 months) IDDM, and 59% in cases with duration from 1-40 years, compared with a frequency of 5% in non-insulin-dependent diabetes mellitus (NIDDM or Type II) and 0% in healthy controls (Table 2).

Table 2. Antibodies to GAD in Europid controls and individuals with IDDM and NIDDM.

Status	Number of Subjects	Age (years) mean ± SD	Antibodies to GAD* No.	Prevalence (%)
Controls**	55	34 ± 12	0	0
NIDDM	20	62 ± 12	1	5
IDDM - recent onset (<1 yr)	26	30 ± 14	18	69
IDDM - longer duration (3-42 yr)	54	39 ± 13	32	59

*Sera considered positive if they contained an arbitrary 20U of antibody, representing 3 SD above mean of the control group.
**Blood donors.

More recently, in an independent series from Perth, Western Australia, we found that 74% of newly diagnosed children of cases of IDDM were positive for antibodies to GAD.[21] We have also confirmed that, unlike ICA, antibodies to GAD can be detected for many years in patients with IDDM.[20] These results are consistent with those reported by other workers.[18,19]

GEOGRAPHIC DIFFERENCES IN ANTIBODIES TO GAD

Having established a consistent frequency of antibodies to GAD in Europid subjects with IDDM,[20,21] we were interested to study patterns in Asian patients, as the prevalence and incidence of IDDM is substantially lower in most Asian countries.[1,4] If the frequency of antibodies to GAD was similar to that in Europids, it might imply that the pathogenesis of IDDM was similar. Alternatively, if the frequency was lower, it might suggest that heterogeneity exists, different auto-antigens implicated or alternative aetiologies operative in different ethnic groups.

Sera were available from Asian IDDM patients and controls from the 3rd Asia-Oceania Histocompatibility Workshop.[22] Clinical data for the patients including age at onset, duration of insulin treatment, height and weight, history of ketoacidosis and presence of IDDM or NIDDM in first degree relatives was available, as was HLA-A, B, and DR locus typing.[22] The criteria for IDDM[22] were consistent with those of the National Diabetes Data Group.[23]

The prevalence of antibodies to GAD are shown in Table 3. When compared with Europids with IDDM from Australia (63%), the prevalence antibodies to GAD was much lower in all Asian populations with IDDM, being 31% in Japanese, 29% in Thais, 23% in Chinese and 5% in Koreans.[24] However, there was a similar mean level of antibodies to GAD among those diabetics from each population who gave a positive reaction.

Table 3. Mean age, age at onset, duration of diabetes (±SD) and prevalence of antibodies to glutamic acid decarboxylase in IDDM subjects from the Asian groups compared with Europids.

Population	No. of subjects	Mean age (years)	Mean age at onset (years)	Mean duration (years)	With antibodies to GAD No.	%
Europids	49	38 ± 14	24 ± 12.0	14 ± 10	31	63
Japanese	16	36 ± 10	30 ± 9.6	6 ± 3	5	31
Thai	7	*	< 40	...[1]	2	29
Korean	21	29 ± 7	24 ± 6.8	5 ± 5	1	5
Chinese**	13	30 ± 12	21 ± 10.2	7 ± 8	3	23

*Not recorded in Histocompatibility Workshop Records
**Cantonese (from Hong Kong)

PREDICTABILITY OF ANTIBODIES TO GAD

Several studies have confirmed that antibodies to GAD are more predictive of IDDM in Europids than antibodies to other islet cell antigens such as ICA and IAA.[18,19] In Europids, 20-30% of cases of IDDM do not show antibodies to GAD.[20,21] This suggests that there are other autoantigens involved in these cases or that there may be genetic differences in the autoimmune response to GAD.

Table 4. Prevalence of antibodies to GAD according to age of onset and HLA-DR genotypes in Europid IDDM patients.

HLA-DR Genotype	No. Tested	Antibodies to GAD present No.	%
DR3, DR4 DR4, DR4 DR3, DR3	26	20	76.9**
DR4, DRX* DR3, DRX DRX,DRX	23	11	47.8
Total	49	31	

*X is an HLA-DR antigen other than DR3 or DR4
**p=0.043, when compared with the prevalence (47.8%) in patients with a non-DR3 or non-DR4 antigen.

This latter possibility has been explored by us recently. In Europids with IDDM who possessed the "high risk" alleles HLA-DR3 or DR4, 76.9% were positive for antibodies to GAD.[25] This was significantly higher than the prevalence of 47.8% found in patients with at least one HLA-DR antigen other than DR3 or DR4. A similar picture was seen in Asian IDDM patients in relation to the high and low risk HLA-DR and DQ alleles seen in those populations.[25]

CONCLUSIONS

Our studies in Europids and Asian IDDM patients show wide ethnic differences in the prevalence of antibodies to GAD. These data are not population-based and the number of Asian subjects is presently rather small. IDDM is quite rare in most Asian populations,[1,4] and extensive studies would be necessary to obtain such data. However, if the Asian subjects are pooled, the sample becomes of substantial size. In a subsequent study, we have surveyed a large sample of Korean IDDM patients and the results are in line with present findings of a low prevalence of antibodies to GAD in this population group.[33]

While it is not yet known whether autoimmunity to GAD causes IDDM, at least in Europids it is clearly highly predictive for the subsequent development of IDDM.[18,19] Clare-Salzler et al.[26] have proposed a possible aetiological mechanism implicating GAD through molecular mimicry. There is a striking sequence similarity between the P2-C protein of Coxsackie virus and GAD. Thus, in a Coxsackie virus infection, the P2-C protein may trigger an immune response which then cross-reacts with a self-antigen (in this case GAD) and initiates the beta-cell destruction, eventually resulting in IDDM. This model, like many that depend on molecular mimicry, may be overly simplistic and a true insight may require knowledge of the antigenic epitope that breaks tolerance of the T cell level.

There is another important aspect to the story. The traditional classification of diabetes applicable to Europids may not necessarily be appropriate for other ethnic groups. The classification of newly diagnosed patients with diabetes mellitus is currently based predominantly on clinical assessment,[23,27] and it is usually possible to make a diagnosis of IDDM and NIDDM.[27-29] However, problems arise in the classification of persons, usually non-obese and 30 years and over, who present with what appears to be NIDDM.[25-27] However, within a short period (months or a few years), a failure of control of diabetes with diet and oral hypoglycaemic agents occurs. Progression to insulin-dependency is then quite rapid. This unusual type of diabetes appears to occur among Asians as it does among Europids,[29] and may be frequent among Koreans.[29] Among Europids, this form of diabetes is now believed to be a late-onset and slowly evolving form of IDDM, since studies have shown markers of autoimmunity, including ICA,[29,31] and in our own limited experience, these subjects have antibodies to GAD.[20] Among Koreans, likewise, this slow-onset adult type is regarded as a variant of IDDM.[30] We have shown that antibodies to GAD were present in three such subjects,[20] and if this is confirmed in larger studies, a routine test for antibodies to GAD becomes a likely scenario in the future for all subjects presenting with the non-obese form of NIDDM. This would provide physicians with valuable information about the possibility of future insulin dependency in these patients.

The difficulty in performing assays for antibodies to GAD has placed a strict limitation on the extent of epidemiological studies to date. The development of a simple and reliable radioimmunoprecipitation assay[20] should allow larger studies in various geographic and ethnic populations. Through these studies, a wider appreciation of the role of GAD in the aetiology of IDDM, and the predictive value of antibodies to GAD for the early detection of IDDM, should be possible. The availability of this assay for antibodies to GAD provides new impetus to IDDM research and a feasible approach to population screening for early cases of potential IDDM. The American Diabetes Association, in a recent policy statement, has proposed that immune intervention should be restricted to a high risk category, effectively first-degree relatives of IDDM cases in whom, as a result of screening, it can be predicted with near certainty that they will develop IDDM.[32] The true potential of the assay for antibodies to GAD to detect cases that constitute the high risk group needs to be assessed urgently.

Acknowledgements

This study was supported by a grant from Miles Diagnostics USA and the Leon Mow Research Trust. We thank Dr. T. Tuomi for advice on the manuscript and Mrs. Sue Fournel for preparation of the manuscript.

REFERENCES

1. M. Rewers, R.E. LaPorte, H. King, J. Tuomilehto, Trends in the prevalence and incidence of diabetes: Insulin-dependent diabetes mellitus in childhood, *World Hlth Stat Q.* 41:179 (1988).
2. A. Green, E.A.M. Gale, C.C. Patterson, Incidence of childhood onset insulin-dependent diabetes mellitus: The EURODIAB ACE study, *Lancet* 339:905 (1992).
3. J. Tuomilehto, M. Rewers, A. Reunanen, P. Lounamaa, R. Lounamaa, E. Tuomilehto-Wolf, H.K. Åkerblom, Increasing trends in Type I (insulin-dependent) diabetes mellitus in childhood in Finland. Analysis of age, calendar time and birth cohort effects during 1965 to 1984, *Diabetologia* 34:282 (1991).
4. M-Z. Bao, J.-X. Wang, J.S. Dorman, M. Trucco, HLA-DQβ non-asp-57 allele and incidence of diabetes in China and the USA, *Lancet* 2:497 (1989).
5. J.S. Dorman, R.E. LaPorte, R.A. Stone, M. Trucco, Worldwide differences in the incidence of Type I diabetes are associated with amino acid variation at position 57 of the HLA-DQβ chain, *Proc Natl Acad Sci. USA* 87:7370 (1990).
6. G.F. Bottazzo, A. Florin-Christensen, D. Doniach, Islet cell antibodies in diabetes mellitus with autoimmune polyendocrine deficiencies, *Lancet* 2:1279 (1974).
7. J.P. Palmer, C.M. Asplin, P. Clemons, K. Lyon, O. Iatpati, P. Raghu, T.L. Paquette, Insulin antibodies in insulin-dependent diabetes before insulin treatment, *Science* 222:1337 (1983).
8. S. Baekkeskov, J.H. Nielsen, B. Marner, T. Bilde, J. Ludvigsson, A. Lernmark, Autoantibodies in newly diagnosed diabetic children immunoprecipitate human pancreatic islet cell proteins, *Nature*, 298:167 (1982).
9. S. Baekkeskov, H-J. Aanstoot, S. Christgau, A. Reetz, M. Solimena, M. Cascalho, F. Folli, H. Richter-Olesen, P-D. Camilli, Identification of the 64K autoantigen in insulin-dependent diabetes as the GABA-synthesizing enzyme glutamic acid decarboxylase, *Nature* 347:151 (1990).
10. W. Hagopian A. Lernmark, Autoimmune diabetes mellitus, *in*: "The Autoimmune Diseases II," N.R. Rose, and I.R. Mackay, eds., Academic Press, Orlando, (1992).
11. R. Lipton R.E. LaPorte, Epidemiology of islet cell antibodies, *Epidemiol Rev.* 11:182 (1989).
12. J.K. Karjalainen, Islet cell antibodies as predictive markers for IDDM in children with high background incidence of disease, *Diabetes* 39:1144 (1990).
13. D.G. McLarty, I. Athaide, G.F. Bottazzo, A.M.B. Swai, K.G.M.M. Alberti, Islet cell antibodies are not specifically associated with insulin-dependent diabetes in Tanzanian Africans, *Diabetes Res Clin Practice* 9:219 (1990).
14. A. Elamin, M.I.A. Omer, T. Tuvemo, Islet cell antibodies and endogenous insulin secretion in Sudanese diabetic children, *Diabetes Res Clin Practice* 16:91 (1992).
15. J.P. Palmer D.K. McCulloch, Prediction and Prevention of IDDM - 1991, *Diabetes* 40:943 (1991).
16. R. Lendrum, G. Walker, D.R. Gamble, Islet cell antibodies in juvenile diabetes mellitus of recent onset, *Lancet* 1:880 (1975).

17. A.G. Ziegler, R.D. Herskowitz, R.A. Jackson, J.S. Soeldner, G.S. Eisenbarth, Predicting Type I diabetes, *Diabetes Care* 13:762 (1990).
18. S. Baekkeskov, J.K. Kristensen, S. Srikanta, G.J. Bruining, T. Mandrup-Poulsen, C. DeBeaufort, J.S. Soeldner, G. Eisenbarth, F. Lindgren, A. Lernmark, Antibodies to a M 64,000 human islet cell antigen precede the clinical onset of insulin-dependent diabetes, *J Clin Invest.* 4 79:926 (1987).
19. M.A. Atkinson, N.K. Maclaren, D.W. Scharp, P.E. Lacy, W.J. Riley, 64,000 M^1 autoantibodies as predictors of insulin-dependent diabetes, *Lancet* 335:1357 (1990).
20. M.J. Rowley, I.R. Mackay, Q-Y. Chen, W.J. Knowles, P. Zimmet, Antibodies to glutamic acid decarboxylase discriminate major types of diabetes mellitus, *Diabetes* 41:548 (1992).
21. Q-Y. Chen, G.C. Byrne, T. Jones, T. Tuomi, I.R. Mackay, M.J. Rowley, W.J. Knowles, P.Z. Zimmet, Antibodies to glutamic acid decarboxylase in Australian children with insulin-dependent diabetes mellitus and their first degree relatives, *Pediatr Res* 34:785 (1993).
22. S.W. Serjeantson, B.D. Tait, P. Chiewslip, E. Gazit, B.R. Hawkins, M.R.J. Kohonen-Corish, H.K. Lee, D. Propert, T.O. Yoshida, Insulin-dependent diabetes mellitus workshop: Joint Report, *in*: "HLA in Asia-Oceania 1986," M. Aizawa, T. Natori, A. Wakisaka, Y.Konoeda, eds., Hokkaido University Press, Sapporo (1986).
23. National Diabetes Data Group, Classification and diagnosis of diabetes mellitus and other categories of glucose intolerance, *Diabetes* 28:1039 (1979).
24. W. Knowles, P. Zimmet, M. Rowley, I. Mackay, Q-Y. Chen, L. Chapman, G. Senator, S. Serjeantson, The epidemiology of antibodies to glutamic acid decarboxylase: Presence and levels in insulin-dependent diabetes mellitus in Europid and Asian subjects, *Diabetes* 41(suppl. 1): 127A (Abstract), (1992).
25. S.W. Serjeantson, M.J. Kohonen-Corish, M.J. Rowley, I.R. Mackay, P. Zimmet, Antibodies to glutamic acid decarboxylase are associated with HLA-DR genotypes in both Australian and Asians with insulin-dependent diabetes mellitus, *Diabetologia* 35:996 (1992).
26. M.J. Clare-Salzler, A.J. Tobin, D.L. Kaufman, Glutamate decarboxylase: An autoantigen in IDDM, *Diabetes Care* 15:132 (1992).
27. M.I. Harris, P. Zimmet, Classification of diabetes mellitus and other categories of glucose intolerance, *in*: "The International Textbook of Diabetes Mellitus," H. Keen, R. DeFronzo, K.G.M.M. Alberti, P. Zimmet, eds., John Wiley, London (1992).
28. P. Zimmet, H. King, Classification and diagnosis of diabetes mellitus, *in:* "The Diabetes Annual/3," K.G.M.M. Alberti, L.P. Krall, eds., Elsevier, Amsterdam (1987).
29. L. Groop, A. Miettinen, P.H. Groop, S. Meri, S. Koskimies, G.F. Bottazzo, Organ-specific auto-immunity and HLA-DR antigens as markers for ß-cell destruction in patients with Type II diabetes, *Diabetes* 37:99 (1988).
30. B.D. Rhee, K-U. Lee, S.Y. Kim, B.Y. Cho, H.K. Lee, C-S. Koh, H.K. Min, Some immunogenetic and clinical aspects of Type I diabetes in Korea, *in*: "Recent Advances in Insulin Therapy," T.H. Lee, K.B. Huh, S. Baba, eds., Elsevier, Amsterdam (1990).
31. L.C. Groop, G.F. Bottazzo, D. Doniach, Islet cell antibodies identify latent Type I diabetes in patients aged 35-75 years at diagnosis, *Diabetes* 35:237 (1986).
32. American Diabetes Association Position statement: Prevention of Type I diabetes mellitus, *Diabetes* 39:1151 (1990).

33. P.Z. Zimmet, M.J. Rowley, I.R. Mackay, W.J. Knowles, Q.-Y. Chen, L.H. Chapman, S.W. Serjeantson, The ethnic distribution of antibodies to glutamic acid decarboxylase: Presence and levels in insulin-dependent diabetes mellitus in Europid and Asian Subjects, *J Diab Comp* 7:1 (1993).

DOES BETA CELL DEATH RESULT EXCLUSIVELY FROM GENETICALLY-MEDIATED AUTOIMMUNE MECHANISMS? A POLEMIC - THE CASE FOR ENVIRONMENTAL FACTORS IN THE ETIOLOGY OF INSULIN-DEPENDENT DIABETES MELLITUS

Allan L. Drash

Professor of Pediatrics
Director of Research, Division of Endocrinology
Metabolism & Diabetes Mellitus
Children's Hospital of Pittsburgh
University of Pittsburgh School of Medicine
Pittsburgh, PA 15213
USA

INTRODUCTION

The past two decades have evidenced a remarkable, almost revolutionary, advance in our understanding of factors surrounding beta-cell destruction, the clinical evolution of IDDM and the early clinical course of the disease. This accumulation of new knowledge is ultimately to be directed toward the cure, or much more likely, the prevention of beta cell destruction and consequently the prevention of insulin-dependent diabetes mellitus. This review attempts to identify the major streams of evidence regarding beta cell destruction and to make a case for expanding collaborative, epidemiological studies with increasing focus on environmental factors that may either cause or sustain beta cell inflammatory processes. The review ends with a series of very personal, biased recommendations as to my view of priorities in the field for the next few years. If my views induce controversy, all the better. We need vigorously designed and executed studies sharply focused on the many unresolved issues regarding the etiology of IDDM. Controversy should push these appropriate studies towards completion with the development of important new knowledge.

THE DOGMA

Insulin-dependent diabetes mellitus is not a genetically transmitted disease. It does

Standardization of Epidemiologic Studies of Host Susceptibility
Edited by J. Dorman, Plenum Press, New York, 1994

not fit classical Mendelian genetics. However, the evidence for a genetic "underpinning" in most, if not all cases of IDDM, is now beyond dispute. Very specific alterations within the major histocompatibility complex (MHC) located on chromosome 6 result in major increases in the likelihood that the beta cell will undergo inflammatory damage. Early observations identified specific HLA antigens which, while present in the general population, were highly correlated with the development of IDDM. More recently, identification of specific amino acid alterations at codon 57 in the DQ beta chain, and in codon 52 in the DQ alpha chain can account for an increased risk ratio for the development of IDDM in excess of 200. Other genetic alterations also undoubtedly exist both on chromosome 6 and possibly other chromosomes that either increase the likelihood of beta cell destruction or in some way protect against its occurrence.[1-6]

It is widely accepted that the mechanism by which beta cell destruction occurs is autoimmune in nature. The scientific evidence supporting this theory is impressive. Gepts originally described inflammatory invasion of islet tissue in histo-pathologic studies carried out in patients who died soon after the diagnosis of IDDM. These changes, referred to as insulitis, represent lymphocytic infiltration which characterizes a number of autoimmune diseases, including Hashimoto's Thyroiditis and Addison's disease. Antibodies to specific target tissue are considered the sine qua non of autoimmunity.[7-9] A number of antibodies to islet tissue have been identified in newly-diagnosed patients with IDDM. The antibodies include cytoplasmic, complement-fixing antibodies to the surface of the beta cell, and more recently antibodies to the enzyme glutamic acid decarboxylase (64k). Depending upon the type of assay employed, investigators report the presence of one or more islet cell antibodies at the time of diagnosis of IDDM in 60-90% of cases. Prospective longitudinal studies of first degree relatives of IDDM patients may identify the development of immunological alterations months or years prior to the eventual conversion to overt diabetes. The presence of islet cell antibodies is agreed to be the single most reliable predictor for the eventual conversion to IDDM. Additional evidence of immunologic alterations that may be identified either in the newly diagnosed patient or in the first degree relative during prospective follow- up studies include antibodies to the insulin molecule (IAA+) or a variety of alterations in the T cell and B cell subsets.[10-19]

The currently accepted theory of beta cell destruction involves a prolonged (months or years) "incubation period" during which time there is progressive beta cell destruction, leading eventually to loss of greater than 90% of the beta cell mass before the expression of carbohydrate intolerance. The "autoimmune cascade" appears to involve macrophage presentation of specific beta cell antigens by genetically altered MHC receptor sites to helper T cells leading to the release of a variety of cytokines and the generation of a family of beta cell antibodies from selected B cells.[20-29] The direct beta cell destructive process is thought to derive from the accumulation of free radicals in the micro-environment of the beta cell with a number of oxides and peroxides being involved in this destructive process. The prime candidate for the major biochemical toxic agent is excessive accumulation of nitric oxide.[30-35]

The accumulation of evidence pointing toward an autoimmune nature of the beta cell destructive process quite naturally led to the development of pharmacological strategies to alter this assumed pathologic mechanism. A large variety of immuno-modulators and immuno-suppressive agents have been tried in the newly diagnosed diabetic patient with variable success. In most cases this intervention therapy has resulted in no apparent alteration in the course of the disease. However, in the collaborative cyclosporin international studies, there was clear evidence of an increase in the frequency, completeness and duration of clinical remission when the cyclosporin therapy was introduced with other appropriate diabetes management early after the clinical diagnosis of disease.[36-50] Recent attempts have been focused on preventing disease in genetically susceptible individuals, primarily first degree relatives of IDDM patients with ICA positivity and declining insulin responsiveness.[51]

CAN AUTOIMMUNITY REALLY BE THE ANSWER?

If IDDM is an autoimmune disease, it is certainly an unusual one. The mean age of diagnosis of IDDM is approximately 8 years with a peak incidence in early adolescence. The disease occurs equally in males and females and the antibodies to islet tissue decline and disappear during the months and years following diagnosis.[52-56] The more usual autoimmune diseases have their clinical appearance in mid to later life, are invariably more common in females than males, and have persistence of high titers of organ specific antibodies for years, if not indefinitely.

The apparent evidence supporting the autoimmune theory cuts in both directions. While the presence of insulitis in the histology of the pancreas in individuals dying shortly after the diagnosis of the disease is strongly suggestive of an autoimmune process, it is important to underline the fact that these findings are identified in only 30-60% of specimens while the remainder may have no evidence of histological alteration. Over the past 40 years, a number of children with newly diagnosed IDDM have died in our institution. Careful histological examination of the pancreatic material from these children reveal that a number of them have had no evidence of histological alterations using currently available techniques. In a recent acute death, minor changes consistent with the diagnosis of diabetes were found in a very small proportion of the islet tissue, while most of the pancreas appeared entirely normal.

Much weight is placed on the presence of islet cell antibodies to buttress the claim that IDDM is an autoimmune disease. It is important to understand that antibodies may or may not develop as a consequence of a variety of beta cell damaging experiences. For example, in those few patients studied following purposeful ingestion of the rodenticide Vacor^R, acute diabetic ketoacidosis resulted. In several of these individuals, islet cell antibodies were identified shortly after the acutely toxic ingestion, indicating that the antibody response was a direct consequence of chemical damage rather than the expression of autoimmunity.[57,58] On the other hand, the interesting experiences of Sutherland, et al. in pancreas transplantation is especially important. A number of their diabetic patients underwent pancreatic transplantation from a healthy non-diabetic identical twin without the use of immunosuppression. In each case, massive beta cell destruction occurred within weeks, leading to recurrence of overt diabetes, but, in most cases, without the presence of circulating islet antibodies. That is, the complete disease process was reactivated as a consequence of the pancreatic transplantation, but without the development of characteristic beta cell antibodies.[59,60]

If IDDM was invariably a consequence of autoimmune beta cell destruction, one would anticipate uniformly identifying evidence of autoimmunity prior to, or at the time of the diagnosis of disease. In the experience of all investigators, a small but variable number of newly diagnosed patients have no evidence of autoimmunity, using currently available investigative techniques, including the search for antibodies to a variety of beta cell antigens. In most studies, 10-20% of patients fit into this category. Does this suggest that these patients have a non-immunologic pathologic mechanism for beta cell destruction? Similarly, large longitudinal follow-up of first degree relatives, while confirming that the presence of islet cell positivity is the most important single predictor of eventual disease, have documented IDDM in small numbers of subjects without ICA positivity, and the absence of disease in individuals followed with ICA positivity for 10 or more years.[61,62]

The general lack of success with immuno-modulatory and immuno-suppressive agents should give one pause in regard to anticipating that this disease is a consequence of immunologic mechanisms gone wrong. The very potent immuno-suppressive agent cyclosporin, used in high dosage in newly diagnosed IDDM patients, resulted in an increased clinical remission in 15-30% of patients followed over the first year following diagnosis. Continued cyclosporin therapy in the second year after diagnosis of diabetes resulted almost invariably in loss of remission, and the necessity to re-institute insulin therapy in almost all such patients.[63-66] If the disease is immunologic in nature, why were there primary "cyclosporin-failures" in approximately 75% of patients, and why was not continued cyclosporin therapy effective after initial induction of clinical remission? Do these findings suggest that a large number of patients have beta cell destruction by a mechanism not involving immunologic alterations? Most investigators would reject this position and explain the current failures as inevitable because of the extensive beta cell destruction already present at the time of clinical diagnosis of diabetes. This has led to a refocusing of immunosuppressive techniques to the "pre-diabetic population". We await the results of these new intervention trials with great interest, but little optimism for success.

EVIDENCE FOR ENVIRONMENTAL FACTORS IN THE ETIOLOGY OF INSULIN-DEPENDENT DIABETES MELLITUS

Environmental factors may play one of several possible roles in beta cell destructive processes. At one extreme, there are undoubtedly cases of diabetes resulting directly from ingestion of beta-cytotropic agents with no intervening genetic susceptibility or autoimmune mechanisms necessary, to the other extreme in which beta cell destruction results exclusively from genetically-mediated autoimmune processes which are internally triggered without interphase with specific environmental stimulants. It seems likely, however, that most cases of IDDM fall in a middle ground where environmental factors, rather than being directly causative, act as stimulants or provocateurs of the autoimmune system. Further, it is my belief that in the great majority of cases, IDDM does not result from a single environmental insult leading to a relentless autoimmune destructive process, but rather there are "multiple-hits" from the environment, resulting in a waxing and waning of the inflammatory process. What is the evidence?[67,68]

Animal Studies

In medicine, we have traditionally looked to animal models for insights and understanding of disease processes that cannot be fully elucidated in the human subject. Animal research has played a principal role in the study of diabetes for well over 100 years. Strangely, those stringent advocates of an exclusively internally-mediated, autoimmune disease process accept the evidence of disease mechanisms from genetically susceptible animals, while rejecting the overwhelming evidence of environmental manipulation on such models as important to the human condition. To summarize the animal experience: 1) Toxic agents, for example alloxan and streptozotocin, can regularly induce beta cell destruction and diabetes in both genetically susceptible and normal animals. Overt diabetes can be rapidly induced by a large dose of streptozotocin by direct toxic action of the drug or probably by autoimmune mechanisms utilizing small doses of the toxin repeatedly (this is reminiscent of the human situation with the drug Vacor[R] in which acute large dose ingestion results in rapid beta cell destruction while small dose chronic ingestion appears to induce autoimmune beta cell destruction).[69] 2) A variety of infectious agents can induce beta cell destruction in diabetes susceptible and non-susceptible animal models. The identified agents include Encephalomyocarditis (EMC) Virus, Mengo Virus and Coxsackie B Virus variants. The beta cell destructive process may be either directly cytolytic or the induction of beta cell damage leading to eventual autoimmune destruction.[70] 3) Environmental manipulation will alter the expression of diabetes in genetically susceptible animal strains. The BB rat and the NOD mouse strains are immunologically defective animals which are at markedly increased risk for the development of beta cell destruction and overt diabetes, mimicking human IDDM in many respects. Environmental manipulation, particularly dietary alteration, can result in marked variation in the expression of disease in both of these animal models.[71-73] This is particularly true in relationship to cow's milk protein and/or beef in the animal chow. Adding bovine serum albumin or whole milk or beef protein to the chow increases the incidence of diabetes while replacing the animal chow with synthetic proteins and amino acids reduces the expression of diabetes to near zero. This evidence strongly suggests that cow's milk protein or beef has a specific provocative effect on the immune system of these animals, resulting in progressive beta cell destruction. 4) A variety of immune suppressive or modulating strategies introduced well before beta cell damage will reduce the expression of diabetes in the genetically susceptible animal models. These observations provide some optimism that immunologic intervention prior to diabetes may prevent disease in the human situation.[74-77]

Human Observations

Twin Studies. The famous diabetes twin studies have provided important insights into the genetic components to both NIDDM and IDDM. The investigators found that the ultimate development of diabetes in the second of a diabetes twin pair, when the first twin had non-insulin-dependent diabetes mellitus, occurred in greater than 90% of cases: that is, the disease was concordant in almost 100% of cases.[78] However, in IDDM the concordance rate was less than 50%.[79,80] In traditional genetic explanation, this would be interpreted to mean that genetic susceptibility, present equally in the twin pairs, resulted in clinical expression of disease based in large part on environmental experiences. This obvious explanation has been rejected by those denying the importance of environmental aspects of the disease process by invoking unique immunologic identity resulting from chance mutations or formation of unique rearrangement of cell lines within the immune system, depending upon early immunologic exposures. This would explain immunologically distinct individuals who are otherwise genetically identical. While possible, such explanations to avoid obvious environmental mechanisms seem strained. The evidence of disease development in HLA identical, haploidentical and non-identical siblings while confirmatory of the experience in identical twins is obviously less impressive, but substantiates the importance of the interaction between genetic susceptibility and environmental provocation.

Geographic Distribution Of IDDM. The conference on the epidemiology of IDDM held in 1983 identified clearly, and for the first time, the remarkable variation in expression of IDDM in various parts of the world.[81] Many follow-up studies have confirmed this, with the finding of high incidence of IDDM in the Scandinavian countries, intermediate levels in much of the West, and very low levels in the Far Eastern countries, including Japan, China and Korea.[82,83] This distribution is extraordinarily similar to that seen for the worldwide distribution of atherosclerosis and morbidity and mortality from coronary artery disease. The latter is closely tied with the ingestion of saturated fat and cholesterol and tightly linked with dietary dairy product ingestion. Conversely, both atherosclerotic risk and IDDM incidence are inversely correlated with the content of naturally occurring anti-oxidants in the diet, with anti-oxidant ingestion traditionally high in the diet of the Orient while quite low in that of the typical diet ingested by individuals living in Western civilization.[84-91]

Highly important genetic studies carried out by our laboratories, in collaboration with many other investigators, have documented a remarkable parallelism between the population distribution of the homozygous non-aspartic acid status at codon 57 of the HLA-DQ beta chain and the incidence of IDDM. That is, in those few countries studied to date, the incidence of diabetes, either high or low, is directly correlated with this diabetes susceptibility gene distribution. However, it appears that probably something less than one individual in 20 carrying the gene alteration that increases diabetes susceptibility will, in fact, develop overt diabetes, continuing to implicate environmental factors as responsible for actual disease induction.[92-94]

Secular Trends In IDDM Incidence. If the incidence of IDDM were exclusively dependent upon the frequency of the diabetes susceptibility genes within the population, one would expect that the annual incidence would be remarkably similar year after year, reflecting the absence or very slow change in the distribution of diabetes-related genes in the population over time. That, in fact, is not the case. There is now substantial evidence based on carefully constructed registry data extending in many countries over the last 20-40 years, that there have been substantial increases in IDDM incidence in many locations, particularly in the Scandinavian countries. The incidence of diabetes among Finnish children has almost doubled during the past 20 years. Similar observations have been made in Norway, Denmark and Sweden. In our own experience in Allegheny County, Pennsylvania we have observed an increase of approximately 1% per year in the incidence of IDDM until a more recent acute acceleration.[95-100]

Probably the most dramatic evidence of a secular increase is the experience in the Italian island of Sardinia. The incidence of IDDM currently in Sardinia approaches 30/100,000 per year, just below that in Finland, and approximately five times greater than the incidence of diabetes in Italian children living on the mainland of Italy. This high incidence is "explained" by the finding of a very high frequency of homozygous non-aspartic acid genetic status in the population of Sardinia.[92,93,101] However, recent review of the earlier incidence of diabetes in Sardinia, not yet published or verified, suggests that the attack rate for this disease among Sardinia school children 20-30 years ago was in the neighborhood of 4-5/100,000 per year, the incidence rate now seen in the mainland of Italy. Obviously, there has been no significant increase in the genetic susceptibility to diabetes in Sardinia over the past quarter century. The increase in attack rate, if it is accurate, must reflect significant environmental changes that have occurred during that time period, almost certainly associated with industrialization, changes in life style patterns and changes in dietary habits.

Diabetes Epidemics. While secular changes in IDDM incidences over a period of several years strongly stress the validity of a totally genetic-immunologic explanation for IDDM, acute increases in incidence, referred to by some as diabetes epidemics, truly defy genetic explanation, and must be viewed as a consequence of acute environmental changes. There are now several well documented, abrupt increases in IDDM incidence followed by decreases back to the base line in several countries including Poland, Latvia and England.[102-106]

Our own recent experience in Pittsburgh is particularly important in this regard. In both our Children's Hospital and Allegheny County registry we have continued surveillance both retrospectively and prospectively covering approximately 40 years. While we have documented a slight, just statistically significant incidence increase, of about 1% per year from about 1965 through 1985,we have recently observed a major increase in incidence during the interval 1985-1989. This increase in incidence is seen predominantly in males, in younger children and in blacks. It is statistically correlated with chicken pox epidemics with a lag phase of between two and three years. While we do not suggest that chicken pox is a primary etiologic factor in IDDM, we strongly feel that our recent observations of a highly statistically significant increase in incidence among children in Western Pennsylvania is indicative of changing environmental stressors. Our documentation of increasing frequency of IDDM in younger children is consistent with the clinical experience of many pediatric diabetologists in various parts of the world, and again suggest environmental stressors appearing at earlier ages.[107]

Migrant Studies. While classical prospective migrant studies have not been carried out in IDDM, there are a number of important and valuable observations that strongly suggest that susceptibility to IDDM may be significantly altered by changes in geography and lifestyle. These conclusions result from the study of individuals of a particular geographic or nationality group living at a distance from their natural home, within a population that has an IDDM incidence significantly different than the study population. Such observations have documented that pure Japanese living in Hawaii have an IDDM incidence approximately five times greater than do Japanese children living in their home land. Similarly there is an approximate doubling of the incidence of pure French and Italian children in Montreal when compared with the incidence in France and Italy.[108,109] The incidence of IDDM in Indian children following migration from South Africa to England showed a dramatic increase from very low levels to that compatible with children in England.[110]

These studies and others strongly support the thesis that environmental factors may either increase or decrease the expression in diabetes in susceptible individuals. The "migrant study" observations have been appropriately criticized by the absence of experiences documenting reduced risk in certain populations. However, the natural migrant drift in the past several decades is from East to West, from Third World countries to Western civilization, from poverty to affluence, and from the tropics to temperate zones. All of these moves in general involve movement of individuals from countries at lesser risk for the development of diabetes to countries where the incidence rate is higher.

CANDIDATES FOR ENVIRONMENTAL FACTORS IN THE INDUCTION OF IDDM

There are four general environmental categories that have been proposed as potential causative or provocative agents in IDDM expression. These include infectious agents, environmental toxins, nutrient factors, and physical and emotional stress. While it is likely that genetically susceptible individuals actually develop diabetes after multiple environmental insults, possibly involving adverse experiences from each of the four general areas, a discussion of the present available evidence for causation will be presented separately.

Infectious Diseases. A number of viral infections have been associated with the development of IDDM, either based on epidemiologic surveys or individual case reports. The association between mumps infection and IDDM has a history of approximately 100 years duration with episodic case reports appearing in the literature associating diabetes in individual cases to mumps occurring a few weeks previously. Coxsackie infections, particularly B3 and B4, have been associated with IDDM both in disease surveillance statistics and in a few isolated, but highly important cases. Cytomegalo virus (CMV) has also been associated with carbohydrate intolerance and diabetes, and insulitis has been identified in the pancreas of infants dying with disseminated CMV infection. The best documented association between a specific viral infection and IDDM is that associated with the congenital rubella syndrome. Rubella infection, acquired by the fetus as a result of trans-placental passage results in a generalized rubella infection which may have devastating effects on the fetus. In the great majority of cases the infection "burns out" soon after delivery. However, it leaves behind an autoimmune destructive process which, over a period of 5 to 20 years, will result in beta cell destruction and overt diabetes. HLA studies indicate that diabetes eventually develops in those rubella syndrome children who have the genetic predisposition to disease. In fact, the eventual development of diabetes may approach 100% in those individuals who are HLA-DR3 and/or DR4. No DQ alpha or beta chain studies have been reported on this group of patients, but this should be accomplished. The congenital rubella syndrome-diabetes association is strong evidence of a viral-induced autoimmune phenomenon leading to beta cell destruction. The recent studies of Yoon et al. documenting an increased frequency of CMV viral fragments in the genome of children with IDDM suggests that such patients either acquired CMV infections very early in life, and that this initiated a beta cell destructive process or that IDDM may be associated with persistency of the infection.[111-120]

Environmental Toxins. It is well documented that a number of chemical agents have beta cell destructive properties. Streptozotocin, the agent used to induce diabetes in laboratory animals, has been used to destroy the pancreas in humans with severe hypoglycemia resulting from inoperable islet cell malignancy. The rodenticide Vacor[R] is highly beta cell toxic, resulting in an acute induction of diabetic ketoacidosis following large dose oral ingestion.[57,58] Inadvertent chronic small dose ingestion, as apparently occurred in Korea when the rat poison was mixed with feed, is strongly reminiscent of the low dose streptozotocin induced-autoimmune destruction in laboratory animal models. Nitrosomines, produced as a consequence of food curing by smoking techniques, has been implicated in cases of IDDM in Iceland.[121,122] The possibility that the widely used grilling or barbecuing of beef may be a factor needs investigation. Further, almost nothing is known about the health hazards of a large number of industrial environmental contaminants as they enter our food and water supply. The recent finding that the apparently benign, inert silicone breast implants may be associated with autoimmune disease should raise increasing concern about the health hazards of "biochemical advances" in our society.

Nutrients. Much recent attention has focused on the relationship between dietary factors and the incidence of IDDM. Much of this attention has focused on the ingestion of milk and dairy products. There is strong epidemiological evidence between the per capita intake of dairy products and the incidence of IDDM. Early studies suggested a possible protective effect of breast feeding while some studies have disputed this.[123-132] Proper studies focusing on neonates who are genetically susceptible to IDDM are in the planning stage.

Repeated studies by Scandinavian investigators have documented that newly diagnosed IDDM children have an increased frequency of antibodies to a variety of milk proteins.[133] The recent studies of Dosch et al. have sharply focused attention on the possible relationship between early ingestion of bovine serum albumin,and the later development of diabetes in genetically susceptible individuals. They document that an antibody to a specific fragment of bovine serum albumin, referred to as Abbos, is present in 100% of newly diagnosed diabetic children, while essentially never present in non-diabetic children or normal adults. The Abbos epitope is immunologically cross reactive with a beta cell autoantigen, P69 which may explain, by virtue of molecular biological mimicry, why ingestion of this compound may induce beta cell inflammatory responses.[134] Of considerable interest is the recent finding that IgA deficiency may be linked to the same amino acid alteration on HLA-DQ beta as is the diabetes susceptibility. Since IgA deficiency would decrease the integrity of the gastrointestinal mucosa, thereby increasing the likelihood of absorption of large foreign proteins, the set-up for nutritionally induced autoimmunity would be vastly increased in the child with the genetics susceptibility for diabetes.[135,136]

It is well documented that the antioxidant potential of the islet tissue of the pancreas is inherently deficient when compared with most of the other tissues and organs of the body. This means that the islet tissue is at increased risk for free radical damage. The balance between oxidant and antioxidant concentrations within the body fluids and tissues is a delicate one. Free radical excess or antioxidant deficiency will result in disease. Nutritional intake is an important factor in this balance with antioxidant ingestion becoming increasingly important as a preventive approach to human disease. It is possible that the lower content of naturally occurring antioxidants that characterizes the diet of Western Man may be inadequate to neutralize free radical accumulation, thus increasing the likelihood of beta cell damage and clinical diabetes mellitus. Conversely, the considerably lower incidence of diabetes in the Orient may be at least partially related to their high dietary intake of antioxidant compounds.

Stress. The relationship between acute emotional stress and the induction of thyrotoxicosis has been widely accepted among endocrinologists for many years. The anecdotal association between the two is convincing. Only recently has research begun to explore the complexities of the inter-relationship between the endocrine and the immunologic systems. It is now quite clear that alterations in the hypothalamic-pituitary-adrenal axis, through the release of both ACTH and adrenal steroids, can result in major alterations in the effectiveness of the immunologic surveillance system. While the anecdotal association between emotional trauma and the onset of IDDM is not as strong in diabetes as it is in thyrotoxicosis, there are, none the less, a number of studies suggesting such a cause and effect relationship.[139,140]

CONCLUSIONS

Insulin-dependent diabetes mellitus results from beta cell destruction. The potential mechanisms of beta cell destruction are numerous, but in most cases, it appears likely that this is a genetically-mediated, autoimmune process which in most cases are triggered by single or multiple environmental events. Viral infectious agents seem to be particularly likely candidates while ingested toxins, either industrial in nature or apparently nutritional in source, almost certainly play an important role. The place of emotions in the susceptibility to beta cell damage requires additional studies.

RECOMMENDATIONS

Epidemiological Studies

There are now in place several large multicenter study groups using epidemiological principles and techniques to study various aspects of the problems of diabetes mellitus.

Recommendation 1. Centers of excellence in diabetes epidemiology should continue to be developed with the development of regional, national, and international networks to enhance communication and collaboration. The epidemiological study groups currently in place must develop increasingly effective means for communication and collaboration, and, where appropriate, development of sharply focused international studies.

Recommendation 2. The Sardinian experience. The incidence in IDDM in children and adolescence in Sardinia approaches 30/100,000 year, one of the highest regional incidence rates in the world. Their "cousins" on the mainland of Italy have much lower IDDM incidence rates, in the neighborhood of 4-6/100,000 per year. If the preliminary and somewhat fragmentary evidence that the incidence in Sardinia 20 to 30 years ago was comparable to that of mainland Italy, then there has been a remarkable increase in incidence unparalleled by any other experience in the world. A "blue ribbon" panel of experts should be appointed to review as completely as possible the data on IDDM incidence in Sardinia over the last 50 years. If such a review confirms the presence of a major increase in incidence rate (four to six-fold), then a very intensive research program should be focused on identifying the lifestyle changes that have occurred in the island of Sardinia during the past two or three decades, including attention to infectious disease, industrialization, changes in dietary habits, including breast feeding frequency, and changes in stress level. The secret to environmental factors in the etiology of IDDM may be waiting for us on the island of Sardinia.

ENVIRONMENTAL STUDIES

Virus Infections

The research of Yoon and colleagues has opened a new door into the study of the relationship between genetic alterations and viral infections as they relate to the etiology of diabetes mellitus. While classical infectious disease studies should go forward, the broader application of the viral fragment methodology should be encouraged.

Recommendation 3. Family studies, with a proper population-based focus, should be undertaken with prospective study of the development of viral fragments and its relationship to the expression of IDDM. I would recommend initiating studies in multiplex families in which specimens are obtained from cord blood at birth in order to determine the frequency of viral fragments in the genome at birth in diabetes and control families. Periodic re-study, carried out in conjunction with other studies related to the prediction of IDDM onset, should be organized, in order to determine whether the incorporation of viral fragments may be associated with the beginning of pancreatic pathology.

Environmental Toxins

Essentially no systematic studies have been carried out attempting to determine whether environmental contaminants may play a significant role in beta cell damage.

Recommendation 4. Large scale, international, collaborative studies, focusing on environmental contamination, should be organized in the near future. The development of such research activities must involve close collaboration with experts in the field of environmental contamination, working with geneticists, immunologists and clinical diabetologists.

Nutrition

I am personally convinced that nutritional alterations are responsible for much of the worldwide variation in the expression of IDDM. Celiac disease or gluten-induced enteropathy is, in some sense, a classical autoimmune disease, with extraordinary similarities to IDDM. On the other hand, it is a disease resulting from a very specific toxic agent, causing local inflammatory damage. The agent gluten is a ubiquitous nutrient, which for most of the world's population, is of nutritional value. However, in the very small number of genetically determined individuals, this normal nutrient becomes a toxin. We need to mount a vigorous search for the "gluten" of IDDM.

Recommendation 5. The evolving evidence that there is a special relationship between ingestion of cow's milk protein and the later development of IDDM in genetically susceptible individuals, requires prompt and vigorous pursuit. International studies are currently in the developmental stage which will focus on newborn infants of diabetic parents, comparing the later development of IDDM in those offspring with high genetic susceptibility and those with decreased genetic susceptibility, and further, comparing exclusive breast milk feedings for the first nine to twelve months with early introduction of cow's milk protein. In order to achieve adequate numbers of subjects, these studies must involve multiple centers with long term follow-up. However, the potential importance of these studies can not be overemphasized, as the outcome may reflect directly on recommendations toward use of cow's milk in general childhood nutritional practices.

Recommendation 6. A collaborative, international study to carefully compare nutritional differences between countries including micronutrients and trace metals should be organized. These clinical studies should be complemented by animal studies in appropriate, genetically susceptible models.

Stress

It is well documented that physical and emotional stress can lead to acute and chronic alterations in the immunologic response system. It seems highly likely that such alterations in individuals who are genetically predisposed to beta cell damage may result in the initiation of the process that leads eventually to IDDM.

Recommendation 7. Carefully controlled studies of the effects of quantitatible stress on the immunologic response of individuals with IDDM and relatives at high-risk for diabetes and appropriate controls should be developed.

Genetics

Molecular biology has overtaken the study of the genetics of diabetes. Direct study of genomic DNA has provided a quantitative leap forward in the capacity to accurately identify small biochemical alterations (single amino acid changes) that may dramatically alter the susceptibility to IDDM. The techniques are now available and in place to "immortalize" lymphocyte cell lines from patients with IDDM and their families, thus making these valuable specimen resources indefinitely available to many investigators.

Recommendation 8. Establishment of a diabetes biological resource bank that would provide a variety of biological materials such as lymphocyte cell lines to qualified investigators. The National Institutes of Health is currently seriously considering such a recommendation that would probably initially be applied only to selected specimens obtained within the U.S. It seems imperative that other national groups should quickly become active collaborators and participants in such an undertaking. The genetic comparison of IDDM patients and their family members from Finland, for example, compared to Japan should immediately reveal important genetic differences. These observations, of course, must be placed within the context of the control population from which the specimens are derived. Consequently, appropriate controls must also be placed into the biological resource bank.

Recommendation 9. Special patient populations remain of considerable interest. Multiplex families containing two or more sibs with IDDM, or parent/child IDDM (or NIDDM/IDDM) family clusters are particularly informative. A large, international, collaborative study should also be undertaken, focusing on MODY patients. The infant with diabetes mellitus not only presents very special therapeutic challenges, but also unique opportunities to study etiology. A collaborative study focusing on the child under one year of age at diabetes-onset should be organized. The "incubation period" is short in such patients and one should be able to detect environmental factors and the interaction between the infant's unique genetic make-up and such things as viral infections, if, in fact, this relationship exists.

Immunology/Pathology

While classical immunological studies, both in vitro and in vivo, are moving rapidly forward, the tissue of most importance to the disease process is rarely available for examination. The "insulitis hypothesis" is built literally on a handful of postmortem examinations. From these limited studies, we already know that there is considerable diversity of histological changes with some pancreatic tissue showing classical and intense lymphocytic infiltration, while others appear histologically normal, at least by standard H and E preparation. The histology of the pancreas undoubtedly holds important secrets that should provide specific direction and understanding as to pathologic events. No self-respecting physician would consider undertaking the treatment of a patient with malignancy without a histological diagnosis. We should do no less.

Recommendation 10. Safe and effective techniques for pancreatic biopsy should be developed and applied routinely to the patient with newly diagnosed IDDM. Because of the importance of this resource, the handling of these tissue specimens should be limited to one or two selected pathological centers that can bring state-of-the-art techniques for complete analysis including histological, immunological, biochemical and hormonal methodology. It is likely that such studies will allow for specific classification into one of several pathologic processes including, for example, pure autoimmunity, direct viral infection, chemical toxin, congenital malformation, etc. With such information in hand, rational therapeutic strategies, such as the selection of immunosuppressants, antiviral agents, antioxidants, etc. can be made.

The Prediction of IDDM and Intervention Trials

There is considerable excitement of the possibility of preventing IDDM by identifying individuals at high-risk and introducing some pharmacologic agent to prevent progressive beta cell loss. Our enthusiasm "to do good" must be tempered by the equally important command "to do no harm". The introduction of potentially harmful pharmacologic agents to healthy children who may (but may not) eventually develop diabetes mellitus is of very considerable concern. While the methodology for predicting IDDM in first degree relatives is certainly improving, it continues to lack precision with something less than 70% of the individuals considered at highest risk by most investigators developing IDDM within five years of onset in the index case.

Recommendation 11. Continued search for IDDM markers in order to improve the prediction of diabetes must be encouraged.

Recommendation 12. Techniques must be developed that will allow quantitation of insulin secretory capacity over time. Present methods are crude, at best, and detect decline in insulin secretion only when a major portion of beta cell mass has been destroyed. A primary objective should be the development of sophisticated, highly sensitive techniques to study maximum insulin secretory capacity with methods that are safe and effective.

The challenge ahead of us is to develop accurate and reliable techniques to predict disease in the general population as well as the families of IDDM patients. This is but the necessary first step before the subsequent development of safe and effective intervention strategies to prevent, or possibly cure, this disease. While the primary focus has been on immunomodulatory agents, it is essential that these research horizons be expanded to include approaches to altering the various environmental factors which may either initiate or sustain beta-cell inflammatory processes.

There is much work left to be done. Increasingly, advances in the field of diabetes will depend upon collaborative, multi-discipline efforts involving scientific input from the molecular biologist to the diabetes clinician. With this sort of effort, I am confident that the coming decade will see the unraveling of the mysteries of beta cell pathology with development of effective therapeutic strategies for both prevention of disease and healing of the diseased pancreas. We and our patients can then face the future with confidence.

REFERENCES

1. J.A. Todd, J.I. Bell, H.O. McDevitt, HLA-DQ beta gene contributes to susceptibility and resistance to insulin-dependent diabetes mellitus, *Nature* 329:599 (1987).
2. J.A. Todd, J.I. Bell, H.O. McDevitt, A molecular basis for genetic susceptibility in insulin-dependent diabetes mellitus, *Trends Genet*. 4:129 (1988).
3. P.A. Morel, J.S. Dorman, J.A. Todd, et al., Aspartic acid at position 57 of the DQ beta chain protects against Type I diabetes: A family study, *Proc Natl Acad Sci. USA* 85:8111 (1988).
4. M. Trucco, J.S. Dorman, Immunogenetics of insulin-dependent diabetes mellitus in humans, *Crit Rev Immunol*. 9:201 (1989).
5. A.L. Drash, Insulin-dependent diabetes mellitus in children and adolescents: Genetics and etiology, *Curr Opinions Pediatr*. 1:61 (1989).
6. M. Trucco, To be or not to be ASP 57, that is the question, *Diabetes Care* 15:712 (1992).
7. W. Gepts, Pathologic anatomy of the pancreas in juvenile diabetes mellitus, *Diabetes* 14:619 (1965).
8. W. Gepts, Islet cell survival determined by morphology, *Diabetes* 27:251 (1978).
9. W. Gepts, Islet cell morphology in Type I diabetes, virus and autoimmunity in diabetes, *in:*"Virus and Autoimmunity in Diabetes Mellitus", Behring Inst Mitt 75:39 (1984).
10. G.F. Bottazzo, A. Florin-Christensen, D. Donjach, Islet cell antibodies in diabetes mellitus with autoimmune polyendocrine deficiency, *Lancet* 2:1270 (1974).
11. S.A. Arslanian, D.J. Becker, B. Rabin, et al., Correlates of insulin antibodies in newly diagnosed children with insulin-dependent diabetes before insulin therapy, *Diabetes* 34:926 (1985).
12. D.W. Drell, A.L. Notkins, Multiple immunological abnormalities in patients with Type I (insulin-dependent) diabetes mellitus, *Diabetologia* 30:132 (1987).
13. S. Baekkeskov, M. Landin, J.K. Kristensen, et al., Antibodies to a 64,000 MW human islet antigen precede the clinical onset of diabetes mellitus, *J Clin Invest*. 79:926 (1987).
14. G.J. Bruining, J.L. Molenaar, D.E. Grobbee, et al., Ten-year follow-up study of islet cell antibodies and childhood diabetes mellitus, *Lancet*. 1:1100 (1989).
15. T.J. Wilkin, Antibody markers for Type I diabetes: The issues, *Trends Endocrinol Metab*. 1:204 (1990).
16. J. Karjalainen, M. Knip, H. Hyoty, et al., Relationship between serum insulin autoantibodies, islet cell antibodies and coxsackie-B4 and mumps virus specific antibodies at the clinical manifestation of Type I (insulin-dependent) diabetes, *Diabetologia* 31:146 (1988).
17. N.K. Maclaren, W.J. Riley, Thyroid gastric and adrenal autoimmunities associated with insulin-dependent diabetes mellitus, *Diabetes Care* 8 (suppl. 1):34 (1985).
18. N.K. Maclaren, Perspectives in diabetes: How, when, and why to predict IDDM, *Diabetes* 37:1591 (1988).
19. A.G. Ziegler, R. Ziegler, and P. Vardi, Life-table analysis of progression to diabetes of anti-insulin autoantibody-positive relatives of individuals with Type I diabetes, *Diabetes* 38:1320 (1989).
20. G.F. Bottazzo, Death of a beta cell: Homicide or suicide?, *Diabetic Med*. 3:119 (1986).
21. G.S. Eisenbarth, Type I diabetes mellitus. Autoimmune disease, *N Eng J Med*. 314:1360 (1986).

22. A.C. Tarn, C.P. Smith, K. M. Spencer, et al., Type I (insulin-dependent) diabetes: A disease of slow clinical onset?, *Br Med J.* 294:392 (1987).
23. J. F. Bach, Mechanisms of autoimmunity in insulin-dependent diabetes mellitus, *J Clin Exp Immunol.* 72:108 (1988).
24. L. C. Harrison, I.L. Campbell, J. Allison, J.F.A.P Miller, MHC molecules and beta-cell destruction immune and nonimmune mechanisms, *Diabetes* 7:815 (1989).
25. N. MacLaren, D. Schatz, A. Drash, et al., Initial pathogenic events in IDDM, *Diabetes* 38:534 (1989).
26. J. Nerup, T. Mandrup-Poulsen, J. Molvig, et al., On the pathogenesis of insulin-dependent diabetes mellitus - A discussion of three recently proposed models, *in:*"Diabetes Mellitus: Pathophysiology and Therapy," W.Creutzfeldt, T. Lefbvre, eds, Heidelberg, Springer Verlag (1989).
27. Y. Wang, L. Hao, K.J. Lafferty, The pathogenesis of immunologically mediated diabetes, *in:* "Autoimmunity and the Pathogenesis of Diabetes. Endocrinology and Metabolism Series," F. Ginsberg-Fellner, R.C. McEvory, eds, Heidelberg, Springer Verlag (1989).
28. A.L. Drash, R. Lipton, J. Dorman, et al., The interface between epidemiology and molecular biology in the search for the causes of insulin-dependent diabetes mellitus, *Ann Med.* 23:463 (1991).
29. R. Lipton, J. Atchison, J.S. Dorman, et al., Genetic immunologic and metabolic determinants of risk for insulin-dependent diabetes mellitus in families: The Pittsburgh Etiology and Epidemiology of Diabetes Study, *Diabetic Med.* 9:224 (1992).
30. S.P. LeDoux, C.R. Hall, P.M. Forbes, et al., Mechanisms of nicotinamide and thymidine protection from alloxan and streptozotocin toxicity, *Diabetes* 37:1015 (1988).
31. I.N. Nomikos, Y. Wang, K.J. Lafferty, Involvement of 0_2 radicals in "autoimmune" diabetes, *Immunol Cell Biol.* 67:85 (1989).
32. J. Mendola, J. Wright, P. Lacey, Oxygen free-radical scavengers: An immune destruction of murine islets in allograft rejection and multiple low dose streptozocin induced insulitis, *Diabetes* 38:379 (1989).
33. B. Kallman, V. Burkart, K.D. Kroncke, et al., Toxicity of chemically generated nitric oxide towards pancreatic islet cells can be prevented by nicotinamide, (In press).
34. H. Kolb, V. Kolb-Bachofen, Nitric oxide: A pathogenic factor in autoimmunity, *Immunol Today* 13:157 (1992).
35. T. Mandrup-Poulsen, S. Helwvist, J. Molvig, et al., Cytokines as immunoeffector molecules in autoimmune endocrine disease with special reference to insulin-dependent diabetes mellitus, *Autoimmunity* 4:191 (1989).
36. R.B. Elliott, J.R. Crossle, C.C. Berryman, et al., Partial preservation of pancreatic ß cell function in children with diabetes, *Lancet* 2:103 (1981).
37. E. Heinze, A. Thon, U. Vetter, et al., Gammaglobulin therapy in six newly diagnosed diabetic children, *Acta Pediatr Scand.* 74:179 (1985).
38. A. Rubenstein, D. Pyke, Immunosuppression in the treatment of insulin dependent (Type I) diabetes, *Lancet* 1:436 (1987).
39. J. S. Skyler, Immune intervention studies in insulin-dependent diabetes mellitus, *Diabetes Metab Rev.* 3:1017 (1987).
40. R. Assan, G. Feutren, J. Sirmai, Cyclosporine trials in diabetes: Updated results of the French experience, *T Transplant Proc.* 20 (suppl.4):178 (1988).
41. P.F. Bougneres, J.C. Carel, L. Castano, et al., Factors associated with early remission of Type I diabetes in children treated with cyclosporine, *N Engl J Med.* 318:633 (1988).

42. The Canadian-European Randomized Control Trial Group, Cyclosporin-induced remission of IDDM after early intervention: Association of 1 yr of cyclosporin treatment with enhanced insulin secretion, *Diabetes* 37:1574 (1988).
43. G.S. Eisenbarth, "Immunotherapy of diabetes and selected autoimmune diseases," Boca Raton, CRC Press (1989).
44. H. Kolb, M. Schmidt, U. Kiesel, Immunomodulatory drugs in Type I diabetes, *in:* "Immunotherapy of Diabetes and Selected Autoimmune Diseases," G.S. Eisenbarth ed., CRC Press Boca Raton (1989).
45. J.L. Mahon, J. Dupre, C.R. Stiller, et al., Immunosuppression in IDDM: Rationale, risks, benefits and strategies {comment}, *Diabetes Care* 13:806 (1990).
46. J. Silverstein, N.K. MacLaren, W. Riley, Immunosuppression with azathioprine and prednisone in recent-onset insulin-dependent diabetes mellitus, *N Engl J Med.* 319:599 (1988).
47. G. Mendola, R. Casamitijana, R. Gomis, Effect of nicotinamide therapy upon beta cell function in newly diagnosed Type I (insulin-dependent) diabetes patients, *Diabetologia* 32:160 (1989).
48. S.C. Shah, J.L. Malone, N.E. Simpson, A randomized trial of intensive insulin therapy in newly diagnosed insulin-dependent diabetes mellitus, *N Engl J Med.* 320:550 (1989).
49. B. Vialettes, R. Picq, M. DuRostu, et al., A preliminary multicentre study of the treatment of recently diagnosed Type I diabetes by combination nicotinamide: Cyclosporine, *Diabetic Med.* 7:731 (1990).
50. R. Lipton, R. LaPorte, D. Becker, et al., Cyclosporine therapy for the prevention and cure of insulin-dependent diabetes: An epidemiologic perspective of benefits and risks, *Diabetes Care* 13(7):776 (1990).
51. A.L. Drash, S.A. Arslanian, Can insulin-dependent diabetes mellitus be cured or prevented? A status report on immunomodulatory strategies and pancreas transplantation, *in:* "Pediatric Clinics of North America", P. Mahoney, ed., Philadelphia, Saunders, (1990).
52. R. LaPorte, H. Fishbein, L. Kuller, et al., The Pittsburgh insulin-dependent (IDDM) registry: The incidence of insulin-dependent diabetes mellitus in Allegheny County, Pennsylvania, *Diabetes* 30:279 (1981).
53. A. Drash, "Clinical Care of the Diabetic Child", Chicago, Il., Yearbook Medical Publishers, Inc., (1987).
54. M. Rewers, R. LaPorte, H. King, et al., Trends in the prevalence and incidence of diabetes: Insulin-dependent diabetes in childhood, *World Health Stat Q.* 41:179 (1988).
55. L. Blom, G. Dahlquist, L. Nystrom, et al., The Swedish childhood diabetes study - Social and perinatal determinants for diabetes in childhood, *Diabetologia* 32: 7 (1989).
56. G. Dahlquist, L. Blom, T. Tuvemo, et al., The Swedish childhood diabetes study - Results from a nine year case register and a one year case-referent study indicating that Type I (insulin-dependent) diabetes mellitus is associated with both Type II (non-insulin-dependent) diabetes mellitus and autoimmune disorders, *Diabetologia*, 32:2 (1989).
57. J. Karam, P. Prosser, P. Lewitt, Islet cell surface antibodies in a patient with diabetes mellitus after rodenticide ingestion, *N Engl J Med.* 299:11 (1979).
58. J. Karam, P. Lewitt, C. Young, et al., Insulinopenic diabetes after rodenticide (Vacor) ingestion: A unique model of acquired diabetes in man, *Diabetes* 29:971 (1980).
59. R. Sibley, R. Sutherland, F. Goetz, et al., Recurrent diabetes mellitus in the pancreas iso - and allograft, *Lab Invest.* 53:132 (1985).

60. D.E.R. Sutherland, K.C. Moudry-Munns, B.A. Elick, Pancreas transplantation, *in:* "Diabetes Mellitus: Theory and Practice", H. Rifkin, D. Porte eds., New York, Elsevier Publishers (1990).
61. B.A. Mijllward, L. Aviggi, P.J. Hoskins, et al., Immune changes associated with insulin-dependent diabetes may remit without causing the disease: A study of identical twins, *Br Med J.* 292:793 (1986).
62. D.K. McCullock, L.J. Klaff, S.E. Kahn, et al., Nonprogression of subclinical ß cell dysfunction among first degree relatives of IDDM patients: Five year follow-up of the Seattle Family Study, *Diabetes* 39:549 (1990).
63. T. Mandrup-Poulsen, J. Molvig, H.U. Andersen, et al., The Canadian-European Randomized Control Trial Group, Lack of predictive value of islet cell antibodies, insulin antibodies, and HLA-DR phenotype for remission in cyclosporin-treated IDDM patients, *Diabetes* 39:204 (1990).
64. P.F. Bougneres, P. Landais, C. Boisson, et al., Limited duration of remission of insulin-dependency in children with recent overt Type I diabetes treated with low-dose cyclosporin, *Diabetes* 39:1264 (1990).
65. P. Chase, N. Simmon-Butler, S. Garg, et al., Cyclosporine A for the treatment of new onset insulin-dependent diabetes mellitus, *Pediatr.* 85:241 (1990).
66. J. Dupre, M.R. Jenner, J.L. Mahon, et al., Endocrine-metabolic function in remission-phase IDDM during administration of cyclosporine, *Diabetes* 40:598 (1991).
67. T. Orchard, J. Dorman, R. LaPorte, et al., Host and environmental interactions in diabetes mellitus, *J Chron Dis.* 36:979 (1986).
68. A. Drash, Reflection on the beta cell and its demise, Diabetes in the young, *Bull ISGD.* 19:233 (1988).
69. H. Kolb, Mouse models of insulin-dependent diabetes: Low dose streptozotocin-induced diabetes in nonobese diabetic (NOB) mice, *Diabetes Metab Rev.* 8:751 (1987).
70. J. Yoon, Role of viruses in the pathogenesis of IDDM, *Ann Med.* 23:437 (1991).
71. R. Elliott, J. Martin, Dietary protein: A trigger of insulin-dependent diabetes in the BB rat?, *Diabetologia* 26:297 (1984).
72. D. Daneman, L. Fishman, C. Clarson, et al., Dietary triggers of insulin-dependent diabetes in the BB rat, *Diabetes* 5:93 (1987).
73. D. Coleman, J. Kuzava, E. Leiter, Effect of diet on incidence of diabetes in nonobese diabetic mice, *Diabetes* 39:432 (1990).
74. W. Sumoski, H. Baquerizo, A. Rabinovitch, Oxygen free radical scavengers protect rat islet cell from damage by cytokines, *Diabetologia* 32:792 (1989).
75. A. L. Drash, W.A. Rudert, S. Borquaye, et al., Effect of probucol on development of diabetes mellitus in BB rats, *Am J Cardiol.* 62(3):27 (1988).
76. A.A. Like, M. Anthony, D.L. Guberski, et al., Spontaneous diabetes mellitus in the BB/W rat: Effect of glucocorticoids, cyclosporin A and antiserum to rat lymphocytes, *Diabetes* 32:326 (1983).
77. N. Murase, I. Lieberman, M. Nalesnk, et al., Prevention of spontaneous diabetes in the BB rat with FK-506, *Lancet* 336:373 (1990).
78. A. H. Barnett, C. Eff, R.D.G. Leslie, et al., Diabetes in identical twins: A study of 200 pairs, *Diabetologia* 20:87 (1981).
79. P. Olmos, R.A. Ahern, D.A. Heaton, et al., The significance of the concordance rate for Type I (insulin-dependent) diabetes in identical twins, *Diabetologia* 31:747 (1988).
80. C. Johnston, B. Millward, P. Hoskins, et al., Islet cell antibodies as predictors of the later development of Type I (insulin-dependent) diabetes: A study in identical twins, *Diabetologia* 32:382 (1989).

81. R.E. LaPorte, N. Tajima, H.R. Akerbloom, et al., Geographic differences in the risk of insulin-dependent diabetes mellitus. The importance of registries, *Diabetes Care* 89:101 (1985).
82. Diabetes Epidemiology Research International, Geographic patterns of childhood insulin-dependent diabetes mellitus, *Diabetes* 37:1113 (1988).
83. WHO DIAMOND Project Group, WHO multinational project for childhood diabetes, *Diabetes Care* 13:1062 (1990).
84. F.K. Gey, P. Puska, J. Paul, et al., Inverse correlation between plasma vitamin E and mortality from ischemic heart disease in cross-cultural epidemiology, *Am J Clin Nutr.* 53:326S (1991).
85. F. Gey, M. Ulrich, P. Jordan, et al., Increased risk of cardiovascular disease at suboptimal plasma levels of essential antioxidants: An epidemiological up-date with special attention to carotene and vitamin C, *Am J Clin Nutr.* In Press.
86. D. Steinberg, S. Parthasarathy, T. Carew, et al., Beyond cholesterol: Modifications of low-density lipoprotein that increase its atherogenicity, *N Eng J Med.* 320:915 (1989).
87. R. Riemersma, D. Wood, C.C.A. Macintyre, et al., Risk of angina pectoris and plasma concentrations of vitamins A, C and E and carotene, *Lancet* 337:1 (1991).
88. B. Grei, Ascorbic acid protects lipids in human plasma and low-density lipoprotein against oxidative damage, *Am J Clin Med.* 54:113S (1991).
89. S. Keizo, E. Nikim, H. Shimasaki, Free radical-mediated chain oxidation of low-density lipoprotein and synergistic inhibition by vitamin E and vitamin C, *Arch Biochem Biophys.* 279:402 (1990).
90. B. Halliwell, J. Gutteridge, C.E. Cross, Free radicals, antioxidants, and human disease: Where are we now?, *J Lab Clin Med.* 119:598 (1992).
91. J. A. Olson and S. Kobayashe, Antioxidants in health and disease: Overview, *Proc Soc Exp Biol Med.* 200:245 (1992).
92. J. Dorman, R. LaPorte, R. Stone, et al., Worldwide differences in the incidence of Type I diabetes are associated with amino acid variation at position 57 of the HLA-DQbeta chain, *Proc Natl Acad Sci. USA* 87:7370 (1990).
93. L. Contu, C. Carcassi, M. Trucco, Diabetes susceptibility in Sardinia, *Lancet* 338:65 (1991).
94. R. Lipton, M. Kocova, R. LaPorte, et al., Autoimmunity and genetics contribute to the risk of insulin-dependent diabetes mellitus in families: Islet cell antibodies and HLA-DQ heterodimers, *Am J Epidemiol.* 136:503 (1992).
95. Diabetes Epidemiology Research International Group, Secular trends in incidence of childhood IDDM in 20 countries, *Diabetes* 39:858 (1990).
96. M. Metcalfe, J. Baum, Incidence of insulin-dependent diabetes in children aged under 15 years in the British Isles during 1988, *Br Med J.* 302:443 (1991).
97. G. Joner, O. Sovik, Increasing incidence of diabetes mellitus in Norwegian children 0-14 years of age (1973-1982), *Diabetologia* 32:79 (1989).
98. L. Nystrom, G. Dahlquist, M Rewers, et al., The Swedish childhood diabetes study. An analysis of the temporal variation in diabetes incidence 1978-1987, *Int J Epidemiol.* 19:141 (1990).
99. J. Tuomilehto, M. Rewers, A. Reunanen, et al., Increasing trend in Type I (insulin-dependent) diabetes mellitus in childhood in Finland. Analysis of age, calendar time and birth cohort effects during 1965-1984, *Diabetologia* 34:282 (1991).
100. A. Green, E.A. Gale, C.C. Patterson, Incidence of childhood-onset insulin-dependent diabetes mellitus, the EURODIAB ACE Study, *Lancet* 339:905 (1992).

101. S. Muntoni, L.M. Songini, High incidence rate of IDDM in Sardinia, *Diabetes Care* 15:1317 (1992).
102. M. Rewers, R. LaPorte, M. Walczakj, et al., Apparent epidemic of insulin-dependent diabetes mellitus in midwestern Poland, *Diabetes* 36:1206 (1987).
103. M. Rewers, R. Stone, R. LaPorte, et al., Poisson regression modeling in temporal variation in incidence of childhood insulin-dependent diabetes mellitus in Allegheny County, Pennsylvania and Wielkopolska, Poland, *Am J Epidemiol.* 129:569 (1989).
104. L. Wagenknecht, J. Roseman, W. Herman, Increased incidence of insulin-dependent diabetes mellitus following an epidemic of coxsackie virus B5, *Am J Epidemiol.* 133:1024 (1991).
105. A.C. Burden, M.L. Burden, E.R. Williams, et al., Evidence of frequent epidemics of childhood diabetes, *Diabetes* 40:373A (1991).
106. World Health Organization DIAMOND Project Group on Epidemics, Childhood diabetes, epidemics and epidemiology: An approach for controlling diabetes, *Am J Epidemiol.* 135:803 (1992).
107. T. Dokheel, The Pittsburgh Diabetes Epidemiology Research Group, An epidemic of childhood diabetes in the United States? Evidence from Allegheny County. (In Press).
108. G. Mimura, Present status and future view of the genetic study of diabetes mellitus, *in:* "Clinico-genetic geneses of diabetes mellitus", G. Mimura, S. Baba, Y. Goto, J. Köbberling, eds, Amsterdam Excepta Medica (1982).
109. E. Colle, J. Siemiatycti, J.R. West, et al., Incidence of juvenile onset diabetes in Montreal: Demonstration of ethnic differences and socio-economic class differences, *J Chronic Dis.* 34:611 (1981).
110. Z. Laron, M. Karp, M. Modan, The incidence of insulin-dependent diabetes mellitus in Israeli children and adolescents 0-20 years of age: A retrospective study 1975-1980, *Diabetes Care* 8:24 (1985).
111. H. F. Harris, A case of diabetes mellitus following mumps, *Boston Med Surg J.* 140:645 (1987).
112. D.R. Gamble, K.W. Taylor, H. Cumming, Coxsackie viruses in diabetes mellitus, *Br Med J.* 4:260 (1984).
113. J. W. Yoon, M. Austin, T. Ondera, et al., Virus induced diabetes mellitus: Isolation of a virus from the pancreas of a child with diabetic ketoacidosis, *N Engl J Med.* 300:1173 (1979).
114. C. W. Pak, H. Meun, R.G. McArthur, et al., Association of cytomegalovirus infection with autoimmune Type I diabetes, *Lancet* ii:1(1988).
115. H.A. Sultz, B.A. Hart, M. Zielezny, et al., Is mumps virus an etiologic factor in juvenile diabetes mellitus?, *J Pediatr.* 86:654 (1975).
116. J. M Forrest, M.A. Menser, J.A. Burgess, High frequency of diabetes mellitus of young adults with congenital rubella, *Lancet* 2:332 (1971).
117. F. Ginsberg-Fellner, L.M.W. Witt, B. Fedun, Diabetes mellitus and autoimmunity in patients with congenital rubella syndrome, *Rev Infect Dis.* (suppl. 1:S170 (1985).
118. A.L. Drash, Evidence for viral infections in the etiology of insulin-dependent diabetes mellitus, *in:* "Diabetologie Pediatrique", P. Czernichow, H. Dorchy eds., Paris (1990).
119. S.S. Roberts, New clues to IDDM origins: IDDM may arise from a case of mistaken identity in which the immune system mistakes a normal beta cell antigen for a virus, *Diabetes Care* 15:137 (1992).

120. J. Yoon, C. Pai, Viral etiology of IDDM, *in:* Genetic and Environmental Factors for Type I Diabetes", Z. Laron , M. Karp eds., Freund Publishing House, Ltd. London, England (1992).
121. T. Helgason, M.R. Jonnasson, Evidence for a food additive as a cause of ketosis-prone diabetes, *Lancet* 2:716 (1981).
122. T. Helgason, S.W.B. Ewen, I.S. Ross, et al., Diabetes produced in mice by smoked/cured mutton, *Lancet* 2:1017 (1982).
123. K. Borch-Johnsen, T. Mandrup-Poulsen, B. Zachau-Christiansen, et al., Relation between breast-feeding and incidence rates of insulin-dependent diabetes mellitus, *Lancet* 2:1083 (1984).
124. F.W. Scott, D. Daneman, J.M. Martin, Evidence for a critical role of diet in the development of insulin-dependent diabetes mellitus, *Diabetes Res.* 7:153 (1988).
125. E.J. Malyer, R.F. Hammkah, E.C. Gay, et al., Reduced risk of IDDM among breast-fed children. The Colorado IDDM Registry, *Diabetes* 37:1625 (1988).
126. G. Dahlquist, L. Blom, L. Persson, et al., Dietary factors and the risk of developing insulin-dependent diabetes in childhood, *Br Med J.* 300:1302 (1990).
127. F.W. Scott, Cow milk and insulin-dependent diabetes mellitus: Is there a relationship?, *Am J Clin Nutr.* 51:489 (1990).
128. K. Dahl-Jorgensen, G. Joner, K. Hanssen, Relationship between cow's milk consumption and incidence of IDDM in childhood, *Diabetes Care* 14:1081 (1991).
129. S.M. Virtanen, L. Rasanen, A. Aro, et al., Childhood Diabetes in Finland Study Group, Infant feeding in Finnish children <7yrs. of age with newly diagnosed IDDM, *Diabetes Care* 14:415 (1991).
130. R.B. Elliot, N. Bibby, S. Reddy, Casein peptide precipitates diabetes in the non-obese diabetic mouse and possibly humans, *in:* "Genetic and Environmental Risk Factors for Type I Diabetes", Z. Laron, M. Karp, eds., Freund Publishing House, Ltd. London, England (1992).
131. S. Virtanen, Dietary risk factors of IDDM in children, *in:* "Genetic and Environmental Risk Factors for Type I Diabetes", Z. Laron, M. Karp, eds., Freund Publishing House, Ltd. London, England (1992).
132. S. Virtanen, L. Rasanen, A. Aro, et al., and the Childhood Diabetes in Finland Study Group, Feeding in infancy and the risk of Type I diabetes mellitus in Finnish Children, *Diabetes Med.* 9:815 (1992).
133. E. Savilahti, H. Akerblom, V-M Tainio, et al., Children with newly diagnosed insulin-dependent diabetes mellitus have increased levels of cow's milk antibodies, *Diabetes Res.* 7:137 (1988).
134. J. Karjalainen, J. Martin, M. Knip, et al., Evidence for a bovine albumin peptide as candidate trigger of Type I diabetes, *N Engl J Med.* 327:302 (1992).
135. H-M Dosch, M.F. Hui, P.J. Doherty, Development of normal and abnormal immune function in man, *Dev Immunol.* 2. (In press).
136. O.Olerup, C.I.E. Smith, L. Hamarstrom, Different amino acids at position 57 of the HLA DQ beta chain associated with susceptibility and resistance to IgA deficiency, *Nature* 347:289 (1990).
137. J. R. Lancaster, Nitric oxide in cell, *Am Sci.* 80:248 (1992).
138. S. H. Snyder, D.S. Bredt, Biological roles of nitric oxide, *Sci Am.* 80:68 (1992).
139. A. Bateman, A. Singh, T. Kral, et al., The immune-hypothalamic-pituitary-adrenal axis, *Endocrinol Rev.* 10:92 (1989).
140. D. Gupta, Pathophysiology of immune-neuroendocrine network, *Network Lett.* 14:1 (1992).

141. Diabetes Epidemiology Research International, Preventing insulin-dependent diabetes mellitus: The environmental challenge, *Br Med J.* 295:479 (1987).
142. F.W. Scott, R.B. Elliott, H.Kolb, Diet and autoimmunity: Prospects of prevention of Type I diabetes, *Diabetes, Nutr Metab.* 2:61 (1989).
143. J. Palmer, D. McCulloch, Prediction of IDDM, *Diabetes* 40:943 (1991).
144. N. Maclaren, M. Atkinsen, Editorial: Is insulin-dependent diabetes environmentally induced?, *N Engl J Med.* 3237:348 (1992).
145. P. Pozzilli, A. Segnore, D. Andreani, What future for therapeutic prevention of Type I diabetes mellitus, *Diabetology* 35:1093 (1992).
146. A. Lemmark, Immune intervention yes, but for what reason, for whom, when and how?, *Diabetologia* 35:1096 (1992).
147. R. LaPorte, S. Baba, Magic bullets, reportable disease and prevention of childhood diabetes, *Diabetes Care* 15:128 (1992).

IDDM EPIDEMIOLOGY IN RUSSIA: CURRENT AND FUTURE DIRECTIONS

I. Dedov, N. Lebedev, U. Sunzov, N. Gubanov

National Endocrinology Centre
Institute of Diabetes
ul. Dm. Uljanova, 11
Moscow, 117036
RUSSIA

INTRODUCTION

Various aspects of IDDM have been extensively studied in different regions of the world during the last decades. IDDM prevalence and incidence data plotted on the map of the world cover vast areas in all populated parts of the globe (Figure 1). Nevertheless, there is an enormous territory that remains a "blank space" or "terra incognita" on the epidemiological map of IDDM. This "terra incognita" is Russia and most of the Republics of the former USSR.

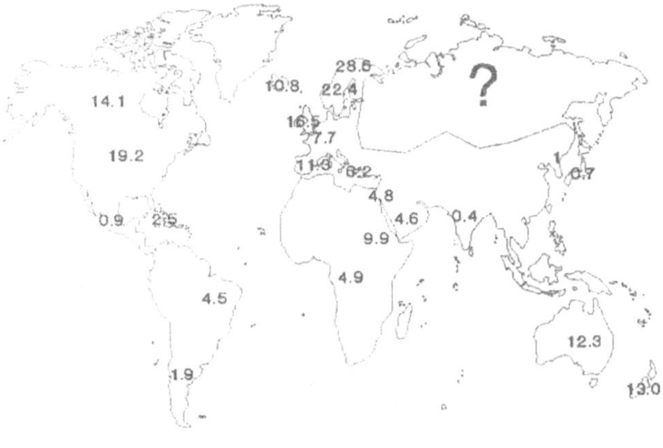

Figure 1. IDDM incidence in different parts of the world.

Standardization of Epidemiologic Studies of Host Susceptibility
Edited by J. Dorman, Plenum Press, New York, 1994

This deplorable lack of information may be attributed to the following:

1) Results of epidemiological surveys obtained in Russia and former USSR were, as a rule, unavailable to diabetologists in other countries due to the information blockade.
2) New classification of diabetes mellitus recommended by WHO in 1979 was not fully adopted in the former USSR until 1987.
3) There was no national program of IDDM epidemiologic investigations, and most of the surveys conducted in the 1970's-1980's were performed by isolated groups of specialists in different regions of the country.

The authors of the present communication hope that it will enable the audience to comprehend the current epidemiologic situation in Russia with respect to IDDM.

Demographic Features of the Country

Russian demographic statistics are based on the results of the national census that were carried out in the former USSR every 10 years. The last one was completed January 1, 1989. Some of its results are shown in Table 1.

Table 1. Population estimates for the former USSR.

Total Population:	147,400,000			
Children:	38,000,000			
Ethnic Composition:	Russians	81.3%		
	Tatars	3.7%	Caucasoids	92%
	Ukrainians	2.9%	Non-Caucasoids	8%
	Chuvashes	1.2%		
	Others	10.9%		

The total population of Russia in the late 1980's was approximately 150,000,000, with Russians exceeding four-fifths of this figure. They were followed by Tatars, Ukrainians, and Chuvashes. The population was dominated by Caucasoids, representing approximately 92% of the total. The urban population was 51.5% of the total. The population of Moscow, the largest city, was 8,800,000 or 6% of the total population of Russia.

Recording Diabetic Patients in Russia

Russia has no national IDDM Registry. The major source of information regarding the morbidity experience in different regions of the country (Table 2) is summarized by annual reports of local general practitioners, such as therapists and pediatricians, who deal with diabetic patients. Minor sources of data include reports from medical facilities in districts, provinces and other territorial units of the country, along with information from the pharmacy service that supplies diabetics with insulin.

Table 2. Sources of information regarding diabetes in Russia.

Available source of information	Importance for morbidity statistic
Reports of general practitioners	+++
Annual hospital reports	++
Annual drug store reports	+
Reports of regional diabetic associations	+/-
Reports of health insurance bodies	-
Screening studies	+/-

These sources operate as follows. Primary information comes to the regional health administration from local endocrinologists, including pediatricians. It is further concentrated in the Departments of Medical Statistics of large cities and provinces, to be finally accumulated in the Russian Ministry of Health.

IDDM Epidemiology in Adults

IDDM prevalence in adult populations from different regions was studied on a limited scale in the 1980's. The results are summarized in Figure 2 which characterizes the epidemiologic situation, not only in Moscow, but also in Siberia and some northern areas of the country (Chucotsk, Kamchatka, Yakutia) with their ethnically diverse populations, including Evenks, Koryaks, Nanaians, Nenets, Yakuts, etc. Russians comprise most of the non-aboriginal group (over 80%).

Figure 2. Prevalence of IDDM in adults (20-69 years of age) in Russia.

The prevalence of diabetes in Moscow was estimated from the results of the 1991 screening study on 99,800 subjects between 20 and 69 years of age in one of the city's communities. A study of the morbidity experience in native populations of the Far North revealed a low prevalence of IDDM (0-1 per 100,000), although its prevalence among immigrants was 15-20.7 per 100,000. The latter figures were not significantly different from those characterizing IDDM prevalence in populations from where the immigrants originated.

IDDM Epidemiology in Children

The number of diabetic children in Russia and throughout the rest of the former Soviet Union, and estimates of the prevalence of childhood IDDM during the last two decades has been the subject of much more accurate and detailed evaluation than studies in the adult population. This was due to the fact that health information at the local, regional and national levels was traditionally separate for adults and children. Moreover, the accuracy of IDDM status for children was greatly promoted by the analysis of annual applications for insulin from local health authorities, indicating both the total number of IDDM patients and the number of newly diagnosed cases.

Data on the IDDM morbidity experience of children in Russia and other Republics of the former USSR normally referred to patients under the age of 15 years. According to the Department of Medical Statistics, Russian Ministry of Health (December 31, 1988), there were 11,180 children of this age group with IDDM or 32.7 diabetics per 100,000 children. Annual incidence in 1985-1988 varied from 5.8 to 6.3/100,000. Comparison with other regions of the world revealed a relatively low incidence of IDDM in Russian children compared to France, Germany, and Poland in the second half of the 1980s.

During the 1980s, we undertook an investigation of IDDM incidence in children living in five large cities of Russia, each with the total population in excess of 1,000,000 (Figure 3). As illustrated, the rates in Novosibirsk and Ekaterinburg (Siberia) were practically identical to those for Moscow and Saint Petersburg (NW Russia): 5.8 to 7.4/100,000. The incidence of IDDM in the city of Ufa (Bashkiria) was significantly lower (4.2/100,000). It is worthwhile to note that the Russian population of Ufa was 52% of the total, whereas in the four other cities, it comprised 73-88%.

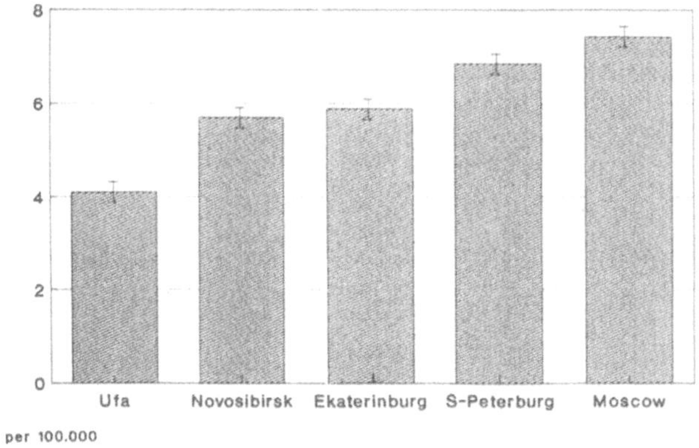

Figure 3. Incidence of IDDM in an urban pediatric population of Russia (1980-1989).

The prevalence of IDDM in children in different regions of Russia during the second half of the 1980's is illustrated by Figure 4. Analysis of these data with respect to the so-called "North-South gradient" (Figure 5) appears to confirm its existence. Eighty-six percent of the regions with a low prevalence of IDDM are located south of 57 N.L., whereas 80% of those with high IDDM prevalence can be found north of 57 N.L.

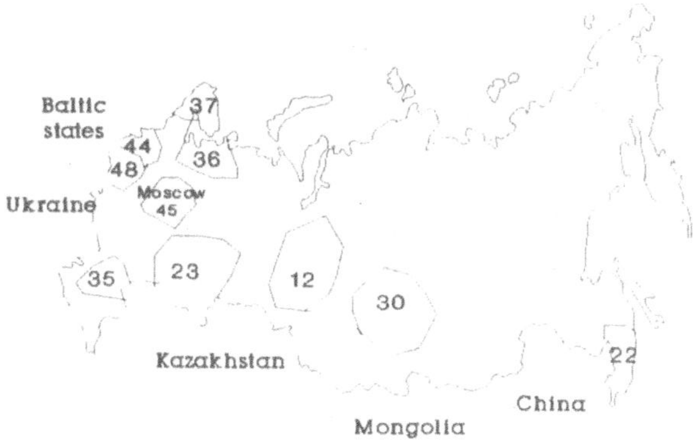

Figure 4. Prevalence of IDDM in Russian children (1985-1989).

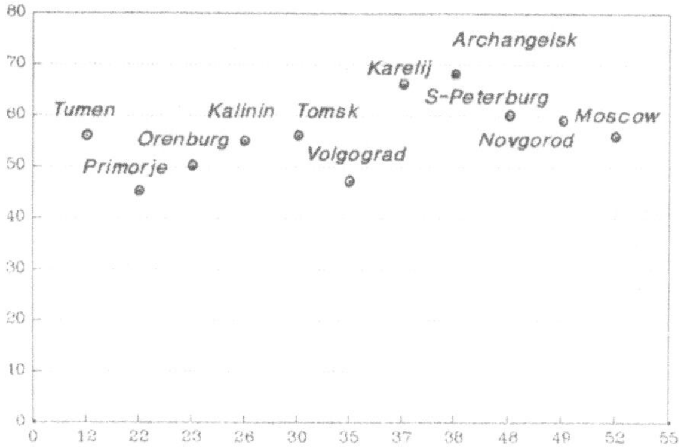

Figure 5. Correlation between IDDM prevalence in various parts of Russia and the latitude of the registry area.

These patterns are even more conspicuous if the analysis is based on the data collected for former Soviet Republics in 1985-1988. It appears that all the Republics having the major part of their territory north of 50 N.L. (Ukraine, Byelaruss, Lithuania, Russia, Estonia), show an IDDM prevalence index above 28, whereas almost all southern Republics with territories lying largely between 35 and 50 N.L. were characterized by lower prevalence of IDDM (5-20 per 100,000). The exception is George, situated much more to the south as compared with Byelaruss or Russia, which shows a similar prevalence of IDDM.

Results obtained in our epidemiologic study in Moscow appear to be of particular importance with respect to selected features of IDDM prevalence in children. It has already been mentioned that the population of the city is 6% of the total. The percentage of Russians in Moscow is also similar to that in the rest of Russia (77% and 81.3%, respectively). Therefore, the epidemiologic situation in Moscow may be considered to reflect the situation in the country.

The annual incidence of IDDM in Moscow children has been recorded since 1967. Figure 6 shows a marked increasing trend in incidence during the last two decades, with a 41% rise between the 1970's and the 1980's (Figure 7).

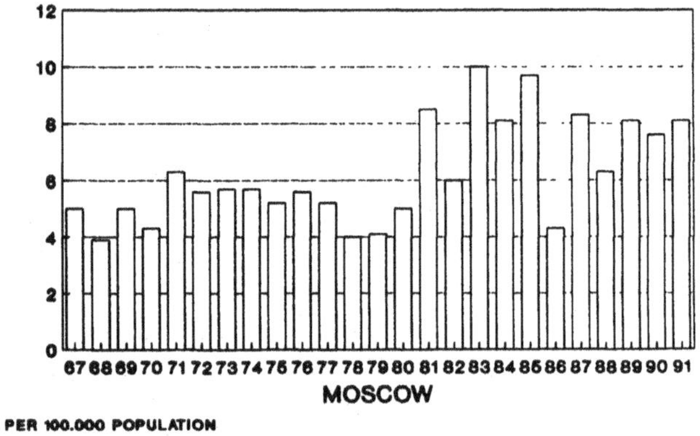

Figure 6. Incidence of IDDM in children age 0-14 years.

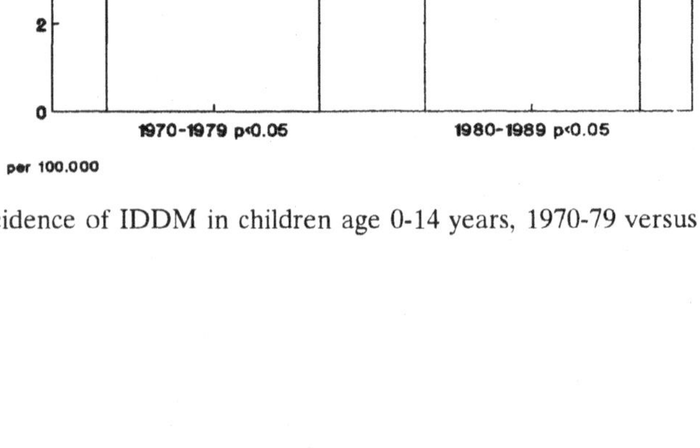

Figure 7. Incidence of IDDM in children age 0-14 years, 1970-79 versus 1980-1989.

The mortality status of Moscow children with IDDM is illustrated by the comparison of our 1975-1989 studies with the mortality experience documented in four other countries (USA, Japan, Finland, and Israel)[1] (Figure 8). The annual mortality of IDDM children in Moscow (per 100,000 subjects) proved to be much higher than in Finland and Israel, but similar to that in USA (Pennsylvania). Seventy-eight percent of deaths among the children in our group were from diabetic coma.

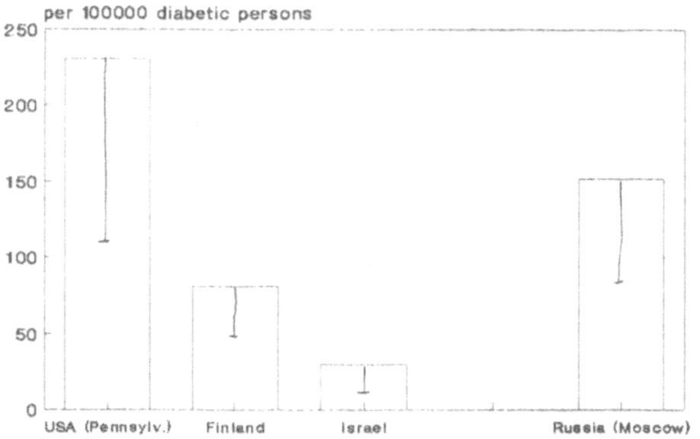

Figure 8. Annual mortality in IDDM children by country.

Prospects of Epidemiologic Studies in Russia

Up-to-date standard methods of epidemiologic investigations need to be developed, taking into account marked differences in IDDM prevalence in ethnic groups within the boundaries of the present day Russia and in numerous regions of the country with specific climatic and geographic conditions. Comparable and statistically significant results can be obtained only in the framework of the national program, stimulating coordination of epidemiologic surveys and application of standard techniques. We represent here the WHO Collaborating Centre of Epidemiology, Prevention and Diabetes Care, recently established in Moscow, and could take responsibility for these activities.

The Russians' program of epidemiology of diabetes is expected to be developing along the following lines:

- advising and contributing to the State program for combatting the most common non-communicable diseases,
- organization of regional IDDM registries (1991-1996) to be crowned by the National IDDM registry (2000),
- studies of epidemiology of IDDM complications,
- immunogenetic studies in relatively isolated ethnic groups,
- IDDM family studies.

To be implemented, this program needs organizational and methodological support by different bodies involved in fundamental and applied studies of diabetes mellitus.

REFERENCES

1. Diabetes Epidemiology Research International Mortality Study Group, Major-Cross country differences in risk of dying for people with IDDM, *Diabetes Care*, 14:49 (1991).

HLA-DQB1, DQA1 AND DPB1 ALLELES IN SARDINIAN PATIENTS WITH INSULIN-DEPENDENT DIABETES MELLITUS

Mauro Congia,[1] Franceso Muntoni,[2] Francesco Cucca,[3] Fulvia Frau,[1] Rosanna Lampis,[1] Susanna Porcu,[2] Marco Songini,[4] and Sergio Muntoni[4]

[1]Institute of Clinic and Biology of Growing Age
Cagliari University
Via Jenner
09134 Cagliari
ITALY

[2]Center for the Study of Thalassemias and Mediterranean Anemias, CNR
Via Jenner
09134 Cagliari
ITALY

[3]Department of Paediatrics, USL 21
St. Michele Hospital
Via Peretti
09134 Cagliari
ITALY

[4]Center for Metabolic Diseases
St. Michele Hospital
09134 Cagliari
ITALY

INTRODUCTION

The incidence of insulin-dependent diabetes mellitus (IDDM) in Caucasians shows remarkable differences, with rates much lower in Central and Southern Europe than in Northern Europe. On the basis of these data, a North-South gradient has been proposed. However, this idea has recently been challenged by the incidence data from Sardinia, an island located in the central part of the Mediterranean Sea. The Sardinian population has an extremely high incidence of IDDM[1,2] (30.2 cases per 100,000 per year; the second highest rate recorded in Europe). It may be considered, due to the high degree of genetic homogeneity, a convenient place for genetic studies. As far as HLA genes are concerned, several studies have shown some peculiarities of our population.[3] For example, the extended haplotypes A30, B18, DR3, DRw52, DQw2[3] and A2, Cw7, B17, 3F31, DR2, DQw1,[4] rare in other Caucasians, are very frequent on our island.

Recently, the DQB1 locus and, in particular, the residue at position 57 of the DQβ chain, was found to be highly correlated with the susceptibility to develop IDDM in Caucasians;[5,6] aspartic acid at this position conferred resistance, while a neutral residue (Val, Ala, Ser) was associated with susceptibility.[5] Subsequent reports indicated that HLA class II genes other than DQB1 alleles have an independent role in IDDM susceptibility.[7] In this respect, several DQA1/DQB1 heterodimers expressed at the cell surface have been implicated in susceptibility to the disease, independently from linkage disequilibrium with DRB1 alleles.[8]

With the aim to find a correlation between high IDDM incidence and genetic susceptibility linked to the HLA system, we performed a molecular study on HLA class II DQA1, DQBl and DPBI genes in healthy subjects and in a group of sporadic IDDM children of Sardinian origin. Our results show that several HLA class II alleles implicated in IDDM susceptibility occur in healthy Sardinians with a higher frequency than in other Caucasian populations. In particular, a very high frequency of Arg-52 positive *and* Asp-57 negative homozygosity was found in the control population. On the basis of our results, we hypothesize that the increased genetic susceptibility to develop IDDM may be due to the particular distribution of HLA class II alleles in Sardinians and it may offer the best explanation for the observed high incidence.

MATERIALS AND METHODS

DNA extraction was carried out by standard methods.[9] Class II typing for DQA1, DQB1, and DPB1 genes were molecularly characterized using allele specific (ASO) and sequence specific (SSO) oligonucleotide probes in a dot blot format on DNA enzymatically amplified from the second exons of the DQA1, DQB1, DPB1 loci with previously reported primers and profiles of amplification.[10,11,12] Nomenclature was assigned according to the official Nomenclature for Factors of the HLA System 1990.[13] Haplotypes were assigned considering the most common allelic combination reported in Caucasians.[14,15] The direct sequencing of single strand amplified DNA using primers specific for different DQA1, DQB1 and DPB1 alleles was performed to confirm the identity of the alleles typed with the oligonucleotide assay.

Statistical Evaluation

We used the chi-square test (with Yate's corrections) for statistical evaluation of the data. Level of significance was set to 0.05 and the p values were adjusted for the number of comparisons, when necessary. The relative risks (RR) were calculated with the Wolf's formula [(a x d)/(b x c)], and when one element of this equation was 0, we used the Haldane's modification.

RESULTS

The distribution of the most frequent haplotypes in Sardinians, compared to the frequency found in other Caucasians[6] is shown in Table 1. The frequency of these haplotypes varies greatly. For example, the DRw16, Dw21 haplotype, a very rare haplotype in Caucasians,[5] accounts for only for 1.5% of all haplotypes, but represents 34% of all Sardinian haplotypes.[4,16] The remaining DR2 haplotypes are very rare in our population, only 3% of all haplotypes, but are common in other Caucasian groups. The B18,DR3,DQw2 haplotype is another haplotype present in Sardinians with an extremely high frequency. As far as DR4 haplotype frequencies are concerned, in Sardinia the DR4,DQB1*0302 haplotype accounts for 10% of all haplotypes, while in Caucasians its prevalence is almost double.

Table 1. Comparison of the frequencies of some haplotypes implicated in IDDM susceptibility among Caucasians and Sardinians.

	Sardinians	Caucasians
DR2,Dw21,DQA1*0102, DQB1*0502	34.0%	1.0%
DR2,Dw2,DQA1*0102, DQB1*0602	3.0%	26.0%
B18,DR3,DQA1*0501, DQB1*0201	30.0%	0.4%
B8,DR3,DQA1*0501, DQB1*0201	0.8%	8.3%
DR4,DQB1*0301	3.4%	5.0%
DR4,DQB1*0302	10.0%	19.0%

HLA Class II Allele Frequencies

The allele frequencies for the various class II loci among the IDDM patients is shown in Table 2.

We found that among DQA1 alleles, only the DQA1*0301 was associated with the disease. Among DQB1 alleles, the strongest association was found with the DQB1*0201 allele, whilst a weaker association was found with the DQB1*0302 allele. It is interesting to note that the DQB1*0502 allele, previously defined as a susceptibility allele,[17] was shown to be significantly decreased in frequency in patients ($p<0.01$) both in our recent study[16] and in a previous study by Carcassi on Sardinian IDDM patients.[4]

As far as the DPB1 locus was concerned, no allele showed an association with the disease. We think that in our population, the DPB1 alleles confer no specific nor any additional risk for IDDM development.

Relative Risk Comparison

The relative risk (RR) for various alleles and allelic combination for Sardinian and Caucasian patients[8] with IDDM is shown in Table 3. The DQA1*0301 allele conferred a RR of 4. This result differs from what has been reported in other Caucasians, where the RR is higher.

Table 2. Allele frequencies for HLA-DQA1, DQB1 and DPB1 loci among 50 IDDM cases and 89 controls of Sardinian descent.

Alleles	IDDMs	Controls	p value
DQA1			
0102	27%	40%	
0101	17%	20%	
0201	2%	9%	
0301	38%	12%	0.003
0501	83%	77%	
DQB1			
0502	24%	34%	0.01
0501	17%	23%	
0201	93%	48%	0.000003
0604	2%	1%	
0302	29%	10%	0.01
DPB1			
0401	32%	55%	
0402	16%	21%	
0201	41%	36%	
0301	49%	48%	
1301	11%	5%	
9-12-17	5%	5%	
1001	2%	3%	

Table 3. Relative risk (RR) for developing IDDM associated with various alleles and genotype combinations in Sardinians and Caucasians.

Alleles	Sardinians RR	Caucasians RR
DQA1*0301	4.0	10.7
DQB1*0201	13.5	1.5
DQB1*0302	3.5	14.1

Genotype Combination	Sardinians RR	Caucasians RR
DQB1*0201/*0302	25.0	20.0
DQB1*0201/*0201	8.7	2.0
DQB1*0302/*0302	-	7.0
DR3/DR4	6.7	12.6
DR4,DQA1*0301, DQB1*0201	17.0	-

Concerning the DQB1 locus, we found several differences between Sardinians and other Caucasians. The greatest relative risk in our patients was conferred by the DQB1*0201 (RR = 13.5) instead of the DQB1*0302 (RR = 3.5). This is the opposite situation compared with the rest of Caucasians, where the greatest risk is conferred by DQB1*0302 allele (RR = 14.1) and not by the DQB1*0201 (RR = 1.5).[8,18,19] As far as genotype combinations were concerned, we found that DQB1*0201/DQB1*0302 conferred the greatest relative risk (RR = 25). This is in accordance to what is found for other Caucasians.[8,19]

When the frequency of homozygosity for the DQB1 alleles was analyzed, we found that Sardinian controls had an excess of DQB1*0201 homozygosity (RR = 8.7), while there were no patients who were homozygous for the DQB1*0302 allele. This is in contrast to what is consistently found in other Caucasian populations where most, if not all, of the homozygous patients have DQB1*0302 on both chromosomes[8,18,19] (RR = 7). On the other hand, only a slight increase in DQB1*0201 homozygosity is generally found in Caucasians. Another interesting observation was finding a rare DR4,DQA1*0301, DQB1*0201 haplotype, present in approximately 10% of our patients. This haplotype was never observed in controls and, thus, conferred a very high relative risk (RR = 17).

DNA Sequencing

All patients analyzed showed the expected sequence for the DQA1 and DQB1 alleles. A sequence microheterogeneity was identified in both controls and patients in about 30% of the DPB1*0401 alleles at position 36 in which an Alanine (encoded by GCG) was substituted by a Valine residue (encoded by GTG).

DQA1*Arg-52 and DQB1*Asp-57 Residues in 50 IDDM Patients and in 89 Sardinian Controls

The role of two amino acid residues has been implicated in conferring protection or susceptibility to the disease, namely position 57 of the DQß chain and position 52 of the DQα chain. As illustrated in Table 4, we found no Asp+ patients. Moreover, the controls showed a relative under-representation of the Asp+ residues in various combinations. No patient homozygous for the Arg-/Arg- combination was found, confirming the susceptibility effect of the arginine residue. In addition, the number of control individuals carrying the Arg+/Arg+ combination was significantly increased ($p < 0.02$) when compared to healthy individuals of other Caucasian population.[29] These data indicate that in healthy Sardinians, not only are the DQB1*Asp-57-negative alleles increased compared to that of other Caucasians, as previously reported,[6] but the high number of DQA1*Arg-52 positive alleles may also contribute to explain a decrease of protection against IDDM.

Table 4. Various DQA1*-Arg-52 and DQB1*-Asp-57 genotypes observed in 50 IDDM patients and 89 controls of Sardinian origin.

Genotype	IDDM	Controls	p	RR
DQα				
R-52/R-52	56%	34%	0.01	2.6
R-52/NR-52	41%	45%	NS	1.1
NR-52/NR-52	0%	21%	0.001	0.035
DQß				
D-57/D-57	0%	17%	0.005	0.047
D-57/ND-57	0%	39%	$<10^{-6}$	0.015
ND-57/ND-57	100%	36%	$<10^{-6}$	178.6

R = Arginine; NR = Non-Arginine; D = Aspartate, ND = Non-Aspartate

DISCUSSION

In Sardinians, a very high incidence of IDDM has recently been reported.[1,2] With the aim to correlate this high incidence with the well known genetic susceptibility linked to the MHC region, we carried out a molecular characterization of HLA class II genes in 89 healthy individuals and 50 sporadic IDDM children from south Sardinia. We found several HLA system peculiarities in our healthy population that may provide an explanation for the high incidence of the disease.

As far as the distribution of relevant haplotypes in the normal population was concerned, a relative lack of an important haplotype of protection, namely the DR2-Dw2-DQB1*0602 haplotype, was observed. This haplotype is virtually absent in Sardinians, where all DR2 positive individuals carry the DR2,Dw21,DQB1*0502 haplotype. This haplotype does not confer a dominant protection against IDDM (as the DQB1*0602 does), but is a neutral or very weak protective haplotype. In fact, this haplotype was found in IDDM patients mainly in association with the highly susceptible DRB1*0301, DQA1*0501, DQB1*0201 haplotype. No patient was homozygous for the DR2,Dw21,DQB1*0502 haplotype.

As far as the pool of DQB1*0201 alleles was concerned, we found an almost complete association with the DRB1*0301, DQA1*0501, DQB1*0201 haplotype, while the DRB1*0701, DQA1*0201, DQB1*0201 haplotype, not a susceptibility haplotype, is very rare in our island (8% vs. 28% of the Italian mainland[20,21]).

The DRB1*0301, DQA1*0501, DQB1*0201 susceptibility haplotype is very frequent in the normal population, amounting to approximately 40% of all haplotypes. In IDDM patients, we found that the DQB1*0201 allele confers the greatest RR in heterozygosity, while the DQB1*0302 was only weakly associated with the disease. The majority of DQB1*0302 positive individuals carried the DQB1*0201 allele in the other chromosome, confirming that the α/β heterodimer confers a high risk for IDDM. The homozygosity for the DQB1*0201 allele was also found to confer a high RR, while a scarcity of patients homozygous for the DQB1*0302 allele was found. Thus, the DRB1*0301, DQA1*0501, DQB1*0201 haplotype is capable of conferring, in cis, the most susceptible α/β heterodimer and is equivalent to the DR4,DQB1*0301 susceptibility haplotype of Caucasians.

Finally, as far as the distribution of the amino acid residues at position 52 and 57 of the DQα and DQβ chains were concerned, we found an increased frequency of Asp-57-negative/Asp-57-negative and Arg-52 positive/Arg-52 positive individuals in the healthy Sardinians in comparison to other Caucasians.[19]

We conclude that an explanation for the observed high incidence of IDDM in Sardinians may be found in the particular HLA genotype distribution of our healthy controls in which a scarcity of protective alleles and a high number of susceptible alleles are present.

REFERENCES

1. A. Green, E.A.M. Gale, C.C. Patterson for the EURODIAB ACE Study Group, Incidence of childhood-onset insulin-dependent diabetes mellitus: The Eurodiab Ace Study, *Lancet*, 339:905 (1992).
2. S. Muntoni, M. Songini, High incidence of Type I diabetes in Sardinia, *Lancet* 337:1047 (1991).
3. L. Contu, C. Carcassi, J. Dausset, The "Sardinian" HLA-A30, B18, DR3, DQw2 Haplotype constantly lacks the 21-DHA and C4B genes. Is it an ancestral Haplotype without duplication?, *Immunogenetics* 30:13 (1989).
4. C. Carcassi, G. Trucco, M. Trucco, L. Contu, A new HLA-extended haplotype is involved in insulin-dependent diabetes mellitus susceptibility, *Hum Immunol.* 31:159 (1991).
5. J.A. Todd, J.I. Bell, H.L. McDevitt, HLA-DQB gene contributes to susceptibility and resistance to insulin-dependent diabetes mellitus, *Nature* 329:599 (1987).
6. J.S. Dorman, R.E. LaPorte, R.A. Stone, M. Trucco, Worldwide differences in the incidence of Type I diabetes are associated with amino acid variation at position 57 of the HLA-DQβ chain, *Proc Natl Acad Sci. USA* 87:7370 (1990).
7. I. Khalil, L. D'Auriol, M. Gobet, et al., A combination of HLA-DQB Asp-57-negative and HLA DQA Arg-52 confers susceptibility to insulin-dependent diabetes mellitus, *J Clin Invest.* 85:1315 (1990).
8. K.S. Ronningen, A. Spurkland, T. Iwe, et al., Distribution of HLA-DRB1, -DQA1 and -DQB1 alleles and DQA1-DQB1 genotypes among Norwegian patients with insulin-dependent diabetes mellitus, *Tissue Antigens* 37:105 (1991).
9. E.M. Southern, Detection of specific sequences among DNA fragments separated by gel electrophoresis, *J Mol Biol.* 98:503 (1975).
10. H.A. Erlich, T.L. Bugawan, HLA class II gene polymorphism: DNA typing, evolution, and relationship to disease susceptibility, *in:* "PCR Technology, Principles and Application for DNA Amplification," H.A. Herlich, ed., Stockton Press, New York (1989).

11. K.S. Ronningen, T. Iwe, T.S. Halstensen, et al., The aminoacid at position 57 of the HLA-DQB chain and susceptibility to develop insulin-dependent diabetes mellitus, *Hum Immunol.* 26:215 (1989).
12. J.I. Bell, J.A. Todd, H.O. McDevitt, The molecular basis of HLA-disease association, *in:* " Advances in human genetics," H. Harris and K. Hirschhorn, eds., Plenum Press, London (1989).
13. J.G. Bodmer, S.G.E. Marsh, E.D. Albert, et al., Nomenclature for factors of the HLA system, 1990, *Tissue Antigens* 37:97 (1991).
14. U.B. Gyllensten, H.A. Erlich, Generation of single-stranded DNA by polymerase chain reaction and its application to direct sequencing of the HLA-DQA locus, *Proc Natl Acad Sci. USA* 85:7652 (1988).
15. G.T. Horn, L. Bugawan, C.M. Long, H.A. Erlich, Allelic sequence variation of the HLA-DQ loci: Relationship to serology and to insulin-dependent diabetes susceptibility, *Proc Natl Acad Sci. USA* 85:6012 (1988).
16. F. Muntoni, M. Congia, F. Cucca, et al., The HLA-DQB1*0502 allele is neutrally associated with insulin-dependent diabetes mellitus in the Sardinian population, *Tissue Antigens* 39:262 (1992).
17. J.I. Bell, J.A. Todd, HLA class II sequences infers mechanism for major histocompatibility complex-associated disease susceptibility, *Mol Biol Med.* 6:43 (1989).
18. H.A. Erlich, T.L. Bugawan, S. Scharf, et al., HLA-DQβ sequence polymorphism and genetic susceptibility to IDDM, *Diabetes* 39:96 (1990).
19. H. Heimberg, Z.P. Nagy, G. Somers, et al., Complementation of HLA-DQA and -DQB genes confers susceptibility and protection to insulin-dependent diabetes mellitus, *Hum Immunol.* 33:10 (1992).
20. M. Morellini, S. Travace, M.C. Mazzilli, et al., A study of HLA class II antigens in an Italian paediatric population with coeliac disease, *Disease Markers* 6:23 (1988).
21. M. Congia, F. Frau, R. Lampis, et al., A high frequency of the A30, B18, DR3, DRw52, DQw2 extended haplotype in Sardinian celiac disease patients: Further evidence that disease susceptibility is conferred by DQA1*0501, B1*0201, *Tissue Antigens* 39:78 (1992).

ANALYSIS OF HLA-DQ ALPHA AND DQ BETA GENES IN TYPE I (INSULIN-DEPENDENT) DIABETES

Raffaella Buzzetti,[1] Lorenza Nisticò,[2] Paolo Pozzilli,[1] and Claudio Giovannini[1]

[1]Cattedra di Endocrinologia
II Clinica Medica
University of Rome "La Sapienza"
ITALY

[2]Istituto di Biologia Cellulare (CNR)
Rome
ITALY

INTRODUCTION

Our knowledge of the genetics of insulin-dependent diabetes mellitus (IDDM), and in particular the HLA system, has gained considerable expansion due to the application of molecular biology. The genetic susceptibility to this disease is linked to the HLA region, particularly the DQ alpha and DQ beta chain genes. The presence of an amino acid other than aspartic acid in position 57 of the DQ beta chain and arginine in position 52 of the DQ alpha chain represent, to date, the best markers to identify subjects at-risk for developing IDDM. The identification of such subjects may be useful for strategies aimed at preventing the disease and may also offer new insight into population screening.

In the study of multifactorial diseases, such as Type I (insulin-dependent) diabetes mellitus, the importance of hereditary factors, and in particular, the role of the HLA system, has gained considerable attention in recent years. This paper presents an analysis of HLA DQ alpha and DQ beta genes as they relate to the development of Type I diabetes.

NEW EVIDENCE FOR ASPARTIC ACID IN POSITION 57 OF THE DQ BETA CHAIN

Comparative analysis of the nucleotide sequences of the DQB1 alleles has shown that the haplotypes which are associated neutrally or negatively with IDDM have aspartic acid in position 57 of the DQ beta chain (Asp-57).[1,2] The susceptibility haplotypes, on the contrary, carry an amino acid other than aspartic acid in the same position (non-Asp-57).

Based on the models suggested for the three dimensional structure of HLA class I molecules, the amino acid residue number 57 of the beta chain of HLA class II antigen should be located on the surface of the molecule in the groove which comes into contact with the antigen and the T lymphocyte receptor. The substitution of Asp for a neutral amino acid could alter the three-dimensional structure of the groove, modifying the interactions between the antigens and the cells of the immune system.[3]

The aspartic acid theory is supported by several studies performed on different Caucasoid populations, where homozygous non-Asp-57 IDDM patients have been reported in percentages which vary between 60% and 90% (Table 1).[4-8] The only exception is the Japanese population, where Asp-57 positive patients were found to be much more frequent than Caucasoid diabetics.[9,10]

Table 1. Amino acid in position 57 of the DQ beta chain in IDDM patients and healthy controls: Comparisons between different studies and different populations.

	DIABETICS			CONTROLS		
	Non-Asp/ Non-Asp (%)	Non-Asp/ Asp (%)	Asp/Asp (%)	Non-Asp/ Non-Asp (%)	Non-Asp/ Asp (%)	Asp/Asp (%)
Americans (Caucasoid) IDDM = 266 Controls = 203	76.3	21.4	0.8	41.9	33	24.1
Norwegians IDDM = 87 Controls = 187	82	16	2	27	51	22
French IDDM = 50 Controls = 73	88	12	-	23.3	54.8	21.9
Spanish IDDM = 102 Controls = 87	82	12	6	29	52	19
Belgian IDDM = 264 Controls = 329	72	25	3	27	44	29
Japanese IDDM = 72 Controls = 85	2.8	48.6	48.6	8.2	34.1	57.6
Central Italy IDDM = 65 Controls = 93	69.2	27.7	3.1	23.7	41.9	34.4

Data pooled from cited references

Our data obtained from IDDM patients from the Lazio region confirm the association between the non-Asp-57 genetic marker and susceptibility to the disease. Approximately 70% of the patients were non-Asp-57 homozygotes vs. 24% of the controls ($p<0.001$). However, 2 of the 65 patients were Asp-57 homozygotes, suggesting that, as already claimed by other authors, this marker can not entirely explain the susceptibility to the disease.

THE ROLE OF THE DQ ALPHA CHAIN

More recent observations confirm that the non-Asp-57 marker cannot be used as the only system to identify susceptibility to IDDM. It is also very likely that the alpha chain of the DQ molecule modulates susceptibility to the disease. One of the first observations was offered by DR7 specificity in Caucasoids and in African Americans. In the former, the haplotype DR7 was not associated with IDDM, although the beta chain (DQB1*0201) had a valine in position 57. In contrast, in African Americans populations, the DR7 haplotype was positively associated with the disease. The only difference between the DR7 haplotypes in the two races was represented by the alpha chain which was DQA1*0201 in Caucasoids and DQA1*0301 in African Americans. Therefore, Todd, et al.[11] hypothesized that the DQA1*0201 alleles play a protective role in the development of the disease. Furthermore, a stronger association between the allele DQA1*0301, rather than non-Asp-57, and IDDM was demonstrated in Japanese subjects (96% of patients compared to 53% of control group)[10].

These previous observations were clarified by Khalil et al. more recently.[7] These authors suggested that the disease also correlates with those alleles of the DQA1 gene carrying an arginine at residue 52 (Arg-52) of the DQ alpha chain. The authors proposed a molecular model of the disease where the susceptibility correlates, quantitatively, with the expression at the cell surface of a heterodimer, composed of a DQ alpha chain bearing an Arg-52 and a DQß chain bearing an amino acid other than Asp in position 57. This finding has been confirmed by a study performed in a Spanish population, where the absolute risk varied from 1.9 to 101.7/100,000 per year according to the possibility of generating 0, 1, 2 or 4 susceptible heterodimers.[5] Our own experience has indeed confirmed this innovative finding and offers useful information. Considering the different combinations of DQ alpha and DQ beta alleles, about 90% of the patients possess at least one susceptibility allele at each locus, and more than one-third of the patients possess all four susceptibility alleles. The highest odds ratio (OR=161) is revealed by the homozygous combination of both susceptibility markers (non-Asp-57 and Arg-52). Furthermore, the two markers considered together seem to provide a graded and reliable evaluation of risk. We conclude that the analysis of both genetic markers allows a better evaluation of the risk to develop IDDM than the consideration of DQA1 or DQB1 gene typing alone.

THE ROLE OF PROTECTIVE GENES

DR2 has been considered the best example of a protective haplotype because of its significant negative association with IDDM in different populations. Because the DQB1 allele most frequently represented in DR2 haplotypes, DQB1*0602, possesses an Asp position 57, this has also been considered strong evidence in support of the importance of the amino acid residue in conferring resistance to IDDM. However, studies performed on patients from central Italy did not confirm the negative association between DR2 and IDDM which was established in other Caucasian populations. These results represent a serious challenge to the hypothesis of Asp-57.

We have tried to resolve this apparent contradiction by evaluating the frequency of DR2-associated DQ beta alleles either possessing (*0601, *0602) or not possessing (*0502) in 19 DR2 positive IDDM patients, identified in central Italy, compared with DR2 positive control subjects (Table 2).[12] This study represents the largest series of DR2 positive IDDM patients analyzed to date. We found the difference between patients and controls highly significant. Thus, 19 out of 21 DR2 haplotypes (90.5%) possess non-Asp-57 DQ beta alleles. In addition, the same non-Asp-57 subtype also has a comparatively high frequency (32.1% of DR2 haplotypes) in the DR2 positive healthy population. Comparisons with other data in various Caucasoid populations show that moving towards southern Europe, the relative frequency of the non-Asp-57 DR2 increases. Concomitantly, the apparent protective effect of the DR2 haplotype disappears. These data strongly support the DQ beta Asp-57 hypothesis and are in line with population data showing a correlation between incidence of IDDM and gene frequency of non-Asp-57 allele.[4]

Table 2. Frequency of DR2 subtypes in different Caucasoid populations.

		IDDM Patients	Controls
	Number	Percent of Non-Asp-57	Percent of Non-Asp-57
North America	21	42%	4.6%
France	32	66%	20%
Central Italy	19	89%	32%
Sardinia	5	100%	75%

Data pooled from cited references

THE IMPORTANCE OF GENETIC MARKERS IN THE PREDICTION OF IDDM

IDDM can be studied at the molecular level for an understanding of its pathogenesis. Recent epidemiological studies suggest that the combined analysis of DQ beta at position 57 and DQ alpha at position 52 represent genetic markers of high predictive value. Further investigations are necessary to identify the genes which confer additional risk. In this respect, the DNA polymorphisms associated with several diabetogenic haplotypes should be analyzed in a large number of families with more than one affected member, as well as in IDDM patients, in order to evaluate the role of the susceptibility to the disease. Another interesting aspect that should be taken into consideration for prediction is the identification of genes which are not linked to the HLA region. This may be obtained by DNA markers in families with more than one affected member. The identification of multiple loci which increase risk of the disease allows the inclusion of such individuals in preventive trials for IDDM. As the concordance for the disease in identical twins is 50%, this suggests that environmental and immunological factors should be taken into consideration for prediction. It is our opinion that the results obtained, to date, advocate the use of more than one marker (genetic, immunological and metabolic) in determining prediction of the development of the disease. Such prediction can open the way for preventive attempts and in this direction trials have already begun.

REFERENCES

1. J.A. Todd, J.I. Bell, H.O. McDevitt, HLA-DQbeta gene contributes to susceptibility and resistance to insulin-dependent diabetes mellitus, *Nature* 329:599 (1987).
2. J.A. Todd, C. Mijovic, J. Fletcher, et al., Identification of susceptibility loci for insulin-dependent diabetes mellitus by trans-racial gene mapping, *Nature* 338:587 (1989).
3. M. Trucco, To be or not to be aspartic 57, that is the question, *Diabetes Care* 15(5):705 (1992).

4. J.M. Baisch, T. Weeks, R. Giles, et al., Analysis of HLA-DQ genotypes and susceptibility in insulin-dependent diabetes mellitus, *N Engl J Med.* 322:1836 (1990).
5. M.D. Gutierrez-Lopez, S. Bertera, M.T. Chantres, et al., Susceptibility to Type I diabetes in Spanish patients correlates quantitatively with expression of HLA-DQ α Arg-52 and HLA-DQ ß non-Asp-57 alleles, *Diabetologia* 35(6):583 (1992).
6. H. Heimberg, Z.P. Nagy, G. Somers, et al., Complementation of HLA-DQA and DQB genes confers susceptibility and protection to insulin-dependent diabetes mellitus, *Hum Immunol.* 33:10 (1992).
7. I. Khalil, L. d'Auriol, M. Gobet, et al., A combination of HLA-DQbeta ASP57-negative and HLA-DQalpha Arg 52 confers susceptibility to insulin-dependent diabetes mellitus, *J Clin Invest.* 85:1315 (1990).
8. K.S. Ronningen, T. Iwe, T.S. Haltsensen, et al., The aminoacid at position 57 of the HLA-DQbeta chain and susceptibility to develop insulin-dependent diabetes mellitus, *Hum Immunol.* 26:215 (1989).
9. T. Awata, T. Kuzuya, A. Matsuda, et al., High frequency of aspartic acid at position 57 of HLA DQß chain in Japanese IDDM patients and non-diabetic subjects, *Diabetes* 39:266 (1990).
10. K. Yamagata, H. Nakajima, T. Hanofusa, et al., Aspartic acid at position 57 of DQbeta chain does not protect against Type I diabetes mellitus in Japanese subjects, *Diabetologia* 32:762 (1989).
11. J.A. Todd, Genetic control of autoimmunity in Type I diabetes, *Immunology Today* 11(4):122 (1990).
12. R. Sorrentino, U. De Grazia, R. Buzzetti, et al., An explanation for the neutral effect of DR2 on IDDM susceptibility in central Italy, *Diabetes* 41:904 (1992).

THE INCIDENCE OF IDDM AND MOLECULAR HLA-DQB1 TYPING IN MACEDONIAN CHILDREN

Mirjana Kocova,[1,3] Janice S. Dorman,[2] Milenka Blagoevska,[1] Milco Bogoev,[1] Martin Ivanovski,[1] Marina Konstantinova[1] and Massimo Trucco[3]

[1]Department of Endocrinology & Genetics
Pediatric Clinic
University of Skopje
Skopje, 91000
MACEDONIA

[2]Department of Epidemiology
Graduate School of Public Health
University of Pittsburgh
Pittsburgh, PA, 15261
USA

[3]Department of Pediatrics
Division of Immunogenetics
Children's Hospital of Pittsburgh
Pittsburgh, PA 15213
USA

INTRODUCTION

Substantial data have been collected concerning the incidence of IDDM in different countries and ethnic groups, showing large geographic variability. The incidence of IDDM is highest in Finland, 42.9 new cases per 100,000 children under 15 years of age per year,[1] whereas the lowest rates are in Far East countries, less than 1/100,000per year in Japan and Korea.[2]

In most of European countries, the incidence is moderate compared to the extremes and appears to be between 4 and 15 new cases per 100,000 per year.[3-10] Many countries, however, have not yet estimated the incidence of IDDM, and the published data cover a very small portion of the world. Several international studies have been conducted to increase the data from different regions. The Balkan Peninsula has been a region with very few published data on the incidence of IDDM.[11,12]

Standardization of Epidemiologic Studies of Host Susceptibility
Edited by J. Dorman, Plenum Press, New York, 1994

The associations between IDDM and certain genes of the HLA-DQ region have been very well documented.[14-17] The data from several studies revealed a strong association between the incidence of IDDM and the presence of HLA-DQB1 alleles encoding an amino acid other than aspartic acid in position 57 of the DQ beta chain of HLA class II molecule (non-Asp-57). Alleles encoding aspartic acid in the position 57 of the DQ beta chain (Asp-57), on the contrary, confer resistance to IDDM.[2,27] These data, however, are not identical in all regions and populations.[2,19,20] This provides an opportunity to explore different environmental factors and additional genetic markers across populations.

The aim of the present study was to evaluate the incidence of IDDM from the registry data in the region of Macedonia, the southern most republic of the former Yugoslavia, and to describe the preliminary results of molecular HLA-DQB1 typing from registered IDDM cases in the region. This is the first molecular epidemiology study of the incidence of IDDM and its association with HLA-DQB1 allele frequencies from the Balkan Peninsula.

MATERIALS AND METHODS

IDDM Incidence Registry

An IDDM incidence registry has been set up for the region of Macedonia for the period of 1986-1991 according the DIAMOND criteria.[3,20] The primary source of data was the medical records from the only university hospital in the region, where almost all children less than 14 years of age are admitted at diagnosis of IDDM. The secondary source was the records from the local hospitals, where the children with IDDM are registered in order to get insulin supplies.

The denominator data for the incidence rates were based on the census, grouped by age, sex and ethnic group. The population of Macedonia is 2,100,000 people and the population younger than age 14 years represents 620,000 children. These data were obtained from the last census in the area in 1980.

The incidence of IDDM and completeness of ascertainment were calculated according to the methods previously described.[21]

Molecular Typing

Molecular HLA-DQB1 typing was performed using genomic DNA extracted from mononuclear blood cells from 35 unrelated diabetic children identified from the Macedonia IDDM registry using the XI International Workshop on HLA typing protocols as previously described.[22,27]

The amplification of the second exon of the DQB1 gene was performed using 0.5 μg of DNA during 30 cycles of the denaturing step at 96°C for 30 seconds, primer annealing at 55°C for 30 seconds and DNA polymerization at 72°C for 30 seconds in the termocycler (Elmer-Perkin). The amplified material was spotted to the Hybond N+ membrane (Amersham). The denaturation and cross-linkage of the amplified material to the membrane was achieved by treatment in 0.4 N NaOH for 20 minutes.

The prehybridization was performed in hybridization buffer containing 3.0 M tetramethylammonium chloride (TMAC), 50mM Tris-HCL(pH 8.0), 2mM EDTA, and 0.1% sodium dodecyl sulfate (SDS), 5X Denhardt's solution, and 100 μg/ml salmon sperm DNA.

Ten ^{32}P labelled oligonucleotide probes were hybridized to the amplified material for at least 6 hours to overnight. The washing of the nonspecific radioactivity was performed, with 2XSSPE/0.1%SDS twice for ten minutes each at room temperature. Two washes with TMAC buffer for ten minutes were performed, one at room temperature, and additional stringent washing at 58°C.

The filters were exposed on X-ray film (Kodak XAR-5) for 2-16 hours.

RESULTS

During the period 1986-1991, a total of 105 children with newly diagnosed IDDM in the age group 0-14 years were registered in the region of Macedonia. The overall incidence of IDDM was 2.46 new cases per 100,000 per year, which is the lowest incidence in Europe reported so far.[28] The completeness of ascertainment of the combined primary and secondary sources for the registry was 96.5%, confirming that almost all children with IDDM from this region have been registered. There was no striking difference between two major ethnic groups, Macedonians representing 70% of the population, and Albanians representing 25% of the entire population. There was no significant difference in incidence between males and females (Table 1.)

The results of the molecular DQB1 typing are presented in Tables 2 and 3.

Table 1. Sex-specific IDDM incidence rates per 100,000/year and 95% confidence intervals (CI) in different ethnic groups in Macedonia.

Ethnic Group	Boys Rate (95% CI)	Girls Rate (95% CI)	Total Rate (95% CI)
Macedonians	2.89 (1.89 - 3.73)	2.47 (1.60 - 3.72)	2.79 (2.21 - 3.40)
Albanians	1.00 (0.11 - 3.27)	1.68 (0.23 - 6.95)	1.36 (0.28 - 3.92)
Others	3.08 (1.50 - 4.63)	2.37 (1.14 - 3.63)	2.60 (1.60 - 3.72)
All	2.48 (1.31 - 3.92)	2.37 (1.62 - 3.57)	2.46 (1.49 - 3.90)

Table 2. DQB1 Allele frequencies in Macedonian diabetics.

DQB1 Allele	Number	Frequency (%)
*0201	21	30.00
*0302	15	21.43
*0502	12	17.14
*0501	8	11.42
*0301	5	7.14
*0602	3	4.29
*0603	3	4.29
*0303	2	2.86
*0503	1	1.43

The most frequent allele was DQB1*0201, followed by *0302 and, interestingly, by *0502. The frequency of *0502 was 17%, and is the highest reported in diabetics. We further studied this group using DR typing, which revealed the association of DQB1*0502 with DR2, i.e., the allele DRB1*1601 in this population.

Table 3. HLA-DQB1 genotype frequencies among diabetics in Macedonia.

Genotype	Number	Frequency (%)
non-Asp/non-Asp	24	68.57
non-Asp/Asp	10	28.57
Asp/Asp	1	2.86

CONCLUSIONS

Our results showed several interesting points. One was the very low incidence of IDDM in the region of Macedonia, which puts it in the range of a "cold spot" for IDDM in Europe and in the world.[28] The data from the region of Balkan Peninsula are rather limited. A study from Croatia revealed yearly variable incidence from 1.03 to 8.3 per 100,000 during the period 1968-1990 for the city of Zagreb.[12] On the contrary, the data from Macedonia were very consistent within the observed period of time.

An older study revealed an incidence in Slovenia of 2.2 new cases per 100,000 per year in 1969-1974,[11] and a three-fold increase was reported for the same region in 1992,[1] as was demonstrated by other studies from Europe.[5,10,23] These two regions are far north from Macedonia and the incidence of IDDM has been shown to decrease from north to south.[10,24] The first data from Greece, however, revealed incidence of 4.6 new cases per 100,000 per year[1], almost double that for Macedonia.

From the molecular HLA-DQB1 typing, there are two interesting points to make. One is the high frequency of DQB1*0502. This allele has been reported with very low frequency in different Caucasian populations, ranging from 0.2 to 4.7%.[25] In the Sardinian population, the frequency of this allele is 7% in diabetics. However, Sardinia has one of the highest incidence rates for IDDM in the world.[26] In the Macedonian diabetics, the frequency of DQB1*0502 is 17%. It would be interesting to explore the frequency of this allele in the non-diabetic control Macedonian population since in Mediterranean populations the frequency of DQB1*0502 seems to be higher than in other parts of the world.[22,26] Our study confirms the association of DQB1*0502 with DR2*1601. The protective role of DR2 in IDDM susceptibility,[22,26,27] as determined by serological typing, does not seem to apply to the population of Macedonia since molecular typing reveals a DR2 haplotype which confers susceptibility to IDDM. This further supports the non-Asp-57 hypothesis[16,17,19] in IDDM susceptibility, because the DQB1*0502 is encoding non-Asp at the position 57 of the beta chain of HLA-DQ molecule, and it is evident that the DQB1 alleles have a stronger influence on IDDM susceptibility than DR. Further analysis of larger number of diabetics, as well as the control nondiabetic population, are needed in this region with a low incidence of IDDM.

REFERENCES

1. A. Green, E.A.M. Gale, C.C. Patterson, for the EURODIAB ACE Study Group, Incidence of childhood-onset insulin-dependent diabetes mellitus: The Eurodiab Ace study, *Lancet* 339:905 (1992).

2. M.Z. Bao, J.X. Wang, J.S. Dorman, M. Trucco, HLA-DQ beta non-Asp-57 allele and incidence of diabetes in China and the U.S.A., *Lancet* 2:297 (1989).
3. R.E. LaPorte, N. Tajima, H.K. Akerblom, Geographic differences in the risk of insulin-dependent diabetes mellitus: The importance of registries, *Diabetes Care* 8 (suppl.):101 (1985).
4. E. Schober, H. Frish, Incidence of childhood diabetes mellitus in Austria 1979-1984, *Acta Paediatr Scand.* 77:299 (1988)
5. G. Soltesz, L. Madacsy, D. Bekefi, I. Danko, The Hungarian Childhood Diabetes Epidemiology Group, Rising incidence of Type I diabetes in Hungarian children (1978-1987), *Diabetic Med.* 7:111 (1990).
6. C. Levy-Marchal, L. Papoz, C. de Beaufort, Incidence of juvenile Type I (insulin-dependent) diabetes mellitus in France, *Diabetologia* 33:465 (1990).
7. Diabetes Epidemiology Research International Group, Geographic patterns of childhood insulin-dependent diabetes mellitus, *Diabetes* 37:1113 (1988).
8. L. Nystrom, G. Dahlquist, M. Rewers, S. Wall, The Swedish Childhood Diabetes study. An analysis of the temporal variation in diabetes incidence 1978-1987, *Int J Epidemiol.* 19:141 (1990).
9. Diabetes Epidemiology Research International (DERI) Study Group, The epidemiology and immunogenetics of IDDM in Italian-heritage populations, *Diabetes Metab Rev.* 6:63 (1990).
10. Diabetes Epidemiology Research International group, Secular trends in incidence of childhood IDDM in 10 countries, *Diabetes* 39:858 (1990).
11. H. Lestardet, J. Besse, Prevalence et incidence du diabete juvenile insulino-dependant en France, *Diabete Metab.* 3:229 (1977).
12. G. Roglic, Z. Metelko, M. Kadrnka-Lovrencic, M. Granic, Z. Skrabalo, Incidence of insulin-dependent diabetes mellitus - The Zagreb registry, 14th International Diabetes Federation Congress, Abstracts 1274 (1991).
13. Diabetes Epidemiology Research International Study Group, Childhood diabetes in Arab countries, *Bull WHO* 68:231 (1990).
14. P.A. Morel, J.S. Dorman, J.A. Todd, H.O. McDevitt, M. Trucco, Aspartic acid at position 57 of the HLA-DQ beta chain protects against Type I diabetes: A family study, *Proc Natl Acad Sci. USA* 85:8111 (1988).
15. H. Reijonen, J. Ilonen, M. Knip, A.K. Hans, HLA-DQB1 alleles and absence of Asp 57 as susceptibility factors of IDDM in Finland, *Diabetes* 40:1640-1644 (1991).
16. M.Trucco, To be or not to be Asp 57, that is the question, *Diabetes Care* 15:705 (1992).
17. M. Trucco, J.S. Dorman, Immunogenetics of insulin-dependent diabetes mellitus in humans, *Critical Rev Immunol.* 9:201 (1989).
18. M. Kocova, W.A. Rudert, E.P.Weisberg, M.Trucco, J.S. Dorman, D.J. Pettit, W.C. Knowler, HLA class II molecular genotypes in Pima Indians: Rarity of IDDM susceptibility alleles, Submitted.
19. J.S. Dorman, R.E. LaPorte, R.A. Stone, M. Trucco, Worldwide differences in the incidence of Type I diabetes are associated with amino acid variation at position 57 of the HLA-DQ beta chain, *Proc Natl Acad Sci. USA* 87:7378 (1990).
20. WHO DIAMOND Project Group: WHO Multinational Project for Childhood Diabetes, *Diabetes Care* 13:1062 (1990).
21. R.J. Vergona, Computer program for IDDM registries, *Diabetes Care* 14:70 (1990).
22. R. Sorrentino, U. DeGrazia, R. Buzzeti, L. Nistico, C. Iannicola, S. Costanzi, R. Tosi, An explanation for the neutral effect of DR2 on IDDM susceptibility in central Italy, *Diabetes* 41:904 (1992).

23. P.J. Bingley, E.A. Gale, Rising incidence of IDDM in Europe, *Diabetes Care* 12:289 (1989).
24. G.J. Vaandrager, G.J. Bruining, F.J. Veenhof, N.M. Drayer, Incidence of childhood diabetes in the Niderlands: A decrease from north to south over North-Western Europe? *Diabetologia* 27:203 (1984).
26. C. Carcassi, G. Trucco, M. Trucco, L.A. Contu, New HLA-DR2 extended haplotype is involved in insulin-dependent diabetes mellitus susceptibility, *Hum Immunol.* 31:159 (1991).
27. D. Zeliszewski, J.M. Tiercy, C. Boitard, X.F. Gu, M. Loche, R. Krishnamoorthy, N. Simmoney, J. Elion, J.F. Bach, B. Mach, G. Sterkers, Extensive study of DRB, DQA, and DQB gene polymorphism in 23 DR2-positive insulin-dependent diabetes mellitus patients, *Hum Immunol.* 33:140 (1992).
28. M. Kocova, M. Trucco, M. Konstantinova, J.S. Dorman, A "cold-spot" of IDDM incidence in Europe, *Diabetes Care* 16:1236 (1993).2

HLA-DR, DQ SYSTEM AND IDDM IN JAPANESE

Nobuo Matsuura,[1,2] Juan M.R. Aparicio,[3] Akemi Wakisaka[3]

[1]Pediatric Clinic
Tonan General Hospital
Sapporo
JAPAN

[2]Diabetes Center
Tokyo Women's Medical College
Tokyo
JAPAN

[3]Department of Pathology
Hokkaido University School of Medicine
Sapporo
JAPAN

INTRODUCTION

The genetic factors involved in the development of insulin-dependent diabetes mellitus (IDDM) have been found to be associated with genes in the major histocompatibility complex (MHC) located on chromosome 6. DQ antigens appear to be the most relevant for the development of IDDM: HLA-DQw2 and DQw3 in Caucasians[1,2] and HLA-DQw4 and DQw9 in Japanese.[3] Moreover, position 57 of the DQß chain and position 52 of the DQα chain were found to be associated with IDDM susceptibility in Caucasians. Homozygosity for non-aspartic acid (non-Asp-57) was observed in 90% of the IDDM patients and none were homozygous for aspartic acid (Asp-57).[2,5]

In the present study of IDDM in the Japanese, HLA antigens were investigated by using the micro-lymphocytotoxicity test and by means of restriction fragment length polymorphisms (RFLP). Although DQ antigens showed the strongest association with IDDM in this study as in Caucasians, some differences were observed in relation to the correlation of the amino acid sequences at the 57th residue of the DQß chain, as proposed by Todd, and at the 52th residue of the DQα chain. Base sequence analysis with allele specific oligonucleotide (ASO) probes and polymerase chain reaction (PCR) were performed in order to clarify these differences.

Standardization of Epidemiologic Studies of Host Susceptibility
Edited by J. Dorman, Plenum Press, New York, 1994

MATERIALS AND METHODS

Patients

A total of 56 unrelated patients with IDDM diagnosed at the Department of Pediatrics of Hokkaido University Hospital were investigated. The age range was 4 to 21 years, with a mean age of 11.70 $\pm$ 3.32. Among the diabetic population, 65 percent were female and 35 percent male.

A random sample of 472 unrelated, healthy Japanese, who were HLA typed at the 3rd Asia Oceania Histocompatibility Workshop, were used for controls of the HLA serotyping. A second group of 24 HLA matched controls was used for RFLP analyses of HLA-DQA1 and DQB1.

RFLP Analyses and Probes

RFLP analyses were performed as previously described.[3] The probes used for HLA are listed in Table 1.

Table 1. Probes used in HLA analysis.

Map Location	Locus Symbol	Probe	Vector Size (kb)	Probe Source
6p21.3	HLA-DQB+	pII-betal	M13/800bp/EcoRI	Sleldon
6p21.3	HLA-DQA+	-	PUC19/1600bp/BamHI	Auffray

PCR Analysis and ASO Probes

PCR was performed for the second exon of the DRB1 and DQB1 genes (1st domain) using genomic DNA to evaluate the 57th amino acid residue. One or two micrograms of DNA was used for amplification as described at the 11th International Histocompatibility Workshop. DRB1 oligonucleotide primers were:

T
5-TTCTTCAA GGGACGGAGCGGGTGTGG-3'/5' - TCGCCGCTGCACTGTCAG-3'
A

DQB1 oligonucleotide primers were: G
5'-TGTGCTACTTCACCAACGGGA-3'/5' - GT TGCAGACACAACTAC-3'
A

These correspond to amino acid residues 17-25, and 87-94 of DRB1 and 14-20 and 78-83 of DQB1 genes, respectively.

Twelve (18-mer length) allele specific oligonucleotide probes, whose sequence contains a codon corresponding to the 57th amino acid residue that reacts with amplified DQB1 and/or DRB1 genes, respectively, were used. After obtaining these results, sixteen (18-mer length) and seven (18-mer length) ASO probes that react specifically with either DQB1 or DRB1 genes from the 11th International Histocompatibility Workshop were applied for reconfirmation of the gene.

RESULTS

HLA Analysis

The antigen frequencies for 56 unrelated patients with IDDM and 472 healthy controls are presented in Table 2. Among the several HLA groups, an increased frequency of B7 (RR = 1.84), B44 (RR = 2.02), Bw54 (RR = 5.32, $p < 0.0001$), Cw1 (RR = 3.25, $p < 0.0001$) in patients was observed along with a decreased frequency of Bw52 (RR = 0.06, $p < 0.0003$). DR2 showed a lower frequency in patients than in controls (RR = 0.15, $p < 0.0001$), whereas, DR1 (RR = 1.94), DRw6 (RR = 1.42), DR9 (RR = 2.64, $p < 0.0005$), DRw53 (RR = 2.82, $p < 0.005$), and DR4 (RR = 2.88, $p < 0.005$) were increased in IDDM, especially DR4.1 (RR = 4.45, $p < 0.0001$), which is in linkage disequilibrium with Bw54.

Among these groups, HLA-DQ antigens showed a decrease in DQw1 (RR = 0.51, $p < 0.02$) and an increase in DQw3 (RR = 2.74, $p < 0.001$). However, the highest association with IDDM was observed for DQw4 (RR = 6.43, $p < 0.0001$).

Table 2. HLA antigen frequencies for IDDM Japan patients and healthy controls.

HLA antigens	IDDM (%)	control (%)	R.R.	P
B7	21.4	12.9	1.84	
B44	19.6	10.8	2.02	
Bw52	1.8	23.5	0.06	<0.0003
Bw54	46.2	14.0	5.32	<0.0001
Cw1	55.4	27.6	3.25	<0.0001
DR1	21.4	12.3	1.94	
DR2	7.1	34.3	0.15	<0.0001
DR4	64.3	38.5	2.88	<0.005
DR4.1(DQw4)	58.9	24.4	4.45	<0.0001
DR4.2(DQw8)	7.1	11.9	0.51	
DRw6	21.4	16.1	1.42	
DRw8	17.9	24.8	0.66	
DR9	48.2	26.1	2.64	<0.0005
DQw1	48.2	64.4	0.51	<0.02
DQw3	76.8**	54.7	2.74	<0.001
DQw4	75.0	31.8	6.43	<0.0001
	(56)*	(472)*		

*Total number tested.
**Mainly DQw9, confirmed by RFLP and PCR analysis.

RFLP Analysis

Associations between the HLA complex and IDDM were also investigated by means of RFLP analyses using both DQA1 and DQB1 probes after digestion with BamHI. Serological results were in accordance with the RFLP evaluation, and Table 3 shows the results. An 8.2Kb BamHI/DQA1 fragment (for which the specificity is DQA1*0301+0201) was increased, reflecting both DQw4 and DQw9 (89.4% in patients against 55.6% of random controls, RR = 6.74). A similar high frequency and relative risk for IDDM was represented by an 11.5Kb BamHII/DQB1 fragment (for which the specificity is DQB1*0401 + 0402 + 0302 + 0303); 89.4% in patients compared to 55.4% in random controls (RR = 6.79).

The 11Kb BamHI/DQA1 band (DQA1*0102 + 0101 + 0103) which corresponds to DQw1 was decreased in IDDM (27.7% vs. 75.6%, RR = 0.12) and supports the decreased frequency of DQw6-DR2, but not DQw5-DR1. The 9.2Kb BamHI/DQB1 fragment (DQB1*0601), which correlates DQw6, was significantly decreased (4.3% vs. 58.9%, $p < 0.01$, RR = 0.03), whereas, the 6.0Kb fragment BamHI/DQB1 (DQB1*0501;2;3), which correlates DQw5, was observed with similar frequency in both populations (21.3% vs. 25.5%)(RR = 0.80).

Table 3. Comparisons between IDDM and both matched and random control frequencies and their specificities.

Fragment Kb specificity	IDDM patients (%)	matched control (%)	random control (%)	R.R.	P
BamHI					
DQA1					
24.0 DQA1*0102a !	23.4	12.4	25.0	0.92	
11.0 DQA1*0101;3 +0102b~	27.7	37.5	75.6	0.12	
8.2 DQA1*0301+0201@	89.4	70.8	55.6	6.74	
DQB1					
11.5 DQB1*0302;3+0401;2	89.4	70.5	55.4	6.79	
9.2 DQB1*0601;2	4.3	29.2	58.9	0.03	<0.01
6.0 DQB1*0501;2;3	21.3	12.5	25.2	0.80	
Number tested	56	24			

@8.2kb DQA1 fragment correspond to DQA1*0301 in Japanese because DQA1*0201 is almost absent.
! DQA1*0102a is associated with DQB1*0604.
~DQA1*0102b is associeted with DQB1*0602.
' To adjust the frequency of DRw53 with IDDM patients.

PCR and ASO Analysis

Table 4 shows the results of the amplified DNA containing the DRB1 and DQB1 genes after twelve allele specific oligonucleotides were applied. Four different amino acids (aspartate, serine, valine and alanine) at the 57th position of both DQB1 and DRB1 chains are identified by these oligonucleotides.

The DRB1 chain homozygous non-Asp/non-Asp amino acid residue was significantly increased ($p < 0.05$) in IDDM compared to controls, whereas the homozygosity for Asp/Asp lines was significantly decreased ($p < 0.03$) (Table 4). For DQB1, no significant deviation of the amino acid combinations were observed. Only 6% of the Japanese IDDM patients were homozygous for non-Asp at the 57th residue.

Table 4. The 57th amino acid residue of DRB1 and DQB1.

Combination	DRB1 IDDM (%)	DRB1 control (%)	DRB1 P	DQB1 IDDM (%)	DQB1 control (%)	DQB1 P
Asp/Asp	12.0	33.3	0.03	40.0	52.2	
Asp/Ser	30.0	16.7		0.0	0.0	
Asp/Val	12.0	25.0		32.0	22.7	
Asp/Ala	0.0	0.0		22.0	18.1	
Ser/Ser	0.0	2.8		0.0	0.0	
Ser/Val	34.0	16.7		0.0	0.0	
Val/Val	12.0	5.6		0.0	2.2	
Val/Ala	0.0	0.0		6.0	4.5	
Ala/Ala	0.0	0.0		0.0	0.0	
Asp/Asp	12.0	33.3	0.03	40.0	52.2	
Asp/NAsp**	42.0	41.6		54.0	40.9	
NAsp/NAsp	46.0	25.0	0.05	6.0	6.8	
	(50)*	(36)*		(50)*	(44)*	

*Total number used.
**NAsp mean aminoacid other rhan aspartic acid.

DISCUSSION

The genetic factors in relation to the HLA antigen associations with IDDM in Japanese were investigated. Several antigens showed deviated frequencies compared to controls; these deviations were grouped into four haplotypes due to their linkage disequilibrium.

Three haplotypes Bw54, Cw1, DR4, DRw53, DQw4; B7, DR1, DQw1 (DQw5) and B44, DRw6 (DRw13), DQw1 (DQw6) were increased in IDDM patients, indicating that those were susceptibility haplotypes, whereas, Bw52, DR2, DQw1 (DQw6) showed a decreased frequency, suggesting it to be resistant for IDDM. Among all the antigens, the DQ antigens showed the strongest positive and negative association with the disease (Table 2). These results were confirmed by both HLA typing and RFLP analysis, and supports that DQ antigens appear to be the most important antigens for the development of IDDM as in Caucasians.[1]

Table 5 summarizes the DR-DQ haplotypes and their relation to IDDM. However, since several haplotypes are lacking in Japanese, the results previously reported in Caucasians and Blacks[1,6,7] were included in order to find common features of both loci. Each antigen was written according to the WHO Nomenclature Committee. For DQA1, DQA1*0101, *0102, and *0301 were thought to be susceptibility alleles, whereas, DQA1*0401, 0501 and *0601 always showed resistance to IDDM. When the amino acid sequence of the alleles were compared, resistant DQA1 alleles have common features at two specific residues: position 56 and 76. The one is the deletion of amino acid at the 56 residue, and the other is the presence of leucine at position 76. However, since there are certain exceptions to this rule, it is reasonable to speculate that a similar contribution other than DQA1 is necessary.

The susceptibility alleles of the DQB1 gene were characterized by sequence for non-aspartic acid in the 57th position of the DQB1 gene. Resistant DQB1 alleles carry Asp at the 57th residue as suggested by Todd. The susceptibility to IDDM appears to be determined only when the susceptibility DQA1 alleles are combined with the susceptible DQB1 alleles. If either one of the DQA1 or DQB1 alleles is considered resistant, IDDM,

Table 5. HLA-DR, DQ haplotypes and IDDM.

DR-DQ haplotypes	DQA1 76th	DQB1 57th	DRB1 57th	IDDM
(Dw1)-DQw5	0101(M)	0501(V)	0101(D)	S
(Dw21)-DQw5	0102(M)	0502(S)	1601(D)	S
(Dw4)-DQw8	0301(V)	0302(A)	0401(D)	S
(Dw10)-DQw8	0301(V)	0302(A)	0402(D)	S
(Dw14)-DQw8	0301(V)	0302(A)	0402(D)	S
(Dw19)-DQw6	0102(M)	0604(V)	1302(S)	S
-DQw2	0301(V)	0201(A)	0701(V)	S
-DQw2	0301(V)	0201(A)	0901(V)	S
(Dw15)-DQw4	0301(V)	0401(D)	0405(S)	S
(Dw23)-DQw9	0301(V)	0303(D)	0901(V)	S
(DRw18)-DQw4	0301(V)	0402(D)	0301(D)	R
(Dw4)-DQw7	0301(V)	0301(D)	0401(D)	R
(DHAG)-DQw6	0103(M)	0603(D)	1303(S)	R
(Dw2)-DQw6	0102(M)	0602(D)	1501(D)	R
(Dw12)-DQw6	0103(M)	0601(D)	1502(D)	R
(Dw9)-DQw6	0101(M)	0503(D)	1401(A)	R
(DRw17)-DQw2	0501(L)	0201(A)	0301(D)	R
(Dw7)-DQw2	0201(L)	0201(A)	0701(V)	R
(DB1)-DQw2	0201(L)	0201(A)	0702(V)	R
(Dw8.1)-DQw4	0401(L)	0402(D)	0801(S)	R
(Dw8.3)-DQw7	0601(L)	0301(D)	0803(S)	R
(DRw12)-DQw7	0501(L)	0301(D)	1201(V)	R

R: resistant
S: susceptible

in general, does not develop. This idea was supported by the crystallographic figure where the 76th residue of the DQα chain is face to face with the 57th residue of the DQβ chain on the tertiary structure of the class II molecule proposed by Brown.[8] These two amino acids might play an important role in constructing a specific structure for the antigen binding site.

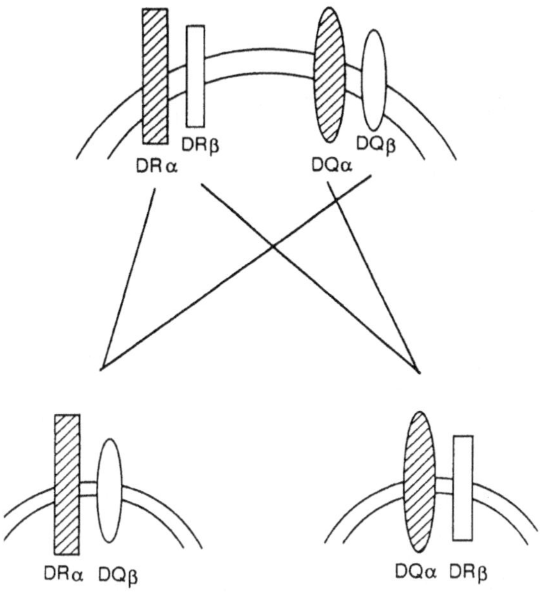

Figure 1. Mixed isotypic class II molecules.

Japanese IDDM patients with DQw4 and DQw9, which have the strongest association with the disease, do not follow this rule. PCR analysis confirmed these results with the exception of three patients who were homozygous for non-Asp-57 of DQB1 gene. Interesting to note, these patients carried non-Asp-57 of DRB1 gene. Our hypothesis is that in the case of Japanese patients with DQw4 and DQw9, DRB1 with non-Asp-57 might substitute in the role of DQB1 with non-Asp-57 found in other ethnic groups. Mixed isotype molecules composed of DQA1 and DRB1 have been demonstrated in vitro on EB transformed B cell lines. Also mouse L cells transfected with DQA1*0301 and DRB1*0405 genes express DQA1-DRB1 mixed isotype molecules on the cell surface. These isotype molecules have been shown to function as stimulators in vitro. Therefore, it is our proposed hypothesis that mixed isotype molecules with susceptible DQA1 and DRB1 alleles might be formed and function in vivo in Japanese IDDM patients with DQw4 and DQw9 (Figure 1). Further studies will be necessary to clarify this point.

Acknowledgements

This work was supported by grant-in-aid for scientific research 02454321 from the Ministry of Education, Science and Culture, Japan.

REFERENCES

1. H. Festenstein, J. Awad, G.A. Hitman, et al., New HLA DNA polymorphisms associated with autoimmune disease, *Nature* 322:64, 1986.
2. J.A. Todd, J.I. Bell, H.O. McDevitt, HLA-DQ gene contributes to susceptibility and resistance to insulin-dependent diabetes mellitus, *Nature* 329:599, 1987.
3. R.J. Aparicio, A. Wakisaka, A. Takeda, et al., HLA-DQ system and insulin-dependent diabetes mellitus in Japanese: Does it contribute to the development of IDDM as it does in Caucasians?, *Immunogenetics* 28:240, 1988.
4. I. Khalil, L. d'Auriol, M. Gobet, et al., A combination of HLA-DQß Asp-57-negative and HLA DQα Arg-52 confers susceptibility to insulin-dependent diabetes mellitus, *J Clin Invest.* 85:1315, 1990.
5. P.A. Morel, J.S. Dorman, J.A. Todd, et al., Aspartic acid at position 57 of the HLA-DQß chain protects against Type I diabetes: A family study, *Proc Natl Acad Sci. USA* 85:8111, 1988.
6. J.A. Todd, C. Mijovic, J. Fletcher, et al., Identification of susceptibility loci for insulin-dependent diabetes mellitus by trans-racial gene mapping, *Nature* 338:587, 1989.
7. J.A. Todd, T. Fukui, T. Kitagawa, et al., The A3 allele of the HLA-DQA1 locus is associated with susceptibility to Type I diabetes in Japanese, *Proc Natl Acad Sci. USA* 87:1094, 1990.
8. H.J. Brown, T. Jardetzky, M.A. Saper, et al., A hypothetical model of the foreign antigen binding site of class II histocompatibility molecules, *Nature* 332:854, 1988.

IMMUNOGENETIC ASPECTS OF ADULT ONSET IDDM IN KOREA

Hong Kyu Lee, Sung-Kwan Hong, Kyong Soo Park, Byoung Doo Rhee, Seong Yeon Kim, Bo Youn Cho, Chang-Soon Koh, and Hun Ki Min

Department of Internal Medicine
Seoul National University Medical College
Seoul
KOREA

INTRODUCTION

Insulin-dependent diabetes mellitus (IDDM) is a clinical diabetes of severe insulin deficiency, resulting from a slowly progressing autoimmune destruction process of pancreatic beta cell. Noninsulin-dependent diabetes mellitus (NIDDM) is a less severe form of diabetes of unknown etiology, although insulin resistance is suspected as the prime abnormality. This distinction is sometimes difficult as there is no clear transition point to apply.[1] Certain HLA class II molecules, particularly DR3 or DR4, have been known to be associated with IDDM susceptibility.[2] With the recent advances of molecular biology, the presence of DNA sequences coding for non-aspartic acid in the 57th position of the DQ beta chain and arginine in the 52nd position of the DQ alpha chain are known to be highly associated with IDDM in Caucasians.[3,4] The associations of HLA alleles with IDDM, but not with NIDDM, was taken as an evidence that IDDM and NIDDM are two distinctively different diseases. Recently, however, several investigators suggest that HLA might also play a role in NIDDM.[5-8]

The incidence rate of typical IDDM in Korea is one of the lowest in the world (0.68/100,000 per year) and many patients with noninsulin-dependent diabetes mellitus become insulin requiring, losing a significant amount of their weight several years after the onset of disease. However, they are not prone to ketoacidosis and do not need insulin treatment for many years. Therefore, the clinical distinction between IDDM and NIDDM among Korean adults is sometimes very difficult. There have been limited data on the immunogenetic characteristics of diabetes in Korea, particularly adult onset IDDM. Thus, we performed HLA typing for the A,B,C and DR loci, and measured islet cell antibody (ICA) and islet cell surface antibody (ICSA). We also tested HLA-DQ associations in IDDM and some NIDDM patients.

Standardization of Epidemiologic Studies of Host Susceptibility
Edited by J. Dorman, Plenum Press, New York, 1994

SUBJECTS AND METHODS

Subjects

Subjects selected for HLA-A, B, C and DR typing, and ICA and ICSA included 61 patients with IDDM and 280 healthy unrelated volunteers. Patients with IDDM were selected by the following criteria: 1) age of onset between 15 and 40 years, 2) body mass index less than 25 kg/m^2, 3) evidence for ketosis-proneness, and 4) current use of insulin. Thirty-three were female and 28 were male. Their age of onset of the disease was 24.3 ± 7.8 years.

Subjects selected for HLA-DQA1 and DQB1 genotyping included 36 adult-onset IDDM patients (onset after age 20 years, previous history of diabetic ketoacidosis or ketonuria, insulin-requiring for the control of blood glucose, and fasting serum C-peptide less than 1.0 ng/ml), 48 NIDDM patients (all other primary diabetes) attending Seoul National University Hospital, and randomly selected control subjects.

HLA Typing for the A, B, C and DR Loci

HLA typing for the A, B, C and DR loci was performed using the standard micro-lymphocytotoxicity technique using the 3rd Asia - Oceania Histocompatibility Workshop typing sera.[9]

Measurement of ICA and ICSA

ICA was measured by indirect immunofluorescence on Group O human pancreas.[10] Complement mediated ICSA was measured by an assay for release of ^{51}Cr using rat insulinoma cells (RINm5F) and rabbit complement.[11]

Oligonucleotide Typing of HLA-DQA1 and DQB1

From 1 μg DNA samples of peripheral lymphocytes, the second exon encoding the first polymorphic domains of HLA-DQA1 and DQB1 genes were amplified by PCR, using specific DNA flanking primers. The primer sequences were:

DQA1: PL 5' GGTGTAAACTTGTACCAC 3',
DQA1: PR 5' CATTGGTAGCAGCGGTAGAG 3',
DQB1: PL 5' CATGTGCTACTTCACCAACG 3',
DQB1: PR 5' GTAGTTGTGTCTGCACAC 3'.

After 30 cycles of amplification, an aliquot of each amplified sample was dot blotted on a nylon membrane (Immobilon-N, Millipore USA) and alleles were identified by color development after hybridization with digoxigenin-alkaline phosphatase 3' end labeled (Boehringer Mannheim, Germany) oligonucleotide probes. We used the 19 sequence specific oligonucleotide probes described by Khalil et al.[4]

Cell Lines

A total of 25 lymphoblastoid cell lines homozygous for the HLA-DQ region fully defined at the Xth International Histocompatibility Workshop were used as reference cells.

RESULTS

HLA-A, B, C antigens did not demonstrate any significant differences in the frequencies between the patients with IDDM and the controls (data not shown). The frequency of HLA-DR4 in IDDM patients was higher than that in controls (57.4% vs. 36.4%, corrected p value < 0.05, Table 1). On the contrary, DR2 was decreased in IDDM patients as compared with the controls (13.1% vs. 32.1%, corrected p value < 0.05).

Table 1. Antigen frequencies of HLA-DR in patients and controls.

DR	Patients (n=61)		Controls (n=280)		
Antigen	No.	Frequency (%)	No.	Frequency (%)	RR
DR1	4	6.5	22	7.9	0.8
DR2*	8	13.1	90	32.1	0.3
DR3	8	13.1	12	4.3	3.4
DR4*	35	57.4	102	36.4	2.3
DR5	2	3.3	14	5.0	0.6
DRw6	3	4.9	47	16.8	0.3
DR7	8	13.1	13	4.6	3.1
DRw8	11	18.0	50	17.9	1.0
DRw9	6	11.4	59	21.1	0.4
DR10	-	-	6	2.1	-

*Corrected p < 0.05 vs. the controls by χ^2test

Prevalence of ICA and ICSA

The overall prevalence of ICA and ICSA in patients with IDDM were 29% and 35%, respectively. As the duration of diabetes increased, the prevalence of ICA and ICSA decreased (Table 2).

Table 2. Prevalence of ICA and ICSA according to the duration of diabetes.

Duration (years)	ICA	ICSA
< 1	43%	50%
1 - 5	31%	32%
> 5	18%	25%

Clinical and Immunogenetic Characteristics of IDDM Patients

IDDM patients were divided into two groups, based on characteristics at diagnosis. Fifty-four percent were initially diagnosed as NIDDM and were treated with diet alone or diet and oral hypoglycemic agents for more than 6 months. However, other clinical parameters, HLA-DR type, prevalence of ICA and ICSA and C-peptide levels showed no significant differences between two groups (Table 3).

Table 3. Clinical and immunogenetic characteristics of the patients with IDDM classified by initial diagnosis.

Initial Diagnosis	IDDM (n=28)	NIDDM (n=33)
Age of onset (years)	23.6 ± 1.4	25.9 ± 1.2
Sex ratio (M:F)	12:16	14:19
History of DKA (%)	44	31
Family history of diabetes (%)	25	15
Current BMI (kg/m^2)	18.5 ± 0.5	17.8 ± 0.4
Current insulin dose (U/kg)	0.7 ± 0.1	1.1 ± 0.3
HLA-DR4 (%)	56	65
HLA-DR2 (%)	12	10
ICA (%)	23	36
ICSA (%)	40	25
Serum C-peptide (nmol/l)	0.14 ± 0.04	0.13 ± 0.03

Values are mean ± S.D.

Determination of DQA1 and DQB1 Alleles

As shown in Table 4, the frequencies of the DQA1 alleles in both IDDM and NIDDM patients was not different from that in control groups. DQB1*0302 and 0201 alleles, which are well known IDDM susceptibility alleles in Caucasians, were not increased in Korean patients with IDDM (Table 5). The frequency of DQB1*0303 was higher in both IDDM and NIDDM patients compared to the control group.

Table 4. Frequencies of HLA-DQA1 alleles in IDDM, NIDDM and normal controls from Korea.

HLA-DQA1	IDDM (n = 36)	NIDDM (n =48)	Controls (n = 238)
	n (%)	n (%)	n (%)
0101	11 (15.3)	12 (12.5)	71 (14.9)
0102	19 (26.4)	20 (20.8)	85 (17.9)
0103	7 (9.7)	14 (14.6)	67 (14.1)
0201	4 (5.6)	5 (5.2)	25 (5.3)
0301	16 (22.2)	29 (30.2)	143 (30.0)
0401	1 (1.4)	2 (2.1)	15 (3.2)
0501	9 (12.5)	8 (8.3)	59 (12.5)
0601	5 (6.9)	6 (6.3)	11 (2.3)

Frequency of Arginine in Position 52 of the DQ∝ Chain and Aspartic Acid in Position 57 of the DQß Chain

When we analyzed our data further, the frequencies of arginine/arginine or non-arginine/non-arginine in the 52nd position of DQα chain in patients with IDDM were not different from those in control subjects. Also, aspartic acid in the 57th position of DQß chain did not emerge as significant (Table 6).

Table 5. Frequencies of HLA-DQB1 alleles in IDDM, NIDDM and normal controls from Korea.

HLA-DQB1	IDDM (n = 36) n (%)	NIDDM (n = 48) n (%)	Controls (n = 174) n (%)
0201	3 (4.2)	3 (3.1)	35 (9.8)
0301	19 (26.4)	21 (21.9)	63 (17.6)
0302	4 (5.6)	10 (10.4)	27 (7.5)
0303	14 (19.4)*	34 (35.4)**	27 (7.5)
04#	6 (8.3)	6 (6.3)	31 (8.7)
0501	3 (4.2)	5 (5.2)	40 (11.2)
0502	3 (4.2)	4 (4.2))	14 (3.9)
0503	5 (6.9)	4 (4.2)	5 (1.4)
0601	1 (1.4)	1 (1.0)	28 (7.8)
0602	2 (2.8)	1 (1.0)	27 (7.5)
0603	6 (8.3)	6 (6.3)	7 (2.0)
0604	6 (8.3)	1 (1.0)	19 (5.3)

04#: 0401, 0402
*: p [χ^2] < 0.05 vs. normal control
**: p[χ^2] < 0.01 vs. normal control

Table 6. Frequency of HLA-DQA1*Arg-52 and HLA-DQB1*non-Asp-57 genotypes.

Genotype	IDDM (n=36) n (%)	Control (n=38) n (%)
HLA-DQA1		
non-Arg/non-Arg	14 (39)	21 (55)
non-Arg/Arg	13 (36)	9 (24)
Arg/Arg	9 (25)	8 (21)
HLA-DQB1		
Asp/Asp	22 (61)	17 (45)
Asp/non-Asp	9 (25)	13 (34)
non-Asp/non-Asp	5 (14)	8 (21)

Asp: Aspartic acid; Arg: Arginine

DISCUSSION

In this study we found that HLA-DR4 had a positive association and DR2 had a negative association with adult onset IDDM in Korea. However, there was no association between DR3/DR4 heterozygosity and IDDM (data not shown). These findings emphasize the differences in HLA associations with IDDM among different ethnic groups. This variability suggests that the primary susceptibility genes for IDDM are not at the HLA-DR locus, but rather, with other genes that are linked to HLA-DR.

Interestingly, we found that over 50% of adult onset IDDM patients initially presented as NIDDM. However, we could not find any differences in clinical and immunogenetic parameters, including HLA frequencies and prevalence of islet cell autoantibodies. The HLA-DQA1*0301 specificity is a well established IDDM susceptibility allele throughout the world and has been confirmed in Caucasians, African Americans and Japanese populations.[12] We could not find, however, a contribution for any DQA1 alleles as susceptibility markers in Korean IDDM patients. HLA-DQB1*0302 and *0201 alleles, which are well known susceptibility alleles for IDDM in Caucasians, were not increased in Korean patients. Because the numbers of our patients were rather small and we studied only adult onset IDDM, this study may not be extended to IDDM in children. We found that the frequency of HLA-DQB1*0303 was significantly increased in our IDDM patients. However, the separation of DQB1*0302 and DQB1*0303 by the method we used was difficult in some cases. We are now repeating HLA-DQB1 typing using specific probes which can separate DQB1*0302 and DQB1*0303 alleles more accurately.

We also found that DQB1*0303 was increased in Korean NIDDM patients. Many diabetologists in the Far East have suspected that patients with NIDDM do not differ from those with adult onset IDDM in terms of their pathogenic mechanisms, since many NIDDM patients are not obese, and lose quite large amounts of body weight soon after the onset of diabetes, even though they do not develop ketoacidosis and become insulin requiring. These findings, thus, provide evidence that some Korean patients with NIDDM share the same susceptibility gene(s) as IDDM patients, suggesting that some individuals with NIDDM might have common pathogenic mechanisms as those with IDDM.

Studies from Finland by Groop et al.[5,6] and by Tuomilehto-Wolf et al.[7] showed that some susceptibility genes for NIDDM may be in the HLA region, supporting our finding of a HLA association in NIDDM. It was already pointed out by Ida et al.[8] that the association between HLA and NIDDM differs among races, and Pontiroli et al.[13] reported no such HLA association in Italian NIDDM patients. However, there is still the possibility that other HLA specificities not tested in these studies, such as DQA1 genes, might be significantly associated in the Caucasian NIDDM population. Most patients with NIDDM in Korea do not have islet cell antibodies in their blood, and anti-GAD antibody positivity in IDDM and NIDDM is much lower than among the Caucasians (data not shown). We are very puzzled as to why a certain HLA specificities would be increased if there is no evidence for autoimmune mechanisms. It is particularly interesting to note that Ida et al.[8] reported a high frequency of the DQw3.2 specificity among Japanese NIDDMs, which is also found to be increased in adult onset IDDM. This study is interesting, especially in that they selected NIDDM patients without anti-islet antibodies.

Non-aspartic acid in position 57 of the DQ beta chain and arginine in position 52 of the DQ alpha chain has drawn intense scrutiny[14,15] after Todd et al.[3] and Khalil et al.[4] reported their importance. Our data do not support these hypotheses, whether analyzed for the presence of individual amino acids, or after alleles were combined to the "high risk" or "low risk" alleles. Because the subjects of the study are from the adult population, our data do not decrease the importance of DQB1*non-Asp-57 or DQA1*Arg-52 found in childhood IDDM.

In conclusion, we did not find a significant contribution for DQB1*non-Asp-57 or DQA1*Arg-52 alleles as disease susceptibility genes. Another important finding was that a group of adult onset IDDMs in Korea presented as NIDDM initially, and did not show any differences in other clinical and immunogenetic characteristics from IDDM. We found a high frequency of HLA-DQB1*0303 in our NIDDM population. These results suggest that many Korean patients with NIDDM might share common autoimmune mechanisms in their pathogenesis.

REFERENCES

1. WHO Expert Committee on Diabetes Mellitus, Second Report, Technical Report Series 646, World Health Organization, Geneva (1980).
2. J.L. Tiwari, P.I. Terasaki, Endocrinology, *in*: "HLA and Disease Associations," J.L. Tiwari, P.I. Terasaki, eds., Springer-Verlag, New York, Berlin, Heidelberg, Tokyo (1985).
3. J.A. Todd, J.I. Bell, H.O. McDevitt, HLA-DQ gene contributes to susceptibility and resistance to insulin-dependent diabetes mellitus, *Nature* 329:599 (1987).
4. I. Khalil, L. d'Auriol, M. Gobet, et al., A combination of HLA DQ-Asp57-negative and HLA-DQ Arg52 confers susceptibility to insulin-dependent diabetes mellitus, *J Clin Invest.* 85:1315(1990).
5. L. Groop, A. Miettinen, P-H. Groop, et al., Organ-specific autoimmunity and HLA-DR antigens as markers for ß-cell destruction in patients with Type II diabetes, *Diabetes* 37:99 (1988).
6. L. Groop, P-H Groop, S. Koskimies, Relationship between ß-cell function and HLA antigens in patients with Type II (non-insulin-dependent) diabetes, *Diabetologia* 29:757 (1986).
7. E. Tuomilehto-Wolf, J. Tuomilehto, A. Nissinen, et al., The search is over: The major genes for NIDDM were also found in the HLA region, *Diabetes* 41:91A (suppl. 1) (1992).
8. T. Ida, H. Kuzuya, M. Hattori, et al., Heterogeneity of non-insulin-dependent diabetes mellitus in HLA types and clinical features: Comparison with insulin-dependent diabetes mellitus, *Endocrinol Japon* 38:9 (1991).
9. P.I. Terasaki, D. Bernoco, M.S. Park, et al., Microdroplet testing for HLA-A, -B, -C and -D antigens, *Am J Clin Pathol.* 69:103 (1978).
10. G.F. Bottazo, A. Florin-Christensen, D. Doniach, Islet cell antibodies in diabetes mellitus with autoimmune polyendocrine deficiencies, *Lancet* ii:1279 (1974).
11. G.S. Eisenbarth, M.A. Morris, R.M. Scearce, Cytotoxic antibodies to cloned rat insulinoma cells in serum of patients with diabetes mellitus, *J Clin Invest.* 67:403 (1981).
12. J.A. Todd, Genetic control of autoimmunity, *Immunology Today* 11:122 (1990).
13. A.E. Pontiroli, A. Calderara, F. Capra, et al., HLA phenotype and secondary failure to oral hypoglycaemic agents, (letter), *Diabetologia* 30:824 (1987).
14. P.A. Morel, J.S. Dorman, J.A. Todd, et al., Aspartic acid at position 57 of HLA DQ-ß chain protects against Type I diabetes: A family study, *Proc Natl Acad Sci. USA* 85:8111 (1988).
15. K. Yamagata, T. Hanafusa, H. Nakajima, et al., HLA A1*1 contributes to resistance and A1*3 confers susceptibility to Type I (insulin-dependent) diabetes mellitus in Japanese subjects, *Diabetologia* 34:133 (1991).

PREDISPOSITION TO INSULIN-DEPENDENT DIABETES MELLITUS: ROLE OF HLA AND NON-HLA GENES

Kjersti S. Rønningen and Erik Thorsby

Institute of Transplantation Immunology
The National Hospital
0027 Oslo
NORWAY

INTRODUCTION

Approximately 0.3-0.4% of European Caucasoids will develop insulin-dependent diabetes mellitus (IDDM).[1] Among Scandinavians, the prevalence is even higher.[2] Among Norwegians, as many as 10-15% of individuals with the particular HLA genotype DR3-DQ2/DR4-DQ8 may develop Type I diabetes (Rønningen et al., unpublished observations). A sibling who shares this genotype with an IDDM patient will have approximately a 25% risk to develop the disease,[3] while the risk reaches 35-50% for a monozygotic twin of a patient.[4]

The concordance rate for monozygotic twins suggests that genetic and environmental factors are approximately equally important for the development of IDDM. Concerning the genetic factors, the figures demonstrate that genes in the HLA complex are by far the most important, but that non-HLA genes contribute.

HLA ASSOCIATIONS IN IDDM

Among Caucasoids, Type I diabetes is strongly associated with DR3 and/or DR4, and the risk is particularly high for DR3/4 heterozygotes. In contrast, IDDM is very rarely found among DR2 positive individuals. Among Japanese IDDM patients, DR4 and DR9 are most prevalent. More recent studies suggested, however, that IDDM susceptibility is more strongly associated with certain DQ than DR specificities. For example, the DR4 association appears to be secondary to an association to DQ8. Todd and co-workers[5] suggested that DQ molecules having serine, alanine, or valine at residue 57 of the DQß chain (non-Asp-57) were associated with susceptibility, while those having aspartic acid (Asp-57) were associated with dominant protection or they were neutral. Interestingly, the non-obese diabetic (NOD) mice, which spontaneously develop an IDDM-like disease, is the only strain of mice having a non-Asp amino acid at residue 57 of its I-Aß chain (the homologue of DQß). However,

Standardization of Epidemiologic Studies of Host Susceptibility
Edited by J. Dorman, Plenum Press, New York, 1994

more recent studies suggest that the picture is more complicated. For example, DQ molecules with non-Asp-57 grossly differ with respect to associated IDDM risk. Some Caucasoid IDDM patients may be Asp-57/non-Asp-57 "heterozygous" or Asp-57/Asp-57 "homozygous"[6,7] and Japanese IDDM patients mostly have DQβ chains of Asp-57 type.[8]

Khalil et al.[9] found that cis or trans encoded DQ molecules consisting of a DQα chain having arginine at position 52 (Arg-52) and a non-Asp-57 DQβ chain (=SS), were most strongly associated with IDDM. Disease susceptibility correlated quantitatively with the expression of such SS DQ molecules.[10] Although DQ molecules being non-Arg-52 and Asp-57 (=PP) were found to be associated with many (but not all) DQ molecules of SS type among Caucasoids, this is not true for the Japanese.

Based on studies of Norwegian patients, we suggested recently[11] that IDDM might be primarily associated to five or six particular cis or trans encoded DQ molecules, which might be common for Blacks and Caucasoids, but partially different for Japanese.

RESULTS OF A RECENT INTERNATIONAL COLLABORATIVE STUDY

As part of the 11th International Histocompatibility Workshop (1991), 981 unrelated IDDM patients collected by 18 different laboratories were genomically typed for HLA class II alleles and compared to 2228 healthy controls.[12] In all three ethnic groups studied (Blacks, Caucasoids, Orientals), susceptibility was found associated to several different combinations of DQA1 and DQB1 genes in cis or trans position. The combinations demonstrating the strongest associations are listed in Table 1 together with the associated relative risks.[13] To a certain extent, the relative risks were additive; i.e., individuals having more than one of these combinations of DQA1 and DQB1 genes were at an even higher risk to develop IDDM.

Table 1. Some combinations of DQA1 and DQB1 genes associated with IDDM susceptibility or protection in Blacks, Caucasoids and Japanese.

DQA1 DQB1	Pos[1]	RR[2]
Susceptibility		
0301,0302 (DQ8)	C	8-12
0501,0201 (DQ2)	C	3-5
0301,0201	C/T	5-20
0501,0302[3]	T	8-35
0301,0402[3] (DQ4)	T	5-15
0301,0401[4] (DQ4)	C	4
0301,0303[4] (DQ9)	C	2
Protection		
0102,0602 (DQ6)	C	0.2
0103,0601[4] (DQ6)	C	0.2

[1]C=cis, T= trans position of the genes; [2]Relative risk (12); [3]Not associated with susceptibility in Japanese; [4]Associated with susceptibility or protection only in Japanese

The three first listed combinations of DQ genes were found associated with IDDM susceptibility in all three ethnic groups. In Black patients, the DQA1*0301 and DQB1*0201 genes were usually found together on the same haplotype (in cis position), together with genes encoding DR4, DR7 or DR9 (Figure 1), while for Caucasoid IDDM patients this was the case to a much lesser extent. In contrast, most Caucasoid (and all Japanese) IDDM patients having this combination of DQ genes carried them on different haplotypes (in trans), the most frequent genotype among Caucasoids being DR3-DQ2/DR4-DQ8. In both cases, the same DQ molecule (DQA1*0301, DQB1*0201) may be encoded[14] (see Figure 1.)

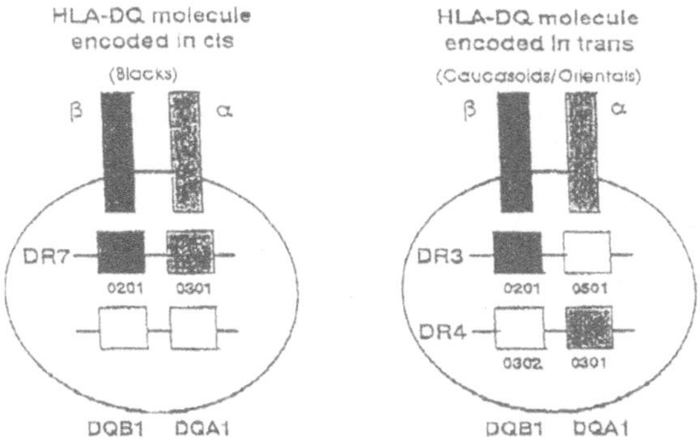

Figure 1. The DQA1*0301 and DQB1*0201 genes are often found on the same haplotype (in cis) among Black IDDM patients, while they most often are found on different haplotypes (in trans) among Caucasoid and Japanese IDDM patients. In both cases, the same DQ molecule is encoded and expressed.

In Japanese, IDDM susceptibility is also associated with other particular combinations of DQ genes, one being DQA1*0301 in combination with DQB1*0401 in the cis position. In Caucasoids, a significant association to the DQA1*0301 and DQB1*0402 genes in the trans position is found[15]. Interestingly, the only difference between the DQβ chains encoded by DQB1*0401 and DQB1*0402 is a leucine to arginine substitution at residue 23, and this minor difference does not seem to be recognized by T cells.[16]

The finding of pronounced IDDM susceptibility associated with some particular combinations of DQA1 and DQB1 genes in cis or trans positions, which in both cases encode the same functional DQ molecule, is strong evidence that the primary HLA association in IDDM is to given DQ molecules as such, and not to products of closely linked genes.[13] An alternative, but very unlikely explanation, would be a primary association to a combination of two other genes in cis or trans positions, one centromeric to DQB1, the other telomeric to DQA1. The strong association to two genes in cis or trans position would also argue against a primary association to given haplotypic combinations of genes; i.e., extended HLA haplotypes.[17] This does not exclude that other HLA genes may contribute. In the 11th Workshop, IDDM susceptibility was not found associated to any DPA1 or DPB1 alleles. However, a contribution by some DRB1 alleles encoding variants of DR4 (e.g. DRB1*0405) could not be excluded.

Protection against development of IDDM was found associated with several other combinations of DQA1 and DQB1 genes. The strongest protection in all three ethnic groups was found for the combination of DQA1*0102 and DQB1*0602, which encodes one of the DQ6 variants. In Japanese, protection was also associated with a variant of DQ6; encoded by DQA1*0103 and DQB1*0601. The DQ6 associated protection was dominant over the DQ associated susceptibility mentioned above.

Until recently, it could not be determined whether protection is primarily associated with the genes encoding DR2 or DQ6, which most often are present on the same haplotype. The Workshop studies revealed that protective DQA1*0102-DQB1*0602 haplotypes may carry different DRB1 genes. Among Blacks, DR2 on this haplotype was encoded by DRB1*1503, while among Caucasoids DR2 was encoded by DRB1*1501. Other recent studies have also demonstrated that protection is more strongly associated with DQ6 than DR2.[18,19]

Taken together, available data suggest that IDDM susceptibility is primarily associated to some particular DQ molecules as such, and not to products of closely linked genes. Similarly, protection appears to be primarily associated with the DQ6 molecule, and is dominant over the DQ associated susceptibility.

THE ROLE OF NON-HLA GENES

Based on the difference in concordance rate between monozygotic twins and HLA-identical siblings, non-HLA genes also appear to be of importance in determining IDDM susceptibility. In the NOD mouse, several such genes (non-H-2) have been identified.[20-22] Most recent studies indicate an interaction of at least 8-10 different genes in determining the risk for IDDM development, and some of these "new genes" may also account for disease heterogeneity (Dr. John A. Todd, personal communication).

In man, a polymorphism in the insulin (INS) gene region on chromosome 11p has been found associated with IDDM susceptibility.[23] However, it was recently claimed that the INS association is significant only in HLA-DR4 positive diabetics.[24]

We analyzed INS polymorphisms in 244 Norwegian sporadic Type I diabetics and 186 Norwegian randomly selected healthy controls. In addition, 75 HLA-DR3/4 and 45 HLA-DR4/X (X = not DR3) controls were selected from the Norwegian bone marrow register to facilitate comparison of diabetics and controls matched for HLA-type. The INS polymorphisms, 1,428/FokI, 1,404/Fnu4HI and 1,127/PstI, were characterized by polymerase chain reaction amplification followed by restriction enzyme digestion.[25] The common allele, which lacks the polymorphic restriction site using FokI and PstI but carries it for Fnu4HI, has been designated '+' and the other allele '-'.

We found a significant association between INS and Type I diabetes; 81% of the patients were INS +/+ compared to 64% of controls ($RR=2.4$, $p<0.0001$). HLA-DR and DQ matching of patients and controls gave, however, no evidence for any DRB1 or DQB1 effect on INS associated susceptibility. The relative risk for IDDM for the homozygote status +/+ at the 1,127/PstI INS site for the different DR genotypes is shown in Table 2. The 1,428/FokI and 1,404/Fnu4HI data did not differ from that generated using PstI digestion due to the strong linkage disequilibrium across the INS region.[26] For this reason only the 1,127/PstI data are presented in Table 2.

Since the function of HLA-DQ molecules is to present peptide-fragments of antigens to CD4+ T cells, which recognize the peptide/HLA complex by their antigen specific T cell receptor (TCR), it is perhaps surprising that no strong or reproducible associations of IDDM have been found to TCR genes.[27] Neither has any strong association to polymorphic genes involved in peptide generation or transport (LMP, TAP) been detected.[28]

POPULATION SCREENING

Using genomic HLA typing, as well as typing for non-HLA genes associated with IDDM susceptibility (INS, etc.), it is now, to a large extent, possible to identify individuals having a strong genetic predisposition to develop IDDM. These individuals can then be followed by yearly tests for detecting early signs of ß-cell destruction (i.e., islet cell antibodies etc.). Prediabetics may in this way get an opportunity for intervention therapy (nicotinamide, immunosuppressive drugs, etc.) trying to prevent the clinical onset of the disease.

Table 2. Relative risk by HLA type for the +/+ homozygote status at the 1,127/PstI INS site in Norwegian sporadic diabetics versus HLA-matched Norwegian controls.

	No. of diabetics with 1,127/PstI INS status		No. of controls with 1,127/PstI INS status		Relative risk (95% confidence intervals)
HLA	+/+	+/- -/-[@]	+/+	+/- -/-[@]	
DR3/4	53	13	51	34	2.7 (1.3-5.7)*
DR4/4 or X	51	8	62	32	3.3 (1.4-7.8)**
DR3/3 or X	16	4	21	14	2.7 (0.7-9.7)
DRX/X	5	1	45	28	3.1 (0.3-28)

[@] Since only 10 of the 438 individuals included in this table were typed as -/- homozygotes, they have been included in the +/- group.
* = $p<0.01$, ** = $p<0.05$

Acknowledgements

We would like to express our gratitude to all the participants in the IDDM part of the 11th International Histocompatibility Workshop (whom will appear from ref. 12). We thank Drs. John A. Todd and Steve C. Bain for helpful suggestions regarding the studies of the INS-gene region.

REFERENCES

1. Diabetes Epidemiology Research International Group, Geographic patterns of childhood insulin-dependent diabetes mellitus, *Diabetes* 37:1113 (1988).
2. G. Joner, O. Sovik, Increasing incidence of diabetes mellitus in Norwegian children 0-14 years of age, *Diabetologia* 32:79 (1989).

3. I. Deschamps, H. Lestradet, M. Busson, J. Hors, Evaluation of recurrence risk in siblings of diabetic children: Importance of age and birth order in relation to HLA genotypes, *Diabetes Res.* 1:125 (1984).
4. A.H. Barnett, C. Eff, R.D.G. Leslie, D.A. Pyke, Diabetes in identical twins, *Diabetologia* 20:87 (1981).
5. J.A. Todd, J.L. Bell, H.L. McDevitt, HLA-DQβ gene contributes to susceptibility and resistance to insulin-dependent diabetes mellitus, *Nature* 329:559 (1987).
6. K.S. Rønningen, T. Iwe, T.S. Halstensen, A. Spurkland, E. Thorsby, The amino acid at position 57 of the HLA-DQβ chain and susceptibility to develop insulin-dependent diabetes mellitus (IDDM), *Hum Immunol.* 26:215 (1989).
7. J.M. Baisch, E.S.T. Weeks, R. Giles, Analysis of HLA-DQ genotypes and susceptibility in insulin-dependent diabetes mellitus, *N Eng J Med.* 26:1836 (1990).
8. K.E.A. Lundin, K.S. Rønningen, A. Aono, et al., HLA-DQ antigens in Japanese patients with insulin-dependent diabetes mellitus, *N Eng J Med.* 26:1836 (1990).
9. I. Khalil, L. d'Auriol, M. Gobet, et al., A combination of HLA-DQβ Asp-57-negative and HLA-DQα Arg-52 confers susceptibility to insulin-dependent diabetes mellitus, *J Clin Invest.* 85:1315 (1990).
10. I. Khalil, I. Deschamps, V. Lepage, et al., Dose effect of cis- and trans-encoded HLA-DQαβ heterodimers in IDDM susceptibility, *Diabetes* 41:378 (1992).
11. K.S. Rønningen, A. Spurkland, T. Iwe, F. Vartdai, E. Thorsby, Distribution of HLA-DRB1, -DQA1 and -DQB1 alleles and DQA1-DQB1 genotypes among Norwegian patients with insulin-dependent diabetes mellitus, *Tissue Antigens* 37:105 (1991).
12. K.S. Rønningen, A. Spurkland, B.D. Tait, et al., "Joint Report: HLA class II associations in insulin-dependent diabetes mellitus among Blacks, Caucasoids and Japanese, 11th International Histocompatibility Workshop", Oxford Publisher, in press (1992).
13. E. Thorsby, K.S. Rønningen, Role of HLA genes in predisposition to develop insulin-dependent diabetes mellitus, *Ann Med.* in press (1992).
14. W.W. Kwok, D. Schwartz, B. Nepom, et al., HLA-DQ molecules form αβ heterodimers of mixed allotype, *J Immunol.* 141:3123 (1988).
15. K.S. Rønningen, H.A. Gjertsen, T. Iwe, et al., Particular HLA-DQ αβ heterodimer associated with IDDM susceptibility in both DR4-DQw4 Japanese and DR4-DQw8/DRw8-DQw4 Whites, *Diabetes* 40:759 (1991).
16. H.A. Gjersten, K.E.A. Lundin, K.S. Rønningen, G. Gaudernack, E. Thorsby, A T-lymphocyte clone recognizing a DQ αβ heterodimer encoded in cis by the DR4-DQw4 haplotype and in trans by DR4-DQw8/DR8-DQw4 heterozygous cells, *Hum Immunol.* 30:262 (1991).
17. R.L. Dawkins, F.T. Christiansen, P.H. Kay, M. Garlepp, et al., Disease associations with complotypes, supratypes and haplotypes, *Immunol Rev.* 70:5 (1983).
18. H.A. Erlich, R.L. Griffith, T.L. Bugawan, et al., Implication of specific DQB1 alleles in genetic susceptibility and resistance by identification of IDDM siblings with novel HLA-DQB1 allele and unusual DR2 and DR1 haplotypes, *Diabetes* 40:478 (1991).
19. D. Zellszewski, J.M. Tiercy, C. Boitard, et al., Extensive study of DRB, DQA, and DQB gene polymorphism in 23 DR2-positive, insulin-dependent diabetes mellitus patients, *Hum Immunol.* 33:140 (1992).
20. J.A. Todd, T.J. Altman, R.J. Cornall, et al., Genetic analysis of autoimmune Type I diabetes mellitus in mice, *Nature* 351:542 (1991).
21. R.J. Cornall, J.B. Prins, J.A. Todd, et al., Type I diabetes in mice is linked to the interleukin-1 receptor and Lsh/Ity/Bcg genes on chromosome 1, *Nature* 353:262 (1991).
22. H.J. Garchon, P. Bedossa, L. Eloy, J.F. Bach, Identification and mapping to chromosome 1 of a susceptibility locus for perinsulitis in non-obese diabetic mice, *Nature* 353:260 (1991).

23. G.I. Bell, S. Horita, J.H. Karam, A polymorphic locus near the human insulin gene is associated with insulin-dependent diabetes mellitus, *Diabetes* 33:176 (1984).
24. C. Julier, R.N. Hyer, J. Davies, et al., Insulin-IGF2 region on chromosome 11p encodes a gene implicated in HLA-DR4-dependent diabetes susceptibility, *Nature* 354:155 (1991).
25. S.C. Bain, J.B. Prins, C.M. Hearne, et al., Insulin gene region-encoded susceptibility to Type I diabetes is not restricted to HLA-DR4 positive individuals, *Nature Genetics* in press (1992).
26. N.J. Cox, G.I. Bell, K. Xiang, Linkage disequilibrium in the human insulin-like growth factor II region of human chromosome 11, *Am J Hum Genet.* 43:495 (1988).
27. J. Hillert, O. Olerup, Germ-line polymorphism of TCR genes and disease susceptibility fact of hypothesis?, *Immunology Today* 13:47 (1992).
28. M. Colonna, M. Bresnahan, S, Bahram, et al., Allelic variants of the human putative peptide transporter involved in antigen processing, *Proc Natl Acad Sci. USA* 89:3932 (1992).

EVIDENCE FOR THE "GENETIC ADMIXTURE HYPOTHESIS" THAT THE INCREASED INCIDENCE OF IDDM AMONG AFRICAN AMERICANS COMPARED TO AFRICANS IS DUE IN LARGE PART TO SUSCEPTIBILITY GENE(S) WHICH ENTERED THROUGH ADMIXTURE WITH THE EUROPEAN AMERICAN GENE POOL

Jeffrey M. Roseman,[1] Rodney C.P. Go,[1] and Ronald T. Acton[1,2]

[1]Department of Epidemiology
[2]Department of Microbiology
University of Alabama at Birmingham
Birmingham, AL 35294-0008
USA

INTRODUCTION

There are approximately 30 million African Americans in the U.S., constituting approximately 12 percent of the population.[1] Most of this subpopulation can trace their ancestry to Africans who were brought over as slaves beginning in 1619 to the Jamestown settlement of the British American colonies and later to the United States. During the major period of slave importation from 1700 to 1808, approximately 400,000 Africans entered, almost all from the area of West and West-Central Africa, predominantly from the present countries of Nigeria and Angola. Through admixture with European Americans, African Americans now have an average of 20 percent European American genes.[2]

Beginning in 1931, there were a number of clinical studies pointing out that the proportion of diabetics among hospitalized African American children was much lower than that among European American children.[3,4,5,6] In 1975, commenting on the lower proportion and the apparent rarity of juvenile diabetes in Africa, Dr. Michael MacDonald speculated, "...that the rate of juvenile diabetes among American Negroes would be even lower, if it had not been for significant racial admixture".[6] This has been called the "Genetic Admixture Hypothesis".[7] There are now four lines of evidence that suggest that admixture with European American genes is in large part the cause of the increased frequency of IDDM in African Americans compared to Africans:

1) lower frequency of IDDM in African Americans than European Americans, but higher than Africans,

Standardization of Epidemiologic Studies of Host Susceptibility
Edited by J. Dorman, Plenum Press, New York, 1994

2) similarities in Major Histocompatibility Complex (MHC) loci associations in European Americans and African Americans with IDDM compared to race-matched controls,
3) a positive relationship between European American genetic admixture proportion and incidence rate of IDDM among U.S. subpopulations of African Americans,
4) increased European American genetic admixture in African American IDDM cases compared to African American controls.

The remainder of this paper will discuss in detail these lines of evidence for the "Genetic Admixture Hypothesis".

EVIDENCE THAT THE FREQUENCY OF IDDM IS LOWER IN AFRICAN AMERICANS THAN EUROPEAN AMERICANS, BUT HIGHER THAN AFRICANS

As noted above, the "Genetic Admixture Hypothesis" was generated from observations of hospital- or clinic-based samples. Such samples have significant threats to their validity. There is not only considerable opportunity for subject selection into health care facilities, but increases in hospitalizations for other diseases could also lead to a lower proportion of IDDM among hospitalized African Americans. One way to overcome this problem is to use population-based samples. One of the earliest population-based studies of the prevalence of diabetes among young African Americans was conducted by the U.S. Army of recruits during WWII.[8] This study found that the prevalence of diabetes at the time of recruitment was 2.9 - 3.0 per 1000 among European Americans, while only 0.8 - 1.9 per 1,000 among African Americans. Other population-based studies also found that the prevalence was lower in young African Americans.[9,10]

All prevalence studies, however, have a major threat to the interpretation that they reflect the underlying incidence. Since prevalence is a product of incidence and mean duration, if African Americans have higher mortality rates (as has recently been demonstrated),[11] and therefore shorter mean duration, then African Americans would have lower prevalence even if the incidences were identical.

Recently incidence estimates have become available from population-based registries. Figure 1 shows the incidences of IDDM from those U.S. registries which have reported both African American and European American rates.[13,14,15,16] In all such registries the African American rates were lower than the European American rates, with incidence rate ratios ranging from approximately 0.2 to 0.9.

The evidence that the incidence rate of IDDM is lower in Africans than among African Americans is scarce. All of the reports from West Africa have been prevalence studies, mostly hospital or clinic-based.[17,18,19] As noted above, such studies have serious threats to their validity with respect to estimating incidence. The problem of shorter mean duration may be even more of an issue in West Africa where, because of poor access to medical care in some locations, more cases might die before coming to medical attention. Another problem with studies of the frequency of IDDM among West Africans is that many with insulin-requiring juvenile onset diabetes have pancreatic calcification, not IDDM.[18,19] Any study of the incidence of IDDM in West Africa would have to exclude such cases. Also, "J-type" or malnutrition-related diabetes has also been reported in Africa.[20] There is debate as to whether this is a form of IDDM, so incidence estimates from Africa would need to take this uncertainty into consideration.

EVIDENCE OF SIMILARITIES IN MAJOR HISTOCOMPATIBILITY COMPLEX (MHC) LOCI ASSOCIATIONS IN EUROPEAN AMERICANS AND AFRICAN AMERICANS WITH IDDM COMPARED TO RACE-MATCHED CONTROLS

Most studies have found that genetic markers at the MHC show similar allelic associations with IDDM in African Americans and European Americans. HLA-B8, B15, DR3 and DR4 are increased significantly, while DR2 is decreased in both African American and European American IDDM samples.[21,22,23,24,25] Consistent with this, complement components Bf and C4 (also in the MHC region) have associations with IDDM in African

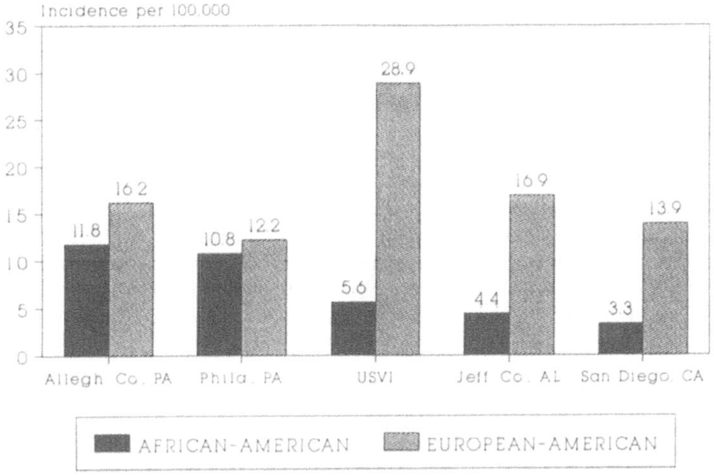

Figure 1. Age-adjusted (0-14) incidence of insulin-dependent diabetes.

Americans similar to those in European Americans.[26,27] A recent study examining HLA-DQA1 and DQB1 alleles found that the A3, and DQw2 alleles, which have been reported to be positively associated with IDDM in European Americans, North Indian Asians and Japanese, were positively associated with IDDM in African Americans, and the DQw6 allele, which was negatively associated in these other groups, was also negatively associated among African Americans.[28] Finally, non-Asp homozygous African Americans and European Americans appear to be at higher risk for IDDM.[29] Although similar associations could be seen as strong support for the "Genetic Admixture Hypothesis", they could also reflect a primary susceptibility allele for IDDM, unrelated to admixture.

A comparison of HLA-DR gene frequencies among controls from Alabama shows that the gene frequencies for HLA-DR3 and DR4 were lower among the African Americans (13% and 4%, respectively) compared to the European Americans (18% and 7%, respectively), suggesting that admixture of these could be a source of the susceptibility gene.[30] There was little difference in frequency of DR2 (18.5% among African Americans, 17% among European Americans), suggesting that increased prevalence in African Americans of an IDDM resistant DR2 linked gene is not the explanation. One piece of evidence indicating that the susceptibility determinant derived from genetic admixture with European Americans is DR4-associated is the report that there is an increased frequency of HLA-DR3 among Black Nigerian IDDM children, but not increased DR4.[31] This finding remains to be replicated. These reports provide tantalizing evidence that if the "Genetic Admixture Hypothesis" is valid, it is likely that the increase is due to HLA-linked susceptibility genes, possibly in linkage disequilibrium with DR4.

EVIDENCE THAT EUROPEAN AMERICAN GENETIC ADMIXTURE PROPORTION IS POSITIVELY ASSOCIATED WITH HIGHER INCIDENCE RATE OF IDDM AMONG U.S. SUBPOPULATIONS OF AFRICAN AMERICANS

Registry results suggest that over the U.S., there is almost a four-fold variation in the incidence of IDDM; much greater than the variation among European Americans. Degree of admixture is one possible source of this variation. One study of admixture showed that there is considerable variation within the U.S., with admixture estimates ranging from 3.7 to 26.5%.[2] There is a trend for more European American genetic admixture (25.2% ± 2.7) in Allegheny Co., Pennsylvania[32] where the incidence of IDDM in African Americans is higher (9.6 per 100,000)[12] than in Jefferson Co., Alabama, where genetic admixture is 17.9% ± 2.3,[33] and the incidence of IDDM is lower (7.0 per 100,000).[14]

EVIDENCE THAT THERE IS INCREASED EUROPEAN AMERICAN GENETIC ADMIXTURE IN AFRICAN AMERICAN IDDM CASES COMPARED TO AFRICAN AMERICAN CONTROLS

Two studies have determined directly whether genetic admixture was a factor in the increased occurrence. Reitnauer et al. did three analyses of admixture.[33] In the first, nine (non-chromosome 6) genetic markers were used and demonstrated that there was significantly greater than zero genetic contribution (9.6 ± 2.3%, $p < 0.01$) from the European American gene pool in a sample of African American IDDM patients from Alabama when a sample of African Americans without IDDM from the same region was used as one "parental" population. In the two other analyses, the amounts of genetic contribution from the European American gene pool to African Americans with and without IDDM were estimated using reported gene frequencies for West Africans for four genetic markers. The estimate of admixture (21.4 ± 2.8%) for the African American IDDM sample was greater than that for the African American controls (17.9 ± 2.3%), although the difference was not significant.

A case-control study of racial admixture as a determinant of IDDM in African Americans in the US Virgin Islands (VI), using a different approach, also provides evidence for the "Genetic Admixture Hypothesis".[15] The race of maternal and paternal grandparents was obtained by interview from diabetic and control subjects. Scores were assigned to each grandparent identified. A grandparent who was identified as White (i.e., the offspring of 2 Whites [White/White]) was assigned a score of 1. A grandparent who was the offspring of any combination of White and another race (White/Other race) was said to be mixed and was assigned a score of 2.5. A grandparent who was the offspring of any combination of races other than White (e.g., Black/Black, Black/Indian, Indian/Indian) received a score of 4. A mean racial score was then computed for both diabetic and control subjects. A lower value for the score would suggest a greater percentage of European American admixture. VI African American IDDM cases had a significantly lower grandparental race score (3.56 ± 0.53) than VI African American controls (3.85 ± 0.36).

CONCLUSIONS

Taken together, the above evidence provides a strong case for the "Genetic Admixture Hypothesis". There are a number of studies, however, which would be necessary to more conclusively determine its validity. First, it would be important to demonstrate that the incidence of IDDM among the African Americans in the U.S. registries is positively associated with the degree of genetic admixture of the African American population from which the registry's cases arise and to estimate the amount of variance in IDDM incidence which is explained by degree of genetic admixture. Such a detailed ecologic analysis remains to be done. It has been argued that because African Americans have more similar environmental exposures to European Americans than they do to Africans, genetic differences must account for the majority of the difference.[6] Second, it is necessary to replicate the results of some studies such as that of MacDonald et al. to see if the finding that HLA-DR4 is not associated with IDDM among West Africans[31] and that of Reitnauer et al. demonstrating that those African Americans with IDDM have a greater degree of European American genetic admixture than race-matched controls[33] are reproducible. Finally, studies of the incidence of IDDM among West Africans would be very useful in evaluating the hypothesis. What remains to be conclusively demonstrated is that the incidence of IDDM is indeed lower in the ancestral African populations which gave rise to African Americans. Such studies should be a high priority.

REFERENCES

1. 1990 Census of Population and Housing, Summary Population and Housing Characteristics, United States, Bureau of the Census (1992).
2. E.T. Reed, Caucasian genes in American Negroes, *Science* 165:762 (1969).
3. E.J. Leopold, Diabetes in the Negro race, *Am Int Med.* 5:285 (1931).
4. A. Altschul, A. Nathan, Diabetes mellitus in Harlem Hospital outpatient department in New York, *JAMA* 119:248 (1942).
5. P. Rosenbaum, Juvenile diabetes at Charity Hospital, *J La State Med Soc.* 199:389 (1967).
6. M.J. MacDonald, Lower frequency of diabetes among hospitalized Negro than children: Theoretical implications, *Acta Genet Med Gemellol.* 24:119 (1975).
7. J.M. Roseman, Chapter VIII, Diabetes in Black Americans *in:* "Diabetes in America, Diabetes Data - Compiled 1984", National Diabetes Data Group, NIH Publication No. 85-1468, 1 (1985).
8. A. Marble, Diabetes mellitus in the U.S. Army in World War II, *The Military Surgeon* 105:357 (1949).
9. H.A. Sultz, E.R. Schlesinger, W.E. Mosher, The Erie County Survey of Long-term Childhood Illness: II. Incidence and Prevalence, *Am J Public Health* 58:491 (1978).
10. K. Gorwitz, G.G. Howen, T.T. Thompson, Prevalence of diabetes in Michigan school-aged children, *Diabetes* 25:122 (1976).

11. J.S. Dorman, R.E. LaPorte, L.H. Kuller, et al., The Pittsburgh insulin-dependent diabetes mellitus (IDDM) morbidity and mortality study: Mortality results, *Diabetes* 33:271 (1984).
12. R.E. LaPorte, N. Tajima, J.S. Dorman, et al., Differences between Blacks and Whites in the epidemiology of insulin-dependent diabetes mellitus in Allegheny County, Pennsylvania, *Am J Epidemiol.* 123:592 (1986).
13. M. Lorenzi, E. Cagliero, N.J. Schmidt, Racial differences in incidence of juvenile-onset Type I diabetes: Epidemiologic studies in southern California, *Diabetelogia* 28:734 (1985).
14. L.E. Wagenknecht, J.M. Roseman, W.J. Alexander, Epidemiology of IDDM in Black and White children in Jefferson County, Alabama 1979-1985, *Diabetes* 38:734 (1989).
15. E.S. Tull, J.M. Roseman, C.L.E. Christian, Epidemiology of Childhood Insulin-Dependent Diabetes Mellitus in the US Virgin Islands from 1979-1988: Evidence of an Epidemic in the Early 1980's and Variation by Degree of Racial Admixture, *Diabetes Care* 14:558 (1991).
16. T.H. Lipman, The epidemiology of Type I diabetes in children 0-14 years of age in Philadelphia, Doctoral Dissertation, University of Pennsylvania (1991).
17. S.R.A. Dodu, The incidence of diabetes mellitus in Accra (Ghana), *West Afr Med J.* 7:129.
18. T.W. Kinnear, The pattern of diabetes mellitus in Nigerian teaching hospital, *East Afr Med J.* 40:288 (1963).
19. B.O. Osuntokun, F.M. Ankinkugbe, T.I. Francis, et al., Diabetes mellitus in Nigerians: A study of 832 patients, *West Afr Med.* 20:295 (1971).
20. World Health Organization, Diabetes Mellitus: Report of a WHO Study Group, Geneva, WHO, 10-24 (Tech. Rep. Ser. no. 727) (1985).
21. R. Patel, A. Ansari, C.L.P. Covarrubias, Leukocyte antigens and disease, III. Association of HLA-B8 and HLA-Bw15 with insulin-dependent diabetes in three different population groups, *Metabolism* 26:487 (1977).
22. R.J. Duquesnoy, M.J. MacDonald, P. Mullins, et al., Increased frequency of HLA-Dw3 in North-American Black patients with juvenile-onset diabetes, *Tissue Antigens* 13:369 (1979).
23. G.E. Rodey, N. White, T.E. Frazer, et al., HLA-DR specificities among Black Americans with juvenile-onset diabetes, *N Engl J Med.* 301:810 (1979).
24. A. Zeidler, J. Loon, S.D. Frasier, et al., HLA-DRw antigens in Mexican-American and Black-American diabetic patients, *Diabetes* 29:247 (1980).
25. R.T. Acton, B.O. Barger, B.R. Boshell, et al., Epidemiology and genetics of insulin-dependent diabetes mellitus in the White and Black Populations from the Southeastern United States *in*: "Frontiers in Immunogenetics", W.H. Hildemann, ed., Elsevier-North, Holland, New York, 239 (1980).
26. B. Budowle, P.J. Reitnauer, B.O. Barger, et al., Properdin Factor B in Black Type I (insulin-dependent) diabetic patients, *Diabetologia* 22:483 (1982).
27. B. Budowle, J.M. Roseman, R.C.P. Go, et al., The complement component C4 in Black Americans with Type I (insulin-dependent) diabetes mellitus, *Diabetologia* 26:166 (1984).
28. C.H. Mijovic, D. Jenkins, K.H. Jacobs, et al., HLA-DQA1 and -DQB1 alleles associated with genetic susceptibility to IDDM in a Black population, *Diabetes* 40:748 (1991).
29. J.S. Dorman, R.E. LaPorte, R.A. Stone, M. Trucco, Worldwide differences in the incidence of Type I diabetes are associated with amino acid variation at position 57 of the HLA-DQ β-chain, *Proc Natl Acad Sci. USA* 7370 (1990).

30. P.J. Reitnauer, J.M. Roseman, B.O. Barger, et al., HLA associations with insulin-dependent diabetes mellitus in a sample of the American Black population, *Tissue Antigens* 17:286 (1981).
31. M.J. MacDonald, O.O. Famuyiwa, I.A. Nwabuelo, et al., HLA-DR associations in Black Type I diabetics in Nigeria: Further support for models of inheritance, *Diabetes* 35:583 (1986).
32. R. Chakraborty, K.I. Mohammad, M. Nwankwo, R.E. Ferrell, Caucasian Genes in American Blacks: New data, *Am J Hum Genet.* 50:145 (1992).
33. P.J. Reitnauer, R.C.P. Go, R.T. Acton, et al., Evidence for genetic admixture as determinant in the occurrence of insulin-dependent diabetes mellitus in U.S. Blacks, *Diabetes* 31(6):532 (1982).

HLA STUDIES AMONG NORTH INDIAN TYPE I DIABETES

E. Bhatia,[1] S. Srikanta,[2] N.K. Mehra,[3] M.M.S. Ahuja[4]

[1]Department of Endocrinology
Sanjay Gandhi Post Graduate Institute
Lucknow
INDIA

[2]Department of Endocrinology
All India Institute of Medical Sciences
New Delhi
INDIA

[3]Department of Anatomy
All India Institute of Medical Sciences
New Delhi
INDIA

[4]Diabetes Foundation
New Delhi
INDIA

INTRODUCTION

The incidence of Type I diabetes varies widely among different racial groups. For example, the incidence among the Japanese and Chinese population is markedly decreased when compared with Caucasians.[1] Even within the same racial group, the incidence of diabetes can vary greatly. Among Caucasians, the incidence of Type I diabetes varies from a high of 28.6/100,000/year in Finland, to a much lower incidence in Central and Southern Europe (such as 4.4/100,000/year in France).[2] The reason for these differences is not clear, but it is thought that both genetic and environmental factors are contributory.

TYPE I DIABETES IN NORTH INDIANS

The exact incidence of Type I diabetes among North Indian Caucasians is not known. However, most observers feel that the rates are considerably less than in the West. In a

Standardization of Epidemiologic Studies of Host Susceptibility
Edited by J. Dorman, Plenum Press, New York, 1994

study from a diabetes clinic in Birmingham, United Kingdom, Odugbesan et al. (1985) reported that the frequency of insulin-dependent diabetes was only 4% compared with 18% in White patients.[3] There are no population-based studies of prevalence/incidence of Type I diabetes from the Indian subcontinent. A multi-center study from different parts of India estimated that 0.5% of the population between 15-30 years of age were diabetic.[4] However, it is not clear from this report how many of these patients were actually Type I diabetics. The reasons for the reduced incidence of Type I diabetes among North Indians is not clear. It is possible that environmental factors, such as a poor protein intake and an increased frequency of infections, may be of importance. However, as discussed in this article, differences in genetic background may also play a role.

MALNUTRITION-RELATED DIABETES

In contrast to the patients in the West, IDDM in the Indian subcontinent is a heterogenous entity. In addition to classical Type I diabetes, young patients with "malnutrition-related" diabetes may also be insulin-dependent. Malnutrition-related diabetes has been subdivided into two major categories: protein deficient pancreatic diabetes (PDPD) and fibrocalculous pancreatic diabetes (FCPD).[5] In the former group, patients are severely insulin-dependent, but do not develop ketosis even when insulin is withdrawn for long periods of time. Patients with FCPD have evidence of chronic pancreatic disease, developing at a young age and without any history of alcohol intake. Since both these entities may have a different genetic background from Type I diabetes, it is important to exclude such patients from any studies of Type I diabetes.[6,7]

GENETICS OF THE NORTH INDIAN POPULATION

The North Indian population consists of Caucasians of Indo-Aryan stock. They constitute a sub-group of Indo-Europeans who invaded Europe, the Middle East, Iran, and India around the second millennium B.C. Over the ages, the Indo-Aryans intermingled with the original inhabitants of the land, and with invading races such as the Mongols and Turks. This intermixing probably accounts for the observed variation in the frequency of some HLA antigens in North Indians, as compared with their Western counterparts.[8] While certain unique disease associations have been observed in this population, eg., HLA-Bw21 with myasthenia gravis,[9] most associations are similar to those seen in the West.

Few reports are available on the HLA associations among North Indian Type I diabetics. We will now summarize the results of our own studies, as well as recent studies in this field by other investigators.

HLA CLASS I AND III ANTIGENS

Initial population studies by our group revealed an increased prevalence of HLA-Bw21 among Type I diabetics when compared with controls (RR = 12.3, p = 5.9 x 10^{-6}).[10,11] This strong positive association seems to be unique to the North Indian population, and is found despite the fact that the control population had a similar prevalence of Bw21 as in Whites. In this study, the frequency of HLA-B8 and B15 were not increased among the diabetic subjects. HLA-B7 was protective (RR = 0.06, p = 3.3 x 10^{-3}). In another study, the same population was studied for various properdin alleles. The rare allele, BfS1, was significantly associated with Type I diabetes in this population.[12] It is of interest to note that BfS1 is in linkage disequilibrium with Bw21.

HLA CLASS II ANTIGENS

Eighty-eight unrelated Type I diabetics and 113 controls were serologically typed for the HLA antigens DR1 to DR7.[13] The frequency of HLA-DR3 was significantly increased (78.4% versus 25.7% in controls, corrected $p = 1.7 \times 10^{-12}$). The relative risk of 10.5 was much higher than that found in the Western IDDM population. Interestingly, DR4 did not show a positive association in this study (RR = 1.1, $p = 0.12$). DR2 showed a significant negative association (RR = 0.2, $p = 1.0 \times 10^{-5}$). HLA-DR associations have also been studied in an immigrant Punjabi North Indian population in the United Kingdom.[14] Again, the strongest association was with HLA-DR3 (RR = 7.7, $p < 10^{-5}$). In contrast to the results of the previous study, DR4 was also increased among the diabetics (RR = 5.7, $p = 0.003$), although the association was weaker than for DR3. DR2 and DR7 were negatively associated.

HLA-DQ gene polymorphisms were studied by Fletcher et al. in North Indians residing in UK.[15] Similar DQ beta gene polymorphisms were associated with DR3 and DR4 as in White Caucasians. However, the DQ beta polymorphism associated with DR2 in Whites was associated with DRw6 in North Indians. DQB1*0201 (linked to DR3) was markedly increased among the diabetics (RR = 12.1, $p < 10^{-6}$). DQB1*0302 was also increased, though to a lesser degree. DQB1*0603 was markedly reduced in diabetics when compared to controls (2.3% vs. 24.7%, $P < 0.02$, RR = 0.1). The same group has also found a positive association with DQA1*0301 with diabetes in this population.[16]

It thus appears that as in Whites, the DQ subregion of the HLA gene complex is primarily responsible for determining susceptibility/resistance to Type I diabetes. The lower incidence of Type I diabetes in North India may be partly due to the reduced frequency of DR4-DQB1*0302 in this population.

HLA-DQ BETA NON-ASP 57 ALLELES

There are no studies among North Indians on the frequency of DQB1* non-Asp-57 alleles among the diabetic and control populations. However, in a recent report from South India, Sanjeevi et al. (1992) studied 10 Type I diabetics and 45 control subjects of South Indian origin with sequence specific oligonucleotide probes for the DQB1 alleles.[17] Homozygosity of non-Asp-57 was present in 7 of 10 diabetics compared with 3 of 45 control subjects ($p < 0.005$). Thus, an amino acid other than aspartic acid at position 57 of the DQ beta chain may play a role in the susceptibility to Type I diabetes in South Indians.

FAMILY STUDIES

To determine the HLA haplotype segregation in relation to Type I diabetes, we studied 7 multiplex diabetic families.[18] Of 17 diabetic sibs, 7 shared both haplotypes, and 3 shared one haplotype with the proband. No HLA non-identical sib was present. This distribution of haplotypes was non-random ($p = 0.005$). In contrast, the inheritance of HLA haplotypes among the non-diabetic sibs was random. This study again highlights the importance of HLA-linked genes in the predisposition to Type I diabetes in North Indians.

CONCLUSIONS

The genetic background of North Indians is similar, though not identical, to that in Western Caucasians. Report of a low incidence of Type I diabetes in North Indians is intriguing, since many environmental or genetic factors may be responsible. It is important to differentiate Type I diabetes from insulin-dependent forms of malnutrition-related diabetes since their genetic backgrounds may differ. Unique HLA associations among North Indian Type I diabetics include the positive association with Bw21 and BfS1. DR3 is markedly increased in patients with Type I diabetes. In contrast, the association with DR4 is absent in one study or weaker than DR3 in another study. As in the West, HLA-DQ subregion genes are probably the primary site for determining susceptibility/resistance to IDDM in this population. HLA-DQ associations in North Indians have not been studied in detail, but appear to be similar to that in the West. Studies of DQB1* non-Asp-57 and DQA1* Arg-52 are not available thus far from North India. Such studies will provide valuable information on whether the lower incidence of Type I diabetes is related to a lower frequency of these genes among this population. The low incidence of Type I diabetes in India still needs a satisfactory explanation, but it is likely that both environmental and genetic factors are important.

REFERENCES

1. R.E. LaPorte, N. Tajima, H.K. Akerblom, et al., Geographic differences in the risk of insulin-dependent diabetes mellitus: The importance of registries, *Diabetes Care* 8 (suppl. 1):101 (1985).
2. Diabetes Epidemiology Research International Group, Geographic patterns of childhood insulin-dependent diabetes mellitus, *Diabetes* 37:1113 (1988).
3. O. Odugbesan, A.H. Barnett, Asian patients attending a diabetic clinic, *Br Med J.* 290:1051 (1985).
4. M.M.S. Ahuja, Epidemiological studies on diabetes mellitus in India, *in*: "Epidemiology of Diabetes in Developing Countries," M.M.S. Ahuja, ed., Interprint, New Delhi (1979).
5. WHO Study Group on Diabetes Mellitus, WHO Technical Representative Series 727:20 (1985).
6. P.K. Kambo, G.A. Hitman, V. Mohan, et al., The genetic predisposition to fibrocalculous pancreatic diabetes, *Diabetologia* 32:45 (1989).
7. J. Abdulkadir, Y. Worku, G.M.T. Schreuder, et al., HLA-DR and DQ antigens in malnutrition-related diabetes in Ethiopians: A clue to its etiology?, *Tissue Antigens* 34:284 (1989).
8. K.K. Mittal, S. Naik, N. Sansonetti, et al., The HLA antigens in Indian Hindus, *Tissue* Antigens 20:223 (1982).
9. N.K. Mehra, A.N. Malaviya, HLA system: Application in clinical medicine, *J Assoc Physicians India* 311:167 (1983).
10. S. Srikanta, M.M.S. Ahuja, A.N. Malaviya, et al., Type I (insulin-requiring) diabetes mellitus in North India: HLA and autoimmunity, *N Eng J Med.* 304:1175 (1981).
11. S. Srikanta, N.K. Mehra, M.C. Vaidya, et al., HLA antigens in Type I (insulin-dependent) diabetes mellitus in North India, *Metabolism* 30:992 (1981).
12. R.L. Kirk, P.R. Ranford, J. Theophilus, et al., The rare factor BfS1 of the properdin system strongly associated with insulin-dependent diabetes in North India, *Tissue Antigens* 20:303 (1982).
13. E. Bhatia, N.K. Mehra, V. Taneja, et al., HLA DR antigen frequencies in a North Indian Type I diabetic population, *Diabetes* 34:565 (1985).
14. O. Odugbesan, J. Fletcher, C. Mijovic, et al., The HLA-D associations of Type I (insulin-dependent) diabetes in Punjabi Asians in the United Kingdom, *Diabetologia* 30:618 (1987).

15. J. Fletcher, O. Odugbesan, C. Mijovic, et al., Class II HLA DNA polymorphisms in Type I (insulin-dependent) diabetic patients of North Indian origin, *Diabetologia* 31:343 (1988).
16. D. Jenkins, C. Myovic, K.H. Jacobs, Allele specific gene probing supports the DQ molecule as a determinant of inherited susceptibility to Type I (insulin-dependent) diabetes, *Diabetologia* 34:109 (1991).
17. C.B. Sanjeevi, V. Seshiah, E. Moller, et al., Different genetic backgrounds for malnutrition-related diabetes and Type I (insulin-dependent) diabetes mellitus in South Indians, *Diabetologia* 35:283 (1992).
18. E. Bhatia, N.K. Mehra, A.N. Malaviya, et al., HLA and autoimmunity in North Indian Type I (insulin-dependent) diabetic multiplex families, *Horm Metab Res.* 18:331 (1986).

ADVANCEMENTS IN DNA TECHNOLOGY AS APPLIED TO THE STUDY OF THE HUMAN HISTOCOMPATIBILITY COMPLEX

Massimo Trucco

Department of Pediatrics
Division of Immunogenetics
University of Pittsburgh
Pittsburgh, PA 15213

INTRODUCTION

The human major histocompatibility complex (MHC), located on the short arm of chromosome 6, encodes the various genes of the human leukocyte antigen (HLA) complex (Figure 1). The HLA region contains three classes of molecules (reviewed in 1). The class I molecules are found on the surface of all nucleated cells associated with peptide fragments derived from proteolytic processing of cellular proteins. The class I molecules A, B and C are associated at the cell surface with β2-microglobulin, a short (12 kD) protein encoded by a gene of chromosome 15. The α and β chains that constitute the class II heterodimeric transmembrane glycoproteins are encoded by a family of genes clustered in the HLA-D region. Class II molecules which present to the helper T cell peptides derived from endocytosis are expressed on B lymphocytes, activated T cells, macrophages, and other antigen-presenting cells. The class III genes encode elements of the complement cascade that are not directly involved in immune recognition.

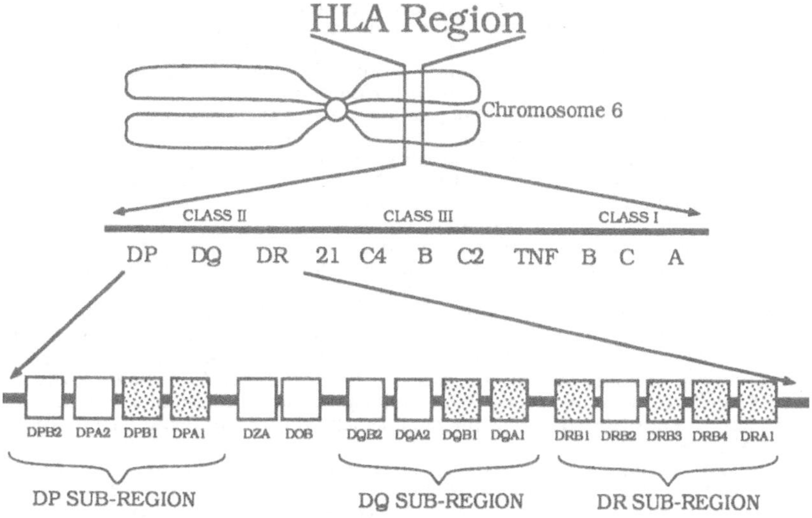

Figure 1. The Major Histocompatibility Complex in humans.

Standardization of Epidemiologic Studies of Host Susceptibility
Edited by J. Dorman, Plenum Press, New York, 1994

The structure of the class II molecules DP, DQ, and DR is similar to the membrane-bound immunoglobulins. Each chain of the dimeric molecule has an extracellular domain, a transmembrane portion, and a short cytoplasmic tail. On the basis of the results obtained from crystallized forms of the class I molecules[2] and the available sequences of class II genes, Brown and collaborators[3] proposed a model for class II molecules in which the two outermost domains of each chain (α_1 and β_1) fold together to form the binding site for the antigenic peptide. A β-pleated sheet, comprised of eight antiparallel strands, forms the floor of the antigen binding site and the sides are formed by an α helical segment from each of the two chains. The amino acid polymorphisms within the class II molecules are clustered into three or four discrete regions of variability[1]. Some of these polymorphic residues are predicted to be located within the cleft, i.e. in an optimal position for interacting with the processed antigen[4]. Other residues are located on the external part of the molecule and constitute the determinants that are recognized by the antibodies generated against foreign HLA molecules by, for example, a *polygravida*, i.e. a woman who has been exposed more than once to paternal HLA molecules expressed on the cells of different fetuses. These determinants are potentially involved in the interaction with the T-cell receptor as well[4].

HLA CLASS II SEROLOGICAL TYPING

HLA class II genes have been typed serologically to acquire the HLA matching necessary for tissue and organ transplantation[1]. Donor/recipient HLA matching is particularly important in bone marrow transplantation because of the inability to use immuno-suppressive measures to decrease the risk of tissue rejection. In addition, HLA typing has been used to assess the risk of developing certain auto-immune diseases such as insulin-dependent diabetes mellitus[5-7]. The development of these diseases appears to be more closely associated with specific HLA class II alleles than class I[8]. For all these purposes, however, the characterization of the alleles of HLA class II molecules by serologic typing has significant limitations.

The most commonly used alloantisera, obtained from *polygravidae*, contain antibodies which recognize both class I and class II molecules. To acquire a class II specific antiserum, blood platelets, which express only class I molecules, are used to adsorb the anti-class I specific antibodies. The adsorption process further reduces the anti-class II titer and thus limits the availability of these antisera. The consistency of the antisera is also a problem since sera from two donors will never have exactly the same specificity, and even the serum obtained from the same donor at different times can frequently change titer and sporadically even its range of reactivity. An additional difficulty is that many allelic variations are not easily recognized by the alloantisera. This is particularly important for the DP locus, since differences in these alleles can only be detected in a mixed lymphocyte culture assay[9, 10]. These and other allelic forms important as autoimmune disease markers, e.g., those involving amino acids present within the antigen binding site[3], are in general not immunogenic [8]. Antibodies specific for these determinants are normally not available either in the human sera or as mouse monoclonal antibodies. These limitations enable only a partial class II typing.

Furthermore, for a successful class II HLA typing, it is very important to have a well purified population of B lymphocytes because class II antigens are preferentially expressed at the surface of these lymphocytes[1]. Class II molecules are also expressed on activated T cells and macrophages, but these cells comprise only a small percentage of the cells in the peripheral blood[1]. In the blood, only 15 - 25% of the cells are B lymphocytes, so a relatively large quantity of blood from each donor must be processed in order to obtain enough B cells to allow typing of all class II allele determinants. The process through which the various lymphocytic subpopulations are normally separated [e.g., centrifugation through a Ficoll-Hypaque gradient, and passage through nylon columns[11]], may damage the cells, generating a background of dead cells that may interfere with the interpretation of the cytotoxic reaction[12]. This is especially true if the assay is read using the conventional eosin stain for the determination of specific killing mediated by the antibodies present in the various sera. In addition, the complement (generally rabbit complement), normally added to the reaction for mediating the specific killing of the target cells by the anti-HLA antibodies, may contain an inherent cytotoxic activity. Thus, even the source of the complement can create some variability in the serological tests.

Serological typing requires then a good preparation of B lymphocytes. In the case of those patients on the list for a bone marrow transplant because of a T cell leukemia, the available quantity of B lymphocytes is extremely low. When B lymphoblastoid cell lines are used as targets instead, the rabbit complement does not work as well. In fact, Epstein-Barr (EB) virus transformation of B lymphocytes has been used frequently to bypass the limiting factor of the paucity of available B cells from peripheral blood.

Although the process of transformation requires much time, expertise, and equipment, and is in general very expensive, it would be more commonly used for the HLA typing if the complement-mediated cytolytic assay was more specific against these types of target cells. Unfortunately, it is the blastoid nature of these cells that makes them much more sensitive to the complement activity. To avoid any non-specific killing of the transformed cell lines, the rabbit serum must be adsorbed prior to use with extremely large quantities of cells from the same line which is to be HLA typed. Thus, in practical terms, each cell line must be tested with its own specifically prepared complement, making this approach difficult for routine use[13].

The advantage of the serological typing, however, is the possibility of obtaining relatively good, though incomplete, results in less than 24 hours. No other conventional procedure is able to match this time schedule.

HLA CLASS II MOLECULAR TYPING

HLA class II molecular typing, although more precise and more complete than the serological typing, is limited in its applications by the complexity of the procedures involved. Restriction fragment length polymorphism (RFLP) analysis, for example, uses specific restriction enzymes to visualize polymorphic DNA sequence elements in the various alleles[14-19]. This procedure requires the isolation of high molecular weight genomic DNA, restriction endonuclease digestion, electrophoretic fractionation of the DNA, and molecular hybridization to visualize the unique digestion patterns. This procedure becomes very laborious, especially if a large number of samples must be processed in a limited amount of time. The conventional typing laboratory does not usually have the necessary equipment and trained personnel to perform these types of molecular analyses.

In consideration of this problem, the X International Histocompatibility Workshop was organized with the specific aim of transferring this technology to the laboratories specialized in HLA serology[20]. The lack of quality controls for the DNA probes, the possible failures in labeling the probes and correctly applying hybridization conditions, together with the difficulties in transferring of the DNA segments from the agarose gels to nitrocellulose filters and reading of autoradiograms, can sometimes make the interpretation of the different band patterns, and the consequent allelic typing, very subjective. For this reason, a large part of the Histocompatibility Testing 1987 was dedicated to the establishment of standards for the RFLP analysis of the class II alleles [20-22].

The introduction of the polymerase chain reaction (PCR)[23] has eliminated many of the problems of molecular typing. Our laboratory was the first to introduce the use of the RFLP analysis on specifically PCR amplified DNA for HLA typing[24, 25]. This method is much more rapid than the RFLP analysis of genomic DNA, and did not require the use of radioactive probes or dot blot hybridizations. Although many laboratories have since adopted and expanded this method to study other gene loci[26], this method still appears too complex for large scale class II typing.

Using PCR amplification of DNA extracted from blood cells and allele specific oligonucleotide probes, we were able to molecularly type hundreds of individuals to demonstrate that variations in the region of codon 57 of the second exon of the HLA-DQB1 gene were reliable markers for susceptibility to insulin-dependent diabetes mellitus[7, 27-29]. Although we were able to generate high quality results using this approach, it became clear that a complete class II molecular typing would be much more demanding. The XI International Histocompatibility Workshop, intended to promote the feasibility of a complete molecular class II typing using PCR, listed 140 unique oligonucleotide probes that would be necessary to achieve a complete class II typing[30]. These probes would be used in separate hybridization reactions and the filters subsequently washed at specific temperatures. The labeling and manipulation of 140 probes and filters becomes a serious limitation when only a few individuals need to be typed. Other limiting factors include the high cost of the necessary reagents (i.e. filter membranes, oligonucleotides, and Taq DNA polymerase), and the need for trained personnel. The time necessary to prepare the samples and the oligonucleotide probes, as well as to interpret the hybridization patterns, makes this approach a very laborious substitute for serological typing. The desired method for molecular typing would incorporate the precision of the molecular approach, and the speed of the serological approach. In addition, there should be a simple procedure to evaluate the reproducibility of the analytical technique as it is performed in a number of laboratories across the country.

The method developed in our laboratory adapts the positive features of these methods into a protocol that allows the rapid and reproducible class II typing of a large number of potential donors. This new procedure involves some automation of the protocols, increasing the productivity and reproducibility of the typing procedure. We had, as an aim, the production of a simple kit that will increase the uniformity of the typing results. This facilitates the subdivision of the extremely large amount of samples to be tested among more typing laboratories, all able to generate comparable results.

Conventional and Reverse Dot Blot Analysis

The most common method for detecting a specific nucleotide sequence within a larger population of DNA molecules is the Southern blot analysis[31]. The target DNA (e.g. genomic DNA) containing the sequence of interest, is immobilized onto a solid support such as a nitrocellulose filter membrane. The sequence of interest is labelled, usually by the incorporation of ^{32}P, and this labelled "probe" is allowed to hybridize to its complementary sequence on the filter. A stringent wash, a combination of high temperature and a low salt buffer, is used to remove probe molecules from partially homologous target sequences. Autoradiography of the filter membrane is used to visualize the presence of the probe hybridized to specific target molecules.

In a conventional dot blot, the target DNA is denatured and spotted directly onto the filter membrane (Figure 2). Thus, numerous target sequences can be assayed in a single blot. A single nucleotide variation from the probe sequence can be resolved by using highly stringent washing conditions. In the case of HLA class II molecular typing, PCR amplified material from a number of individuals can be screened for the presence of a specific allele using synthetic single-stranded oligonucleotides as allele-specific probes. A battery of sequence-specific oligonucleotide probes (SSOPs) are hybridized to replicate filters, and the HLA genotype is determined by the comparison of the hybridization of these SSOPs to the individual filters.

Several recent modifications to this procedure include the use of non-radioactive reporter groups on the probe molecule. The incorporation of stable organic molecules such as biotin and digoxigenin eliminate the problems associated with the handling and disposal of radio-active material, and allow the synthesis of probe molecules with a relatively long shelf life. The detection systems for these molecules depend on the production of a fluorescent compound, working in a fashion similar to that of the radioactive probe, or of a colored precipitate which can be seen directly. Frequently, however, the signal generated by non-radioactive probes is not strong enough to generate films (or filters) that are as easily read as those obtained through the use of radiolabeled probes.

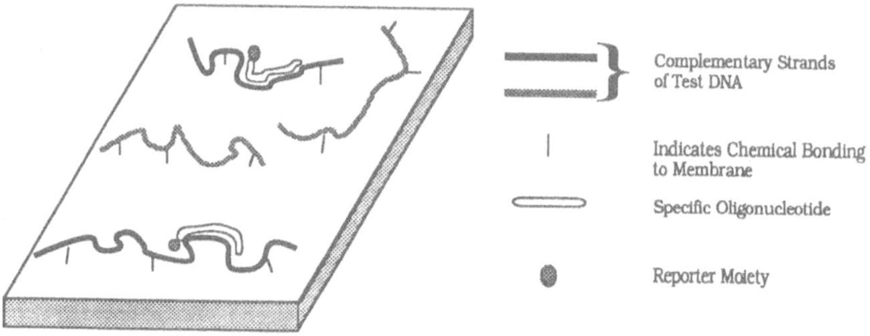

Figure 2. Conventional Dot Blot method.

Reverse dot blots differ from conventional dot blots in that the oligonucleotide probe is anchored to the membrane instead of the target DNA[33]. In addition, the reporter moiety is attached to the target DNA instead of to the probe. For the HLA class II typing, there are several advantages to this approach. The numerous probes used to identify the specific class II alleles can all be spotted onto a single filter. A portion of the labelled PCR product from each amplification is then hybridized simultaneously to the battery of SSOPs. Thus, hybridization is done in a single container, instead of a separate container for each specific probe, and only one labeling per amplification is required.

The sequence-specific oligonucleotide probes are approximately 20 nucleotides in length. This length is sufficient to visualize the specific hybridization, while at the same time discriminating even a single base difference between the probe and target sequences. Oligonucleotides of this length are readily synthesized and purified. For the reverse dot-blot, however, the direct attachment of probes of this size to membranes is not very efficient. In addition, the formation of chemical bonds between the filter and the SSOP will reduce the efficiency of the probe. This difficulty has previously been solved by enzymatically adding a large number of thymidine nucleotides at the end of the probe (poly dT tailing)[33]. The larger size greatly increases the binding efficiency, and, in most instances, the chemical bonds between the filter and the poly dT tail will not limit the availability of the SSOP (Figure 3).

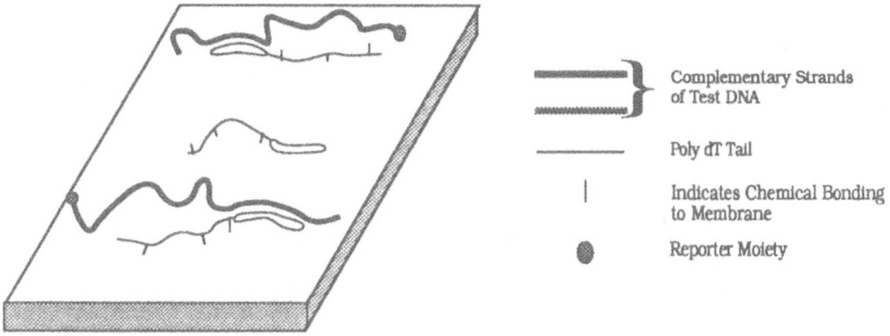

Figure 3. Reverse Dot Blot using Poly dT Tailing.

An alternative method to bind the small probe sequences to the filter membrane was developed in our laboratory[34]. In this procedure, numerous tandem repeats of the probe sequence, from 25-70, are constructed using PCR technology[35]. The 500-1000 base pair molecules bind very efficiently to the filter membrane and the molecules bound do not contain a large amount of non-probe sequence (e.g. the poly dT tail). Thus, a very high concentration of probe sequence can be immobilized onto the filter, increasing the hybridization signal obtained and offering the potential for using shorter hybridization times and/or less sensitive detection methods.This new method, Molecular Typing by Reverse Blotting (MTRB), also differs from the poly dT tailing method in that both complementary strands of the probe sequence are bound to the membrane. Their presence makes it possible to hybridize both labeled strands of the test DNA, further increasing the sensitivity of this assay (Figure 4). Although the complementary strands can anneal to each other, this can be minimized by anchoring the probe to the filter membrane in denaturing conditions.

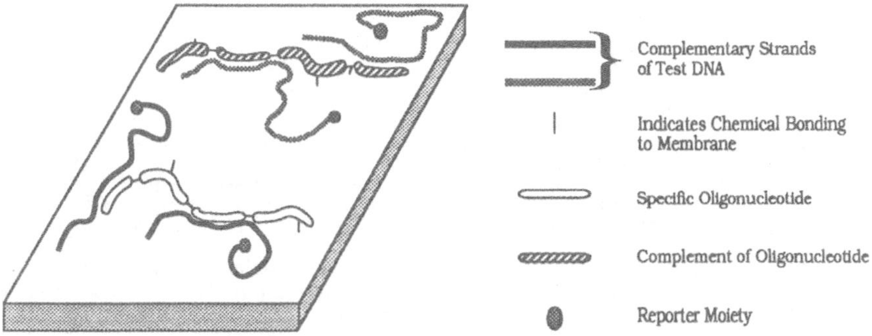

Figure 4. Reverse Dot Blot using probe polymers

Construction of Polymeric Oligonucleotide Probes by PCR

Our method of synthesizing oligonucleotide polymers stems from our studies of the regulation of the HLA class II genes[36]. Gene expression is regulated by trans-acting factors that bind to highly conserved nucleotide sequences present in the promoter region. For studies of DNA-protein interaction, oligomers of the recognition sites of these DNA binding nuclear proteins have been constructed by ligation of annealed complementary synthetic oligonucleotides[37]. This approach involves the production of large quantities of oligonucleotides and generates possibly unstable polymers. In another approach, the stability of the construct in a host plasmid was increased by repetitive head to tail directed cloning of the monomeric recognition sequence[38]. This method, however, is quite time consuming. We have developed a system based on the polymerase chain reaction (PCR) amplification of staggered, complementary oligonucleotides which function as both primer and template[35]. Our approach is rapid, reliable, and relatively inexpensive, and the entire process can be performed using standard PCR equipment and supplies. The procedure is based on a staggered re-annealing of a molecule containing two copies of the probe sequence in tandem (Figure 5).

The dimeric molecule AA can anneal in a staggered fashion with its complement A'A' such that the 3' ends of the duplex are annealed and the 5' ends remain single-stranded. In this configuration, DNA polymerase can extend the molecule by synthesizing the complementary strand, yielding a double-stranded trimer. The pair of dimers can also anneal in two other configurations, one with flush ends, and the other with 5' monomeric sequences annealed. DNA polymerases do not act on these configurations, and so no alternative products of the reaction result. In the next cycle, a trimeric molecule can anneal with the complementary dimeric or trimeric molecules to synthesize tetrameric or pentameric molecules. Successive cycles of denaturation, re-annealing and elongation further increase the size of the multimeric molecules. For a multimeric molecule made up of "n" monomeric subunits, the length of the amplification product can increase by as much as n-1 subunits. Using this approach, we have succeeded in synthesizing polymeric molecules containing 25-70 repeats of the monomeric sequence.

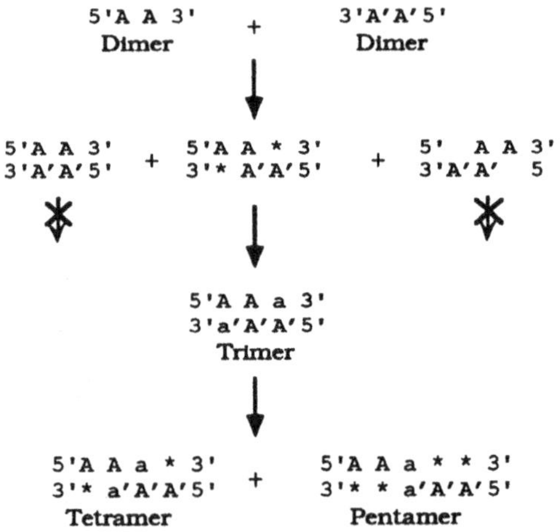

Figure 5. Upper case letters denote any nucleotide sequence which is usually more than 16 and less than 50 nucleotides in length. Letters followed by an apostrophe (') designate the sequence which is complementary to the unmarked letter and also indicates the proper 5'-3' orientation for annealing of the two complementary sequences. Sites available for extension by DNA polymerase are indicated by an asterisk. Lower case letters represent the same nucleotide sequence as the upper case letters, except that these nucleotides have been previously added by DNA polymerase.

Several variations in this basic scheme are also possible. One such variation can be used to facilitate the synthesis of a polymer probe from single-stranded oligonucleotides. Instead of beginning with a dimeric molecule as above, the initial reactant is a single-stranded monomer (ABC). In order to generate the staggered-ended molecules that allow the amplification of the repeat units, a small terminal repeat is contained within the A and C regions. Annealing of these repeats and a single round of primer extension generates the double-stranded dimer as in Figure 5. A second variation can be used to generate a group of polymer probes which differ only in the central portion (Figure 6). As above, the initial reactant is a single-stranded molecule ABC. In order to generate the 5' overhang for the extension reaction, a second single-stranded molecule, complementary to the 5' and 3' ends (A and C) of the monomer sequence, is necessary. Annealing of the 3' ends of the molecules (C and C') forms a duplex molecule with 5' overhanging ends.The first round of amplification adds the 5' portion (A) onto the 3' end of the molecule. In the second round of amplification, annealing by the A and A' regions forms a duplex molecule that can synthesize the additional BCA sequence. As above, additional rounds of amplification further elongate the polymeric molecule.

In the case of the SSOPs used for HLA class II typing, polymers of probes which differ in sequence only in the central "B" portion can be generated using the same A'C' molecule and a unique monomer of the repeat. The most important requirement for this technique is that the complementary regions, A and C, must be sufficiently long so that proper annealing during the early PCR cycles occurs. The position of the monomer A and C regions is important so that undesired duplexes are not formed, resulting in the synthesis of alternate polymer sequences.

The MTRB has the advantage of a consistent source of polymeric probes. This level of reproducibility has not been obtained with the poly dT tailing method and will be an advantage for standardizing the procedure for use by many laboratories. After the long polymers have been synthesized by PCR, the reaction products are ligated into a plasmid vector and used to transform Escherichia coli. Because the products of the synthesis can vary over a considerable size range, it is desirable to select bacterial clones with polymers containing a large number of monomeric repeats. After sequencing the polymers to ascertain that the PCR procedure did not introduce alterations in the desired sequence, bacterial stocks are then prepared and become a permanent source of polymers with uniform size and membrane binding properties. The polymers are then prepared by restriction nuclease digestion of large plasmid preparations from the bacteria. For smaller quantities, the polymers can be amplified from the plasmid by PCR using primers which anneal to the plasmid vector adjacent to the cloning site.

```
5'A B C 3'    +    3'C'A'5'
          ↓
      5'A B C * 3'
      3'* * C'A'5'
          ↓
   5'A B C a * * * 3'
   3'* * * a'b'C'A'5'
```

Figure 6. In this variation, the monomer is 5' A B C 3'. Once the shorter sequences are extended to or beyond the dimer size, the synthesis proceeds as in Figure 5. The B portion can vary considerably in size. When very large, 5' A B C 3' could be derived from a cloned fragment. When B is only a few nucleotides, several slightly different specific sequences could be synthesized using the same common 3' C A' 5' primer-template.

MTRB Method for Typing HLA Class II Alleles

In the case of HLA-DQB1 gene alleles, oligonucleotides of the type described in Figure 6 have been synthesized with our Millipore Cyclone Plus DNA Synthesizer around codon 57[7, 34]. To distinguish between the pairs of alleles characterized by the same oligonucleotide centered on codon 57, another 5 polymers must be added to our filter. These additional oligonucleotides can be chosen to represent another variable segment. For this purpose we used the segment centered on codon 27[7, 34]. Thirteen different polymers are sufficient to properly type all the recognized DQB1 allelic forms. However, certain heterozygous combinations lead to ambiguities which can be resolved with five additional polymer probes. Although in most cases these probes will provide redundant information, their use will verify the accuracy of the typing protocol.

DQA1, DPB1, and DPA1 gene allele specific oligonucleotides can be generated using the same strategy. DRA1 does not constitute a problem since it is not polymorphic. It can be added to our filters as an internal positive control. In the case of DRB allelic variants, the system becomes far too complex to be easily solved as for the alleles at the other loci. The DRB locus may encode one or two expressed genes, e.g., DRB1 and DRB3, or DRB1 and DRB4. Thus, a single individual may express up to 4 different DRB alleles.

Second, different alleles frequently contain the same variable segments but in different combinations. Because of this patchwork of identical, short DNA stretches present in different allelic sequences, one amplification and a small number of oligonucleotide probes are insufficient to type this locus. This problem can be solved, however, by using allele specific primers for the initial PCR amplification of the DNA. Three allele specific amplifications are necessary to recognize all the molecular allelic forms present in the DR1, DR2, and DR4 allelic "families". The DRB1 alleles associated with the DRw52 locus form the fourth group to be amplified with a group specific primer. A fifth primer pair must also be included to distinguish the four DRw52 alleles. The amplified material must still be probed with a large number of SSOPs. Based on the XI International Histocompatibility Workshop Protocols[30], there are at least 60 specific oligonucleotides that must be used to type all of the DRB alleles. However, a partial DRB typing is obtained by the so called "generic" amplification. With this amplification 24 probes are used to assign the sample to one of the five DR groups listed above. If desired, a more specific typing is obtained by a second "group-specific" amplification and hybridization against 7-21 additional SSOPs. The group specific primer pairs are the following:

DR 1 =	A:TTCTTGTGGCAGCTTAAGTT
	B: CCGCTGCACTGTGAAGCTCT
DR 2 =	A: TTCCTGTGGCAGCCTAAGAGG
	B: CCGCTGCACTGTGAAGCTCT
DR 4 =	A: GTTTCTTGGAGCAGGTTAAAC
	B: CCGCTGCACTGTGAAGCTCT
DRw52 associated =	A: CACGTTTCTTGGAGTACTCTAC
	B: CCGCTGCACTGTGAAGCTCT
DRw52 - DRB3 =	A: CCCAGCACGTTTCTTGGAGCT
	B: CCGCTGCACTGTGAAGCTCT

Polymeric oligonucleotide probes for DRB "generic" typing have been generated and successfully used for MTRB typing of 223 Pima Indians[39].

Since, to completely type for DRB, five amplifications are necessary, the number of independent amplifications needed to generate a complete HLA class II typing, becomes 11: 2 for DQA1 and DQB1, 2 for DPA1 and DPB1, 1 for generic DRB, 1 for DRA1, and 5 for DRB allele specific segments. To reduce the number of independent PCR reactions needed for DRB typing, co-amplification conditions were tested. This technique consists of combining primer sets such that two or more amplifications are carried out in the same reaction tube. Pairing sets of primers must be performed in a way that does not compromise the quality and specificity of the reaction. If overlapping segments are generated in the co-amplification, incomplete products can be obtained. We have determined that six separate reactions are sufficient to properly co-amplify all the necessary segments. The permissive pairing of the 11 sets of primers was: tube 1: DPA1 + DRB1 DR1; tube 2: DPB1 + DRB1 DR2; tube 3: DQA1 + DRB1 DR4; tube 4: DQB1 + DRw52 associated DRB1; tube 5: DRA1 + DRB3 DRw52; and tube 6: DRB primers common to all the alleles.

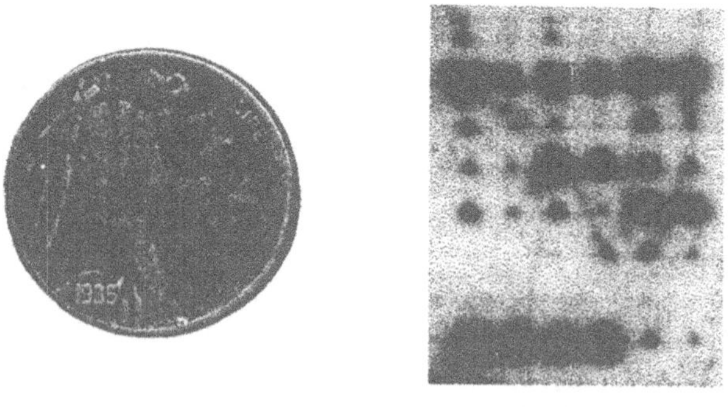

Figure 7. Generic DRB typing of Pima Indian samples by the miniaturized MTRB method.

Efforts to further reduce the number of amplifications necessary to perform the test are in progress. The need to keep separate the DRB1 allele specific amplifications will be, however, a limiting factor. We are also working to determine which primers give the best resolution of the various amplified segments.

In order to prepare a large number of MTRB hybridization filters for DR typing, all of the SSOPs, with their respective controls, are spotted onto a small filter by a robotic arm. The robotic arm can be precise and highly reproducible, allowing miniaturization of the filters and thus utilizing lower amounts of the hybridization reagents.

REFERENCES

1. J.R. Duquesnoy and M. Trucco, Genetic basis of cell surface polymorphisms encoded by the major histocompatibility complex in humans. In: *Critical Reviews in Immunology* 8:103 (1988).
2. P. Bjorkman, M. Saper, W. Samraoui, W. Bennett, J. Strominger, D. Wiley D, Structure of the human HLA class I hisocompatibility antigen. *Nature.* 329: 506 (1987).
3. J. Brown, T. Jardetzky, M. Saper, B. Smaraoui, P. Bjorkman, and D. Wiley, An hypothetical model of the foriegn antigen binding site of class II histocompatibility molecules. *Nature.* 332:845 (1988).
4. P. Bjorkman, M. Saper, B. Samraoui, W. Bennett, J. Strominger, D. Wiley, The foriegn antigen binding site and T-cell recognition regions of class I histocompatyibility antigens. *Nature.* 329: 512 (1987).
5. J. Bertrams, and M. Baur, Insulin dependent diabetes mellitus In: "Histocompatibility Testing 1984." Heidelberg: Springer-Verlag. p. 348 (1984).
6. J. A. Todd, J. I. Bell, H..O. McDevitt, HLA DQβ gene contributes to susceptibility and resistance to insulin dependent diabetes mellitus. *Nature.* 329:559 (1987).
7. P. Morel, J. Dorman, J. Todd, H. McDevitt, and M. Trucco, Aspartic acid at position 57 of the HLA DQ-beta chain protects against type I diabetes: a family study. *Proc. Natl. Acad. Sci. U.S.A.* 85:8111 (1988).
8. M. Trucco and J. Dorman, Immunogenetics of insulin-dependent diabetes mellitus in humans. In: *Critical Reviews in Immunology,* 9:201 (1989).
9. A. Cesbron, P. Moreau, N. Milpied, J. Y. Muller, J.L. Harousseau, J. D. Bignon, Influence of HLA-DP mismatches on primary MLR responses in unrelated HLA-A, B, DR, DQ, Dw identical pairs in allogeneic bone marrow transplantation. *Bone Marrow Transplantation* 6: 337 (1990).
10. T. H. Eiermann, J. Falker, S. F. Goldman, The incidence of DPB1 differences between serological and mixed lymphocyte culture matched unrelated individuals: implications for selection of bone marrow donor. *Bone Marrow Transplantation* 9: 157 (1991).
11. G. Garotta and T. Neri, HLA-DR typing by complement dependent B lymphocyte lysis. In: "Immunological Methods, II," eds. Lefkovits, I. and Pernis, B. p.163 (1981).
12. J. W. Stocker, G. Garotta, B. Hausmann, M. Trucco and R. Ceppellini, Separation of human cells bearing HLA-DR antigens using a monoclonal antibody rosetting method. *Tissue Antigens.* 13:212 (1979).
13. M. Trucco, J. W. Stocker and R. Ceppellini, Monoclonal antibodies against human lymphocyte antigens. *Nature.* 273:666 (1978).
14. S. Rosenshine, I. Cascino, A. Zeevi, R. J. Duquesnoy and M. Trucco, DQ-alpha and DQ-beta RFLP reveals the composition of the DQ molecule recognized by T-cell clones. *Immunogenetics.* 23:187 (1986).
15. M. Trucco, S. Rosenshine, I. Cascino and R. J. Duquesnoy, Analysis of HLA-DQ subregion polymorphisms using sequential restriction endonuclease digestion of genomic DNA. *Immunogenetics.* 24:184 (1986).
16. I. Cascino, S. Rosenshine, E. Turco, M. Marrari, R. J. Duquesnoy and M. Trucco, Relationship between DQ-alpha and DQ-beta RFLP and serologically defined class II HLA antigens. *J.Immunogenet.* 13:387 (1986).
17. A. Zeevi, I. Cascino, L. DiVecchia, M. Marrari, R. J. Duquesnoy and M. Trucco, Alloreactive T-cell recognition of two DQ-beta allelic specificities associated with DQw1. *Human Immunol.* 18:225 (1987).

18. M. Trucco and R. J. Duquesnoy, Polymorphisms of the HLA-DQ subregion. *Immunol. Today*. 7:297 (1986).
19. E. Turco, A. Caré, P. Compagnone-Post, C. Robinson, I. Cascino and M. Trucco, Allelic forms of the alpha- and beta-chain genes encoding DQw1-positive heterodimers. *Immunogenetics*. 26:282 (1987).
20. M. Trucco and E. Ball, RFLP analysis of DQ-beta chain gene: Workshop report. In: "Histocompatibility Testing 1987". Vol.1, p. 860 (1989).
21. E. Turco, R. Fritsch and M. Trucco, First domain encoding sequence mediates human class II beta chain gene cross-hybridization. *Immunogenetics* 28:193 (1988).
22. M. Trucco, I. Cascino, S. Rosenshine, E. Turco, M. Marrari, L. DiVecchia, A. Zeevi, R. J. Duquesnoy, Definition of HLA-DQ polymorphisms by RFLP and in serological and cellular assays. *Transplant. Proc.* 19:848 (1987).
23. R. K. Saiki, S. Scharf, F. Fallona, et al. , Enzymatic amplification of ß-globin geneomic sequences and restriction site analysis for diagnosis of sickle cell anemia. *Science*. 230:1350 (1985).
24. G. Trucco, R. Fritsch, R. Giorda and M. Trucco, Rapid detection of IDDM susceptibility, using amino acid 57 of the HLA-DQ beta chain as a marker. *Diabetes*. 38:1617 (1989).
25. W. A. Rudert and M. Trucco, A rapid test for IDDM susceptibility detection. *Pharmacy Times*. October 1989, p. 38 (1989)
26. J. Y. Tong, S. Hsia, G. L. Parris, D. D. Nghiem, E. M. Cottington, W. A. Rudert, M. Trucco, Molecular compatibility and renal graft survival: the HLA DQB1 genotyping. *Transplantation*, in press (1992).
27. M. Z. Bao, J. X. Wang, J. S. Dorman and M. Trucco, HLA-DQ beta non-Asp-57 allele and incidence of Diabetes in China and the U.S.A. Lancet.2:297 (1989).
28. M. Trucco, To be, or not to be Asp 57, that is the question. *Diabetes Care*. 15:705 (1992).
29. J. Dorman, R. LaPorte, R. Stone and M. Trucco, Worldwide differences in the incidence of Type I diabetes are associated with amino acid variation at position 57 of the HLA-DQ beta chain. *Proc. Natl. Acad. Sci. U.S.A.* 87: 7370 (1990).
30. 11th Intenational HLA Workshop reference protocols, 1990.
31. E. M. Southern, Detection of specific sequences among DNA fragments seperated by gel electrophoresis. *J. Mol. Bio.* 98:503 (1975).
32. E. Turco, R. Fritsch and M. Trucco, Use of immunologic techniques in gene analysis. In: "Immunodiagnosis of Cancer". Second edition. Vol. 53, 205 (1990).
33. R. K. Saiki, P. S. Walsh, C. H. Levenson, H. A. Erlich, Genetic analysis of amplified DNA with immobilized sequence-specifc oligonucleotide probes. *Proc. Natl.Acad.Sci. USA*. 86:8230 (1989).
34. W. A. Rudert and M. Trucco, Rapid detection of sequence variations using polymers of specific oligonucleotides. *Nucl. Acid. Res*. 5:1146 (1992).
35. W. A. Rudert and M. Trucco, DNA polymers of protein binding sequences generated by PCR. *Nucl. Acid. Res.* 18: 6460 (1990).
36. E. Turco, B.J. Manfras, L. Ge, W. Rudert, and M. Trucco, Promoter X boxes from HLA class II beta genes at different loci do not compete for nuclear protein specific binding. *Immunogenetics*. 32: 117 (1990).
37. J. T. Kadonaga and R. Tijian, Affinity purification of sequence-specific DNA binding proteins. *Proc. Natl. Acad. Sci. USA*. 83:5889 (1986).
38. S. Eiseberg, S. C. Francesconi, C. Civalier, S.S. Walker, Purification of DNA-binding proteins by site-specific DNA affinity chromatography. *Methods Enzymol.* 182:521 (1990).
39. M. Kocova, W. Rudert, J. Dorman, M. Phillips, W. Knowler, M. Trucco, Pima Indians' genetic admixture as determined by molecular HLA class II genotyping. Submitted (1992).

MOLECULAR EPIDEMIOLOGY AND DNA TECHNOLOGY TRANSFER: A PROGRAM FOR DEVELOPING COUNTRIES

Janice S. Dorman on behalf of the International Molecular Epidemiology Task Force*

Department of Epidemiology, Graduate School of Public Health
University of Pittsburgh
Pittsburgh, PA 15261
USA

INTRODUCTION

Molecular epidemiology has recently been defined as "a science that focuses on the contribution of potential genetic and environmental risk factors, identified at the molecular level, to the etiology, distribution and prevention of disease within families and across populations".[1] This new field has emerged from the integration of molecular biology into traditional epidemiologic research. The objectives of molecular epidemiology are quite broad and include: 1) descriptive and analytical studies to evaluate host/environmental interactions in disease, 2) the development of strategies for the control of bacterial, parasitic and viral disorders through molecular diagnosis, and 3) the prevention of non-communicable diseases and genetic disorders by assessing risk and identifying susceptible individuals through genetic screening.

As yet, few developing countries have a significant capability in molecular epidemiology due to a lack of trained molecular epidemiologists and a pervasive shortage of appropriate equipment and reagents. Unless these shortcomings are corrected, developing countries will be unable to realize the full benefits of molecular epidemiology. Thus, there is a great need for: 1) adequate training in methods for integrating molecular biology into epidemiologic research and public health, and 2) advanced biotechnology equipment, reagents and supplies for molecular analyses of potential genetic and environmental risk factors.[1] Since training opportunities are currently limited in all nations, educational materials and short courses in molecular epidemiology must be developed across the world. Adequate laboratory facilities and advanced biotechnology equipment and supplies are also essential for molecular epidemiology and are generally unavailable in developing nations. Even when advanced technology is accessible, the problems that frequently arise due to inadequate technology transfer can appear to outweigh its possible benefits.[2,3] Thus, effective plans for the transfer of advanced biotechnology and training in molecular epidemiology, as they apply to international health, are essential.

To help developing countries build up capabilities in molecular epidemiology, we propose the establishment of an International Molecular Epidemiology Task Force. The rationale for the organization of the Task Force is based upon: 1) identification of health problems in various regions of the world which may be ameliorated through molecular epidemiology, 2) establishment of the mission, membership, activities and responsibilities of the Task Force, and 3) justification for support of the Task Force from industrialized and developing countries. Each of these issues will be addressed below.

THE IMPORTANCE OF MOLECULAR EPIDEMIOLOGY IN DEVELOPING COUNTRIES

At an international conference held at Alma-Ata, in the former USSR in 1978, it was concluded that a realistic objective of the World Health Organization would be the provision of primary health care for all the world's people by the year 2000.[4] Strategies for attaining this goal have been formulated, and regional and national plans are currently being implemented across the world. In the industrialized nations, health promotion initiatives have been emphasized, whereas in developing countries, activities are targeted toward the control of communicable diseases, better nutrition and improved maternal and child health.[5] However, with better control of infectious disorders, non-communicable diseases are becoming a major health problem in the Third World.

These efforts can be greatly aided through molecular epidemiology by training in public health, the transfer of advanced biotechnology and the implementation of rapid methods for molecular diagnosis of infectious, inherited and chronic diseases.[6] We outline below specific plans for molecular epidemiology programs in select countries in Asia, the Eastern Mediterranean Region and Latin America. Approaches for effective implementation, which will have a tremendous impact on disease prevention and the future of international public health, are also discussed.

Molecular Epidemiology in Asia

Hepatitis B (HBV) infection is endemic in parts of China, with perinatal and early childhood infections accounting for the high rates of primary hepatocellular carcinoma in older age groups.[7,8] Recombinant DNA technology, such as polymerase chain reaction (PCR), has led to the development of rapid and sensitive methods for the diagnosis of viral infections, such as HBV, that are less time consuming and labor intensive than conventional approaches.[9,10] In addition, molecular methods have been utilized to develop safe, immunogenic and cost-effective recombinant HBV vaccine,[8,11] which is also required for the control of the disease. With the availability of appropriate technology and the implementation of public health approaches, such as the development of vaccination strategies to reduce HBV transmission from infected mothers to their newborns[12], and efficient methods for epidemiologic surveillance,[13] primary and secondary prevention of HBV infection and its consequences in China can be achieved.

In terms of chronic diseases, advanced biotechnology and molecular epidemiology can be used in occupational medicine, which focuses on physical, chemical and biological factors in the workplace and their effect on human health. Molecular methods are now being used to measure exposure to genotoxic and carcinogenic substances, such as DNA and protein adducts, as well as oncogenes and tumor suppressor genes.[14] Screening for these biomarkers in an occupational setting will permit early detection of substances that are potential carcinogens. This may lead to the preventive strategies, including reducing work-related exposure to carcinogenic substances, improving environmental protection measures and initiating early treatment of these disorders. Other non-communicable diseases, such as non-insulin-dependent diabetes (NIDDM), cardiovascular disease and obesity are also increasing in Asia. Recently, progress has been made in identifying potential susceptibility genes for these disorders.[15,16] A number of candidate genes that are involved in control of plasma lipid levels and thrombotic potential have been determined and mutations in some of these genes have been identified.[17] These genetic markers will be used in the future to identify at-risk individuals, ultimately contributing to the prevention of these and other chronic diseases. Thus, molecular epidemiology programs will contribute to the prevention of both infectious and chronic diseases in Asia.

Molecular Epidemiology in the Eastern Mediterranean Region

In the Eastern Mediterranean Region, many countries are undergoing the epidemiologic transition. Despite the high prevalence of communicable disorders in these areas, the burden of chronic diseases is rapidly increasing. In addition, the magnitude of the health problems posed by genetic disorders, such as sickle cell anemia[18] and beta thalassemia,[19] contribute significantly to the psychosocial burden on relevant communities, and imposes additional strain on the existing health care systems. These genetic diseases are very common in the Eastern Mediterranean Region, due to the high degree of consanguinity in this area of the world.

DNA technology is now being utilized for diagnosis of genetic disorders; and genetic screening will permit the detection of the presence of mutant alleles in carriers. Progress with the Human Genome Project will eventually lead to the availability of DNA markers for all disorders with an underlying genetic susceptibility,[20] including non-communicable diseases, which are becoming increasingly prevalent in countries undergoing transition. Thus, in the Eastern Mediterranean Region, there is a great need for the application of advanced biotechnology, with an emphasis on the development of clinical molecular laboratories to provide rapid, accurate and inexpensive methods for diagnosing disorders prenatally, conducting newborn and pre-marital screening, and identifying individuals at-risk or with early stages of disease. Countries, including Tunisia and Saudi Arabia, are in the process of initiating such activities, which will ultimately be incorporated into health care services. However, to effectively utilize relevant technology for improving health care delivery and achieving health goals, planning strategies need to be formulated. The development of national programs using molecular epidemiology will provide models for such advancements in the Eastern Mediterranean Region.

Molecular Epidemiology in Latin America

In Latin America, bacterial, viral, as well as parasitic diseases, such as malaria and leishmaniasis, remain important health problems. Strategies for controlling these disorders include improvements in laboratory facilities for rapid diagnosis and treatment.[9,21] For parasitic disorders, for example, nucleic acid probes are now available for the specific species or sub-species that cause these diseases; they can be used to diagnose disorders, such as malaria, directly from finger-stick blood samples. This permits rapid detection of the organisms at a much more sensitive and specific level than existing clinical methods. Moreover, this technology has been simplified so that large numbers of specimens can be processed quickly, allowing appropriate treatment to be promptly initiated. Continued research into methods for vector control, chemotherapy, and the development of recombinant vaccines are required for lowering the prevalence of parasitic diseases.[22] Molecular epidemiology provides the basis for these efforts in Latin America.

Recombinant DNA technology has also been instrumental in the development of very rapid and specific methods for diagnosing bacterial diseases. With PCR, the detection of Shigella,[23] E. coli and Vibrio cholerae,[24] for example, is greatly simplified and far more accurate than conventional methods. In January, 1991, an outbreak of cholera began in Peru and spread throughout many Latin American countries. By January 1, 1993, over

730,000 cases and 6,300 deaths were reported.[25] PCR has been, and remains an enormously helpful diagnostic tool for the detection of the specific toxin genes associated with this disease.

Acquired immune deficiency syndrome (AIDS) has become a leading international health problem in many regions of the world.[26] Once it was determined that HIV was the cause of AIDS, serological tests to detect HIV antibody were rapidly developed, and HIV can now be detected at the molecular level. In México, for example, HIV studies are being undertaken at the Instituto Nacional de Diagnóstico y Referencia Epidemiológicos (INDRE). The amplification of target DNA sequences permits screening of donated blood, as well as early identification, treatment and possible prevention of these disorders. The control of AIDS and other viral diseases will require active research, the transfer of DNA technology, and the implementation of molecular epidemiology for primary and secondary prevention.

Although communicable diseases are still highly prevalent in Latin America, life expectancy is also increasing, resulting in an aging population and a rise in the occurrence of non-communicable diseases.[27] Moreover, new environmental exposures and changes in lifestyle habits, such as smoking, obesity and hypertension, are contributing to a higher prevalence of potential risk factors. As mentioned previously, molecular methods will soon be available for genetic screening of high-risk individuals. With appropriate technology transfer to Latin America in the future, prevention programs can be approached from a molecular epidemiologic perspective.

Argentina, Brazil and México represent ideal countries for national programs for molecular epidemiology and advanced biotechnology transfer. The basic scientific infrastructure in these countries is sufficiently strong to support the implementation of such programs for national public health. Moreover, Buenos Aires, Argentina, Sao Paulo, Brazil and México City, México were among the initial participating centers in the WHO Multinational Project for Childhood Diabetes, known as the DIAMOND Project.[28] This long-standing collaboration has provided a foundation for the development of studies of the molecular epidemiology of insulin-dependent diabetes mellitus (IDDM), which will serve as the basis for similar efforts for other chronic and infectious diseases.

THE INTERNATIONAL MOLECULAR EPIDEMIOLOGY TASK FORCE

To implement the programs described above, an International Molecular Epidemiology Task Force* will be established in 1993. The mission of the Task Force is to facilitate the development and implementation of national programs in molecular epidemiology and to promote advanced biotechnology transfer for disease prevention in various regions of the world. Members of the Task Force will include internationally recognized scientists and representatives of:

* Ministries of Health
* World Health Organization
* WHO Collaborating Centers
* World Bank
* Pan American Health Organization
* International Centre for Genetic Engineering and Biotechnology
* Fogarty International Center
* National and International Experts on Molecular Biology, Genetics and Epidemiology
* Biotechnology Industry

Within each country, a National Scientific Committee with Government Advisors will also be established to interact with the international representatives and to direct the activities of that nation. The first meeting of members of the Task Force will be in 1993.

We believe that effective advanced biotechnology transfer to developing countries will be based upon: 1) the provision of relevant scientific information to key officials, 2) training in molecular epidemiology and approaches for disease prevention, 3) obtaining financial support for the operation and maintenance of laboratory equipment, 4) the evaluation of advanced biotechnology transfer, and 5) the development of national and international scientific networks. These will represent the primary activities and responsibilities of the members of the Task Force.

Provision of Relevant Scientific Information to Key Officials

The Task Force will provide relevant scientific information to key officials in the national government, academia, industry and the health care profession. Presentations will be made at individual and group meetings, as well as at national and international scientific conferences, such as this NATO Advanced Research Workshop, and an upcoming meeting in Saudi Arabia on the epidemiologic transition and health in developing countries, to be held in 1994. Brochures, journal articles, and technical information concerning the potential contribution of advanced biotechnology transfer to the health, research and economic goals of developing countries will be prepared and distributed.

Initial efforts will also be directed towards conducting a survey of each region to determine the availability, utilization and need for recombinant DNA technology in various areas of the country. This will be based upon the recent international questionnaire on DNA technology and molecular epidemiology that was distributed to scientists from across the world (Annex A). These efforts will contribute to the infrastructure necessary for effective transfer of advanced biotechnology.

Training in Molecular Epidemiology and Approaches for Disease Prevention

The Task Force will develop plans for short regional training courses in molecular epidemiology for 1994. This will provide additional opportunities to local scientists and clinicians in developing countries to receive training in molecular epidemiology and its application to public health. In Latin America, the training program will be based upon an existing course in molecular genetics that is held at INDRE in México City each year. In 1993, the molecular genetics course will, for the first time, include lectures on molecular epidemiology and its applications in México. Full training courses in molecular epidemiology are being planned for 1994 in Saudi Arabia and China, and will include laboratory experiments, instruction in basic epidemiologic methods, such as defining cases of disease, assessing incidence and prevalence, and evaluating disease associations with potential risk factors. Development of specific disease prevention programs using advanced biotechnology will be outlined during the courses, such that they will be ready for immediate implementation. Training in minor repairs on equipment will also be provided. This will enhance effective technology transfer and increase its potential utilization for a variety of research and clinical activities.

Support for the Operation and Maintenance of Laboratory Equipment

One of the major problems with technology transfer to developing countries has been the lack of financial support for the purchase of laboratory equipment and its continued operation and maintenance. Reagents, enzymes and the other chemicals required for molecular analyses are also expensive, limiting effective advanced biotechnology transfer. The Task Force will determine what resources are required to develop molecular epidemiology in these areas. Members of the biotechnology industry will be particular helpful in dealing with these issues. The Task Force will also assess how funds to purchase items can be obtained. Proposals, grants and requests for funding will be submitted by the Task Force to national and international organizations and funding agencies to obtain sufficient financial support for equipment and supplies.

Evaluation of Transfer of Advanced Biotechnology

The Task Force will outline strategies for evaluating advanced biotechnology transfer to developing countries to determine its effectiveness, assess potential barriers, and increase its utilization in the future. Such evaluations will include details about prevention programs, research projects, grants and/or scientific publications that were initiated since the technology became available. The level of funding obtained from various sources will also be determined. Barriers to effective technology transfer, including inefficient utilization and operation of laboratory equipment, the high costs of supporting the technology, difficulties with access to supplies, etc., will also be assessed. The Task Force will review this evaluation with the key officials mentioned above. This will contribute to the formulation of future technology applications in clinical medicine, public health, industry, and agriculture.

Development of National and International Scientific Networks

At the present time, programs sponsored by WHO are targeted towards conditions such as AIDS, malaria, leishmaniasis, leprosy, tuberculosis and other communicable diseases.[29] Emphasis is placed upon the application of research to disease prevention, the development of immunization programs, the delivery of effective chemotherapy, and the utilization of rapid diagnostic methods which can be standardized across populations. In addition, adequate laboratory technology has been promoted and training courses for the prevention and control of disease have been organized in developing nations. International laboratory standards, methods and reference preparations, which have been very effective in stimulating collaborative research, have also been developed. Members of the Task Force are directly involved with these efforts; their experience will be invaluable for the development of programs for molecular epidemiology.

International collaborative studies, such as the WHO DIAMOND Project, are also working towards the achievement of global health. Such international efforts require the development and implementation of standards, which can be applied across populations to assure that accurate comparisons can be made. For example, the identification of HLA-DQ locus polymorphisms, which are highly associated with IDDM,[30] led to the development of a DIAMOND Sub-Project to test the hypothesis that population variation in the frequency of HLA class II alleles is the primary determinant of the worldwide patterns of disease.[1,31] The implementation of standards for this international study in developing countries, for example, where studies of IDDM are urgently needed, is difficult due to the lack of resources and technology. This DIAMOND Sub-Project presents a unique opportunity to promote molecular epidemiology and the transfer of advanced biotechnology and molecular epidemiology through collaborative grants. One such proposal will be submitted to the NIH in 1993 to investigate the molecular epidemiology of IDDM in Latin America, where such studies have been recently initiated.[32-34] The existence of the strong international DIAMOND network and the support of WHO and PAHO will provide an infrastructure for effective transfer of technology and training, and will encourage scientific collaboration between developing and industrialized nations.

CONCLUSIONS

The attainment of "health for all" is a goal which all nations are actively pursuing. While international efforts must continue to be directed towards improving the social and economic conditions of many areas, and providing education and health care to under-served populations, the attainment of international health can also be realized by the development of molecular epidemiology and the transfer of advanced biotechnology. Increasing capabilities in molecular epidemiology, if effective, will greatly contribute to a reduction in the morbidity and mortality associated with the most prevalent diseases in both developing and industrialized countries.

In this manuscript, we outlined plans for molecular epidemiology programs in Asia, the Eastern Mediterranean Region and Latin America. Similar approaches can be adopted by other regions of the world. However, the implementation of molecular epidemiology programs and achieving effective biotechnology transfer to developing countries can only be accomplished with careful planning, coordination and organization at the national and international level. This requires discussions between members of the government, the scientific community, industry and the health professions. Thus, an International Molecular Epidemiology Task Force has been established, with its membership reflecting various areas of expertise and representing many nations across the world. National scientific committees, with government advisors, are also being organized to interact with the international representatives and to direct the programs of individual countries.

The mission of the Task Force is to assist in the development and implementation of national programs in molecular epidemiology and to promote advanced biotechnology transfer for disease prevention in various regions of the world. This will be accomplished by: 1) the provision of relevant scientific information to key officials, 2) training in molecular epidemiology and approaches for disease prevention, 3) obtaining financial support for the operation and maintenance of laboratory equipment, 4) the evaluation of advanced biotechnology transfer, and 5) the development of national and international scientific networks. The development of national molecular epidemiology programs in one country will serve as a basis for additional programs in other countries and regions of the world. Thus, international support is vital to the success of this mission. Through collaboration and communication, the development of molecular epidemiology and effective biotechnology transfer will have a tremendous impact on the prevention of infectious and chronic diseases and the attainment of "health for all by the year 2000".

INTERNATIONAL MOLECULAR EPIDEMIOLOGY TASK FORCE*

Argentina: Dr. Julio Libman, Universidad Nacional de Rosario, Facultad de Ciencias Medicas, Santa Fe 3100, 2000 Rosario, Argentina

Dr. Jorge R. Palazzi, Instituto de Estudios Bioquímicos, Mendoza 1180, 2000 Rosario, Argentina

Brazil: Dr. Laercio Joel Franco, Departamento de Medicina Preventiva, Escola Paulista de Medicina, Rua Botucatu, 740 04023 São Paulo-SP, Brazil

China: Dr. Wang Ke-An, Vice President, Chinese Academy of Preventive Medicine, 10 Tian Tan XiLi, Beijing 100050, People's Republic of China

Dr. Shen Yi-E, Vice Director, Department of Preventive Medicine, WHO Collaborating Center for Occupational Health, Shanghai Medical University, Shanghai 200032, People's Republic of China

Egypt: Dr. Ala'din A.S. Alwan, WHO Regional Office for the Eastern Mediterranean Region, P.O. Box 1517, Alexandria, Egypt 21511

Germany:	Dr. Winfried Albert, Boehringer Mannheim, GMBH Research Center, Tutzing, Bahnhofstrasse 9-15, D-8132, Tutzing, Germany
Italy:	Dr. Arturo Falaschi, Director, International Centre for Genetic Engineering and Biotechnology, Padriciano 99, I-34012 Trieste, Italy
Japan:	Dr. Shigeaki Baba, Hyogo Institute for Research in Adult Diseases, 13-70 Kitaoji-cho, Akashi 673, Japan
	Dr. Naoko Tajima, Jikei University School of Medicine, Third Department of Internal Medicine, 3-25-8 Nishi-shibashi, Minatoku, Tokyo, Japan
México:	Dr. Luis Cabrera Coello, Enfermedades Transmisibles Hpt, Organizacion Pan Americana de la Salud, Organizacion Mundial de la Salud, Av. Paseo de las Palmas 530, Delegacion Miguel Hidalgo, México,, D.F., C.P 11000, México
	Dr. Clara Gorodezky, Chief, Department of Immunogenetics, Instituto Nacionale de Diagnóstico Y Referencia Epidemiológicos, SSA Carpio 470-1, México D.F. 11340, México
	Dr. Juan Manuel Sotelo Figueiredo, Representante Organizacion Pan Americana de la Salud, Organizacion Mundial de la Salud, Av. Paseo de las Palmas 530, Delegacion Miguel Hidalgo, México,, D.F., C.P 11000, México
Saudi Arabia:	Dr. Moshen El-Hazmi, Director, WHO Collaborating Centre for Haemoglobinopathies, Thalassaemias and Enzymopathies, King Saud University, Riyadh, Saudi Arabia
Spain:	Dr. Manuel Serrano Rios, Hospital Clinico San Carlos, Department of Internal Medicine, Ala Norte 6th., c/Martin Lagos s.n.,Madrid 28040, Spain
U.K.:	Dr. Steve Humphries, University College London Medical School, Department of Medicine, Division of Cardiovascular Genetics and Centre for Genetics of Cardiovascular Disorders, Rayne Institute, 5 University Street, London WC1E 6JJ, UK
U.S.A.:	Dr. Paul Basch, Department of Health Research and Policy, School of Medicine, Stanford University, Stanford CA 94305
	Dr. Peter Bennett, Chief, Phoenix Epidemiology and Clinical Research Branch, NIDDK, 1550 East Indian School Road, Building 1, Phoenix, Arizona 85014
	Dr. Janice S. Dorman*, Director, Molecular Epidemiology, WHO Collaborating Center for Diabetes Registries and Training, University of Pittsburgh, Pittsburgh, PA 15261 and Director of IME Task Force

Dr. Josefa Ippolito-Shepherd / Dr. Arlene Fonaroff, Office of the Director, Program Officer for the Americas and WHO, Fogarty International Center, National Institutes of Health, Bethesda, MD 20892

Dr. Thomas Calloway, Director, Scientific Affairs, Roche Molecular Systems, Inc., 1080 US Highway 202, Branchburg, NJ 08876

Dr. Massimo Trucco, Department of Pediatrics, Division of Immunogenetics, Rangos Research Center, Children's Hospital of Pittsburgh, Pittsburgh, PA 15213

Mr. Michael Uhrin, Molecular Diagnostics Reference Laboratory, Central Blood Bank, 812 Fifth Avenue, Pittsburgh, PA 15219

Dr. Raymond A. Zilinskas, Center for Public Issues in Biotechnology, University of Maryland Baltimore County, Catonsville, MD 21228

OBSERVERS: Dr. V. Boulyjenkov, Hereditary Disease Programme, World Health Organization, CH-1211, Geneva 27, Switzerland

Dr. Claus C. Heuck, Health Laboratory Technology and Blood Safety, World Health Organization, CH-1211, Geneva 27, Switzerland

Dr. Gabriel Schmunis, Pan American Health Organization, 525 Twenty-Third Street, NW, Washington, D.C. 20037

REFERENCES

1. J.S. Dorman, Genetic epidemiology of insulin-dependent diabetes mellitus: International comparisons using molecular genetics, *Ann Med.* 24:393 (1992).
2. C.C. Heuck, A. Deom, Health care in the developing world: need for appropriate laboratory technology, *Clin Chem.* 37:490 (1991).
3. R.A. Zilinskas, Biotechnology and the third world: The missing link between research and applications, *Genome* 31:1046 (1989).
4. World Health Organization, "Global Strategy for Health for All by the Year 2000," Geneva (1981).
5. J.M. Last, International Health, *in*: "Public Health and Preventive Medicine," J.M. Last, R.B. Wallace, eds., Appleton and Lange, Norwalk, pp 1127 (1992).
6. J.S. Dorman, C. Gorodezky, "Molecular Epidemiology, DNA Technology Transfer and DiseasePrevention in Developing Countries," Submitted.
7. A. Eddleston, Hepatitis, *Lancet* 335:1142 (1990).
8. H.S. Margolis, Viral hepatitis, *in*: "Public Health and Preventive Medicine," J.M. Last,R.B. Wallace, eds., Appleton and Lange, Norwalk, pp 131 (1992).
9. F.C. Tenover, "DNA Probes for Infectious Diseases," F.C. Tenover, ed., CRC Press, Inc., Boca Raton, FL, (1989).
10. N.G. Stoker, The polymerase chain reaction and infectious diseases: Hopes and realities, *Trans Royal Soc Trop Med Hyg*. 84:755 (1990).

11. Zhi-Yi Xu, H.S. Margolis, Determinants of hepatitis B vaccine efficacy and implications for vaccination strategies, *in*: "Viral Hepatitis in China; Problems and Control Strategies," Monog Virol, Vol 19. Y-M. Wen, Z-Y. Xu, J.L. Melnick, eds., Karger, Basel, pp 87 (1992).
12. A. Bonair, P. Rosenfield, K. Tengvald, Medical technologies in developing countries: Issues of technology development, transfer, diffusion and use, *Soc Sci Med.* 28:769 (1989).
13. R. LaPorte, D.J. McCarty, E.S. Tull, N. Tajima, Counting birds, bees and NCDs, *Lancet* 339:494 (1992).
14. B.S. Hulka, T.C. Wilcosky, J.D. Griffith, *in:* "Biological Markers in Epidemiology, "Oxford University Press, Inc., pp 3 (1990).
15. L.C. Groop, M. Kankuri, C. Schalin-Jäntti, A. Ekstrand, P. Nikula-Ijäs, E. Widén, E.Kuismanen, J. Eriksson, A. Franssila-Kallunki, C. Saloranta, S. Koskimies, Association between polymorphism of the glycogen synthase gene and non-insulin-dependent diabetes mellitus, *New Engl J Med.* 328:10 (1993).
16. F. Cambien, O. Poirier, L. Lecerf, A. Evans, J.P. Cambou, D. Arveiler, G. Luc, J.M.Bard, L. Bara, S. Ricard, L. Tiret, P. Amouyel, F. Alhenc-Gelas, F. Soubrier, Deletion polymorphism in the gene for angiotensin-converting enzyme is a potent risk factor for myocardial infarction, *Nature* 359:6241 (1992).
17. J.W. MacCluer, C.M. Kammerer, Invited Editorial: Dissecting the genetic contribution to coronary heart disease. *Am J Hum Genet.* 49:1139 (1991).
18. R.K. Saiki, S. Scharf, F. Faloona, K.B. Mullis, G.T. Horn, H.A. Erlich, N. Arnheim, Enzymatic amplification of ß-globin genomic sequences and restriction site analysis for diagnosis of sickle cell anemia, *Science* 246:1350 (1985).
19. R.K. Saiki, C.A. Chang, C.H. Levenson, T.C. Warren, C.D. Boehm, H.H. Kazazian, H.A. Erlich, Diagnosis of sickle cell anemia and ß-thalassemia with enzymatically amplified DNA and nonradioactive allele-specific oligonucleotide probes, *New Engl J Med.* 319:537 (1988).
20. J.D. Watson, R.M. Cook Deegan, Human genome project and international health, *J Am Med Assoc.* 364:3322 (1990).
21. D.F. Wirth, W.O. Rogers, R. Barker, H. Dourado, L. Suesebang, B. Alguquerque, Leishmaniasis and malaria: New tools for epidemiologic analyses, *Science* 234:975 (1986).
22. H.K. Webster, Malaria vaccines: Overview and perspective for developing countries, *Asian Pacif J Allergy Immunol.* 7:77 (1989).
23. G. Frankel, L. Riley, J.A. Giron, J. Valmassoi, A. Friedman, N. Streckbine, S. Falkow, G.H. Schoolnik, Detection of Shigella in feces using DNA amplification, *J Infect Dis.* 161:1252 (1990).
24. T. Popovic, C.H.A. Bopp, O. Olsvik, J. Kiehlbauch, Ribotyping in molecular epidemiology, *in:* "Diagnostic Molecular Bicrobiology," D.H. Persing, F.C. Tenover, T.F. Smith, T.J. White, eds., American Society for Microbiology, Washington (1992).
25. Pan American Health Organization and Centers for Disease Control, Update: Cholera - Western Hemisphere, 1992. *Morbidity and Mortality Weekly Report* 42:89 (1993).
26. G.C. Schild, P.D. Minor, Human immunodeficiency virus and AIDS: Challenges and progress, *Lancet* 335:1081 (1990).
27. Pan American Health Organization, "Health Conditions in the Americas, Vol. 1", Pan American Health Organization, Washington, D.C. (1990).
28. WHO DIAMOND Project Group, WHO multinational project for childhood diabetes, *Diabetes Care* 13:1062 (1990).

29. World Health Organization, "Introducing WHO," World Health Organization, Geneva (1976).
30. M. Trucco, To be or not to be Asp 57, that is the question, *Diabetes Care* 15:705 (1991).
31. J.S. Dorman, R.E. LaPorte, R.A. Stone, M. Trucco, Worldwide differences in the incidence of Type I diabetes are associated with amino acid variation at position 57 of the HLA-DQ ß chain, *Proc Natl Acad Sci. USA* 87:7370 (1990).
32. C. Gorodezky, H. Debaz, A. Olivo, M. Moreno, N. Altamirano, C. Robles Valdes, Contribution of DRB, DQA and DQB genes for the susceptibility to Type I diabetes (IDDM) in Mexicans, *HLA 1991* (In Press).
33. M.D. Gutierrez, O. Ramos, M. Calvillán, F. Pérez, M. Martinez, Spanish heritage IDDM group I: Results from Ciudad de Buenos Aires (Argentina). Submitted to the 1993 American Diabetes Association Scientific Meeting.
34. M. Calvillán, M.D. Gutierrez, F. Pérez, E. Carrasco, M. Garcia de los Rios, M. Serrano Rios, Spanish heritage IDDM group II: Results from Santiago de Chile (Chile) study. Submitted to the 1993 American Diabetes Association Scientific Meeting.

ANNEX A

RETURN TO:
Janice S. Dorman, Ph.D.
University of Pittsburgh
Graduate School of Public Health
Department of Epidemiology
Pittsburgh, PA 15261 USA
Phone: 01-412-692-5200 / FAX: 01-412-692-8329

DNA TECHNOLOGY AND MOLECULAR EPIDEMIOLOGY: AN INTERNATIONAL SURVEY

Name: ____________________

Address: ____________________

____

Phone: ____________________ FAX: ____________________

1. Is your institution currently utilizing recombinant DNA technology, such as polymerase chain reaction, DNA hybridization, gene cloning or DNA sequencing, for research and / or clinical practice?

   _____ Yes _____ No

2. If your institution is not currently utilizing recombinant DNA technology, would it be accessible at other institutions in your country?

   _____ Yes _____ No If yes, at what institutions:

   Name: ____________________

   Address: ____________________

   Name: ____________________

   Address: ____________________

3. If your institution is currently utilizing recombinant DNA technology **or** it is available in your country, what are the applications of this technology? Check all that apply:

	Yes	**No**	**Specific Application**
HLA molecular typing	_____	_____	__________
Diagnosis of disease	_____	_____	__________
Genetic studies	_____	_____	__________
Detecting susceptibility genes	_____	_____	__________
Genetic counseling	_____	_____	__________
Gene therapy	_____	_____	__________
Clinical / public health research	_____	_____	__________
Basic science research	_____	_____	__________
Vaccine development	_____	_____	__________
Product development - industry	_____	_____	__________
Agriculture research	_____	_____	__________
Assessing biomarkers of exposure	_____	_____	__________
DNA sequencing	_____	_____	__________
Other	_____	_____	__________

4. How is this technology being financially supported? Check all that apply:

	Yes	**No**	**Name of Source of Support**
Federal government	_____	_____	__________
State / local sources	_____	_____	__________
Universities	_____	_____	__________
Hospitals	_____	_____	__________
Other institutions	_____	_____	__________
Private foundations	_____	_____	__________
National grants	_____	_____	__________
International grants	_____	_____	__________
International agencies (WHO, UNDP)	_____	_____	__________
Industry	_____	_____	__________
Agriculture			
Other	_____	_____	__________

5. If DNA technology is ***NOT*** available at your institution, please list potential barriers to effective technology transfer. Check all that apply, and rank in order of importance (1 = most serious problems):

	Yes	No	Specific Problem
Inadequate financial support from:			
Government	____	____	____________
Academia	____	____	____________
Industry	____	____	____________
Hospitals / institutions	____	____	____________
Inadequate training	____	____	____________
Limited access to needed supplies	____	____	____________
Limited equipment	____	____	____________
Limited opportunities for collaboration	____	____	____________
Inadequate international support	____	____	____________
Technology is not a priority	____	____	____________
Other	____	____	____________

6. How could barriers to effective DNA technology transfer be removed? Please be as specific as possible.

____

____

____

____

7a. If DNA technology is ***NOT*** available, would you be interested in obtaining access to this technology?

____ Yes ____ No If yes, at what institutions would it be most effectively utilized?

Name: ____

Address: ____

Name: ____

Address: ____

7b. What would be the most important applications of this technology? Check all that apply:

	Yes	No	Specific Application
HLA molecular typing	____	____	____________
Diagnosis of disease	____	____	____________
Genetic studies	____	____	____________
Detecting susceptibility genes	____	____	____________
Genetic counseling	____	____	____________
Gene therapy	____	____	____________
Clinical / public health research	____	____	____________
Basic science research	____	____	____________
Vaccine development	____	____	____________
Product development - industry	____	____	____________
Agriculture research	____	____	____________
Assessing biomarkers of exposure	____	____	____________
DNA sequencing	____	____	____________
Other	____	____	____________

7c. How could the technology be financially supported? Check all that apply:

	Yes	No	Name of Source of Support
Federal government	____	____	____________
State / local sources	____	____	____________
Universities	____	____	____________
Hospitals	____	____	____________
Other institutions	____	____	____________
Private foundations	____	____	____________
National grants	____	____	____________
International grants	____	____	____________
International agencies (WHO, UNDP)	____	____	____________
Industry	____	____	____________
Agriculture	____	____	____________
Other	____	____	____________

8. Are you, or is someone at your institution, currently working in the field of molecular epidemiology?

_____ Yes _____ No

Name: ________________________________

Address: ________________________________

If yes, please describe the nature of this work:

	Yes	No	Specific Type of Work
Infectious disease epidemiology	____	____	____________
Parasitic diseases	____	____	____________
Viral diseases	____	____	____________
Other	____	____	____________
Chronic disease epidemiology	____	____	____________
Autoimmune disease	____	____	____________
Cancer	____	____	____________
Cardiovascular disease	____	____	____________
Perinatal / reproductive disorders	____	____	____________
Other	____	____	____________
Pharmacoepidemiology	____	____	____________
Genetic diseases	____	____	____________
Environmental / occupational diseases	____	____	____________
Other	____	____	____________

9. What type of training is available in molecular epidemiology at your institution and/or country? Check all that apply:

	Yes	No	Specific Type of Training and Institution
Course / seminar	____	____	____________
Research opportunities - integration	____	____	____________
of molecular biology and epidemiology	____	____	____________
Clinical training (DNA diagnosis)	____	____	____________
Other	____	____	____________

10. Would someone at your institution be interested in participating in short or long-term training in molecular epidemiology?

	Yes	No	Specific Type of Training
Two-week course	____	____	____________
Extended training	____	____	____________

11. What would be most important to your institution and/or country in terms of the training? Check all that apply:

	Yes	No	Specify Application
Training in Molecular Biology			
Gene cloning	____	____	____________
DNA probe construction	____	____	____________
RFLP analyses	____	____	____________
In situ hybridization	____	____	____________
cDNA libraries	____	____	____________
DNA sequencing	____	____	____________
Polymerase chain reaction	____	____	____________
Southern / northern blots	____	____	____________
HLA typing	____	____	____________
Other	____	____	____________
Training in Epidemiologic Methods			
General epidemiology	____	____	____________
Disease surveillance	____	____	____________
Risk factor assessments	____	____	____________
Family studies	____	____	____________
Evaluating disease aggregation in families	____	____	____________
Developing primary / secondary Prevention programs	____	____	____________
Economic analyses	____	____	____________
Clinical trials	____	____	____________
Manuscript writing	____	____	____________
Grant writing	____	____	____________
Computer analyses	____	____	____________
Other	____	____	____________

12. What applications of training in molecular epidemiology would be most important to the research and public health objectives of your institution and country?

	Yes	No	Specific Type of Work
Infectious disease epidemiology	____	____	____________________
Parasitic diseases	____	____	____________________
Viral diseases	____	____	____________________
Other	____	____	____________________
Chronic disease epidemiology	____	____	____________________
Autoimmune disease	____	____	____________________
Cancer	____	____	____________________
Cardiovascular disease	____	____	____________________
Perinatal / reproductive disorders	____	____	____________________
Other	____	____	____________________
Pharmacoepidemiology	____	____	____________________
Genetic diseases	____	____	____________________
Environmental / occupational diseases	____	____	____________________
Other	____	____	____________________

13. The results of this international survey will be prepared for publication as a collaborative paper. Please list the names and addresses of up to five individuals from your institution and/or country who are interested in molecular epidemiology and DNA technology transfer, and who would be interested in being a co-author on this collaborative paper.

Name: ________________________________

Address: ________________________________

Name: ________________________________

Address: ________________________________

Name: ________________________________

Address: ________________________________

Name: ________________________________

Address: ________________________________

Name: ________________________________

Address: ________________________________

EFFECTIVE DNA TECHNOLOGY TRANSFER TO DEVELOPING COUNTRIES

Clara Gorodezky and Angelica Olivo

Department of Immunogenetics
Instituto Nacional de Diagnóstico y
Referencia Epidemiológicos, SSA
Carpio 470
México D.F. 11340
MÉXICO

INTRODUCTION

In industrialized countries, biotechnology, research and development have given rise to a rapidly growing bioscience-based industry, and to very solid research groups that dedicate their efforts to basic or applied science. In science, the most important contributions to the knowledge base come from basic research. A fraction of this base has the potential for practical application that goes to an advanced research and development laboratory, such as a national, defense or industrial laboratory. The education system produces scientists, technicians and managers that operate the processes involved with the commercial manufacture of the product. The system functions well in a very favorable political-economic-legal environment.[1]

In developing countries, technologies have been traditionally imported and introduced directly into the industrial sector, without involving a research component. Basic research is very expensive and needs financial and administrative support for getting information, materials, equipment and a very good infrastructure. Because of the socio-political and economical problems in developing countries, optimal research conditions are difficult to achieve.

The main differences between these two types of countries are that for poor or developing areas of the world, the following facts are part of the structure:

1. Access to the knowledge base by scientists is often difficult and time consuming. Information is not easily available, and libraries are usually not updated or complete.
2. Materials, reagents and equipment are expensive and many times difficult to acquire.
3. Little of the basic research becomes part of the applied science, although now efforts are addressing this issue.
4. Poor interaction exists between universities and industries regarding applied science.

Standardization of Epidemiologic Studies of Host Susceptibility
Edited by J. Dorman, Plenum Press, New York, 1994

5. Research and development unities hardly exist in the industrial sector.
6. Technology is usually imported.
7. The environment is not always favorable and depends many times on political and economic changes.[1]

Effective transfer of technology can be achieved with the help of international organizations such as the World Health Organization (WHO), the United Nations Children's Fund (UNICEF), the Food and Agriculture Organization (FAO), The United Nations Educational, Scientific and Cultural Organization (UNESCO), The United Nations Industrial Development Organization (UNIDO), the United Nations Development Program (UNDP) or the International Center for Genetic Engineering and Biotechnology (ICGEB). Their assistance falls in three categories: 1) help to policy markers, 2) support for national research and training activities, and 3) assistance for the establishment of international networks. These activities provide training opportunities, which are essential, and allow funds for the purchase of equipment and materials. Finally, with the establishment of international networks, an individual institute can take on projects that it otherwise would have to forsake, perhaps because they would be too expensive, large or complicated.[2]

Several activities have been developed in Mexico that contribute greatly to effective DNA technology transfer. These are described in the following paragraphs.

COURSES AND TRAINING PROGRAMS FOR LATIN AMERICA

The Department of Immunogenetics in Mexico City organizes two annual courses that include lectures and laboratory sessions. One is a current course on histocompatibility and the second one is on molecular genetics. The first one was started in 1982, and students from all over Mexico and various countries in Latin America have been trained. Every year more people attend, and 80 students participated in 1992. The course has continuous financial support from commercial companies. Prominent professors from Mexico and abroad participate every year. Among them are Peter Stastny, Min Sik Park, Edmond Yunis, G.B. Ferrara, Maria Pia Pistillo, Pablo Rubinstein, Fritz Bach, Susan Radka, Zulay Layrisse, Jose Roberto Moraes and others. The aim of this course is to update the knowledge in the field of histocompatibility and immunogenetics, including basic and technological aspects, as well as discuss applications in transplantation, population genetics, disease associations, immune response genes and mechanisms underlying disease susceptibility. Sixty professionals have been trained from Argentina, Bolivia, Costa Rica, Cuba, Colombia, Ecuador, Honduras, Peru, Puerto Rico, Panama, Dominican Republic and Venezuela.

The course on molecular genetics was started in 1986, also with great success. Students from all over Latin America attend. Approximately 40 people from the same countries have been trained, and 85 more registered for the 1992 course. The modern techniques of DNA extraction, southern blot and DNA typing by PCR amplification are done by them in the lab. Modern topics on different DNA technologies are revised, as are their applications in the fields of histocompatibility, bacteriology and parasitology. Financial support from commercial companies is available, and distinguished professors such as Marcelo Fernandez Vina, Edgar Milford, Giovanna Angelini and others, participate. The aim of both courses has been to teach researchers how to work in a developing country laboratory, the mechanisms to purchase materials, and how to get service for equipment maintenance. Most important has been the fact that we started several histocompatibility laboratories in Mexico and in other Latin American countries, and some of them have already incorporated the molecular techniques. Continuous intellectual advice and help to solve technical and practical problems is given. Follow-up of technology transfer has been relevant.

LATIN AMERICAN AND INTERNATIONAL WORKSHOPS

Workshops in which reagents are exchanged and methods are standardized are very important as a source of technology transfer. An example is the V Latin American Workshop of Histocompatibility that was held in Mexico City on December 1989. Latin American histocompatibility labs had, for the first time, the opportunity to assay a method of DNA typing of HLA alleles using the polymerase chain reaction (PCR), and to practice oligonucleotide typing with a nonradioactive method developed by Bugawan et al.[4] The complete protocol, the primers and probes were a contribution of Dr. Henry Erlich (Cetus Corporation). Thirty-three laboratories participated in this Workshop that was chaired by C. Gorodezky (Mexico) and G.B. Ferrara (Italy). The countries involved were Argentina (7 labs); Brazil (4 labs); Chile (2 labs); Colombia (3 labs); Costa-Rica (1 lab); Mexico (3 labs); Uruguay (1 lab); Venezuela (4 labs); Canada (1 lab); Germany (1 lab); Italy (1 lab); and USA (8 labs). Since it is important to communicate what is going on in the HLA field in Latin America, the results were published in a book of Proceedings.[5] The goals of the Workshop are depicted in Table 1.

Table 1. Goals of the V Latin American Workshop of Histocompatibility.

1.	Serological analysis of HLA antigens in Latin American cells from healthy controls, Chagas disease and leishmaniasis patients
2.	Behavior of HLA antisera in non-Latin American cells (from European, Canadian, and USA populations)
3.	Analysis of frequencies of class III molecules in Latin American populations
4.	Molecular analysis by PCR amplification of DNA of native populations and HLA disease associated cells

Looking for HLA associations in leishmaniasis was an important goal, as very little was known about the genetic factors involved in this human disease. There are clear antecedents of the existence of resistance and susceptibility genes located inside and outside the MHC region in mice.[6,7] Moreover, the parasitosis constitutes a real public health problem, since the incidence is 400,000 cases/year and the prevalence is 12,000,000 cases worldwide.[8] The disease is caused by different species and subspecies of a intracellular protozoan parasite belonging to the genus Leishmania. It is transmitted by a sandfly, and the spectrum of clinical manifestations is grouped under several categories: localized cutaneous leishmaniasis (LCL), diffuse cutaneous leishmaniasis (DCL), mucocutaneous (espundia, MCL) and visceral leishmaniasis or Kala-Azar (VL).[9] Understanding of immunological mechanisms has been greatly advanced by the studies in inbred mice, but since very few data were available for the human disease, the Workshop had several main goals: 1) to standardize procedures for immunological and clinical diagnosis, 2) to establish the diagnostic criteria, and 3) to provide the standard protocols for the preparation of leishmanin (the specific antigen to test the cellular immune response). Thus, 447 patients from different countries were studied in the Workshop. The results indicate that the susceptibility to LCL and MCL is associated to HLA-DQw3 in Mexican (Mestizos),[10] Venezuelan (Mestizos),[11] and Brazilian (Caucasians and Mulatos)[12] populations. A resistance gene linked to HLA-DPw4 was also suggested in Mexicans.[10] On the contrary, in the visceral form, no HLA associations were detected.[13]

The results obtained by the different labs using standard procedures and the same reagents, encouraged us to do some molecular typing. As an example, the frequencies of HLA-DQA1 alleles were determined in different populations using the PCR/oligonucleotide typing that was introduced to the Latin American HLA community. Table 2 shows the distribution of these alleles in Colombia, Mexico and Venezuela.[14]

Table 2. Frequencies of HLA-DQA1 alleles (%) in different Latin American populations.

DQA1 Allele	Colombia (Mulatos)	Mexico (Mestizos)	Venezuela (Mestizos)
*0101	18.1	18.0	18.8
*0102	20.2	19.0	16.7
*0103	7.5	10.0	4.2
*0201	7.5	11.1	8.3
*0301	12.8	26.1	22.9
*0401, *0601	34.0	12.0	29.2

The distribution of DQA1 alleles differs among the three groups, showing that DQA1*0401 and DQA1*0601 are the prevalent alleles in Mulatos, while in Mexicans, DQA1*0301 is the most frequent allele, and in Venezuela DQA1*0302, DQA1*0401 and DQA1*0601 are common.

MOLECULAR TECHNOLOGY AND INFECTIOUS DISEASES

Effective DNA technology transfer to nations where infectious diseases are the most important public health problems can contribute to a reduction in morbidity and mortality associated with the prevalent diseases in each place. However, chronic diseases must not be forgotten, since cardiovascular problems, autoimmune diseases, cancer and diabetes, among other illnesses, are also benefiting in terms of having a more accurate diagnosis, analyzing the mechanisms of etiopathogenesis using molecular tools, and in the near future, gene therapy.[15] Table 3 shows how molecular technology and the PCR[16-18] have revolutionized many aspects of medicine including the detection of fungal, bacterial, parasitic and viral pathogens.[17]

MOLECULAR METHODOLOGY APPLIED TO VIRAL DISEASES

The human retroviruses (HIV-1 and HIV-2) and human lymphotropic virus (HTLV-1 and HTLV-2) and those yet to be identified are associated with an increasing number of human diseases. Although serological assays for antibodies to these viruses have played a significant role in the identification of infected individuals, direct detection of the viral pathogen is highly desirable in many cases. Therefore, molecular diagnosis using PCR is very important in several instances: 1) when transcriptional dormancy of the proviral genome exists, 2) if only a small number of infected cells are present within the peripheral blood, 3) when a small number of proviral copies per infected cells are present or if multiple related but distinct viral members exist, 4) or if the infected cells in the host are localized in lymph nodes or other compartments not accessible to monitoring.[19]

Hepatitis B virus (HBV) infection is an important public health problem in many areas of the world.[20] PCR has dramatically increased the sensitivity of detection of HBV, as much as 10,000-fold increase over standard procedures.[21] Another important approach is the use of recombinant DNA technology to produce live vaccine that will be more effective than those currently available. With the new vaccine, strategies are being targeted in Asia towards newborns to reduce transmission from infected mothers to their infants.[22]

Table 3. Applications of DNA technology for research and practical purposes in medicine.

1.	Gene sequencing
2.	Detection of gene expression, using PCR
3.	Fingerprinting for forensic medicine, paternity testing, post-transplant cell population identification, tumor analysis, and twin zygosity determination
4.	HLA DNA typing for analysis of micropolymorphisms, population genetics, evolution of HLA genes, disease associations, and mechanisms of transplant rejection
5.	Diagnosis of viral, bacterial, fungal and parasitic diseases
6.	Diagnosis of genetic diseases such as hemophilia, β-thalassemia, and cystic fibrosis
7.	Diagnosis of new mutation diseases, detection of deletions at the Duchenne muscular dystrophy locus
8.	Development of synthetic vaccines (malaria, leprosy, neurocysticerosis)

Wild polioviruses are endemic to most developing countries.[23] Although the incidence of poliomyelitis declined sharply in the American Continent following intensified immunization programs, the identification of infection caused by wild or vaccinal strains is very important. PCR has been very useful in the laboratory surveillance of poliomyelitis cases in South America, permitting direct and rapid identification of the wild polioviruses.[24]

MOLECULAR TECHNOLOGY FOR BACTERIAL AND PARASITIC DISEASES

Although effective therapy has been available for more than 40 years in the treatment of syphilis, a sexually transmitted disease, two recent developments underscore the need for more effective means of diagnosing the infection: 1) annual increases in new cases are being reported, and 2) a large proportion of patients with HIV infection are co-infected by T. pallidum. Bacterial DNA is efficiently detected in serum, cerebrospinal and amniotic fluids from syphilis patients by using PCR.[25]

Cholera enterotoxin (CT) is a major virulence determinant of Vibrio cholerae 01. However, most 01 strains isolated from the environment do not produce CT nor do they possess the genetic potential to produce CT, although some non-01 strains produce it. These findings make necessary regular examination of isolates for their ability to produce CT in order to assess their clinical significance. PCR is more rapid than the southern blot in detecting the CTX operon because PCR amplifies the target sequence to a detectable level in a short time. The detection limit is 1 pg of chromosomal DNA or broth culture containing three viable cells.[26]

Turning to parasitic diseases, malaria is a very great problem in tropical areas of the world. Assessing the extent of antigenic variation in Plasmodium is relevant for gaining insight into the evolution of the parasite genome and the underlying mechanisms involved in the generation of diversity. It may also be critical for the development of an effective vaccine. The parasite displays genetic variation in natural isolates and there is a high frequency of genetic recombination, resulting from the crossing of two cloned lines of Plasmodium falciparum.[27] Analysis of stage-specific, single copy genes from the parasite at the DNA level and deduced amino acid sequences has given insight into the structural basis for antigenic variation in different strains.[28]

Severity and pathogenicity of leishmaniasis is associated to the species or subspecies of the infecting Leishmania. Identification methods based on phenotypic characteristics are not very efficient. Recombinant DNA methods have proven highly useful. Several probes for the identification of subspecies have been constructed from the kDNA (Kinetoplast DNA) of the minicircles. The use of kDNA of L. mexicana and L. braziliensis as probes has permitted the differential diagnosis of many patients in Brazil. Cloning of minicircle fragments and the analysis of deletions in one minicircle of L. amazoniensis, allowed the obtaining of highly specific probes for two subspecies of L. mexicana, (mexicana and amazoniensis). The same approach has been used to construct probes for L.tropica and L. donovani.[29-31] We have been working with a similar idea, in order to construct probes that may distinguish between the localized and disseminated cutaneous disease, caused by L. mexicana. The analysis by Southern-blot of the isolates that have been screened until now show a very high homology. The fragments obtained of the digestion of kDNA with EcoRI and DraI are already cloned and genomic libraries of kDNA of three different strains are now being screened.[32]

CONCLUSIONS

DNA technology transfer to developing countries is a necessity in order to design prevention strategies for infectious as well as for chronic degenerative diseases. For these purposes, genetic screening to identify high risk individuals and detection of infectious agents at the molecular level will greatly reduce morbidity and mortality associated with the most prevalent diseases in these countries. Mexico has been one of the starting points for transfer of DNA technology, using different approaches: 1) courses and training programs in histocompatibility and molecular genetics, 2) organization of a Latin American Workshop on Histocompatibility, 3) help for the establishment of new laboratories in other cities of Mexico and in other Latin American countries, and 4) help for developing new programs of interest in infectious diseases prevalent in a specific geographical area using DNA technology. This is done by giving continuous technical and intellectual advice and support, and suggesting effective mechanisms for getting financial aid and for purchasing equipment, reagents and materials. Effective transfer requires administrative and political comprehension; authorities must understand the importance of using modern methodologies which may be expensive but very rewarding in terms of results. All of these efforts can be best promoted by international agencies that are absolutely conscious of the health needs in each country.

REFERENCES

1. R.A. Zilinskas, Biotechnology and the Third World: The missing link between research and applications, *Genome* 31:1046 (1989).
2. E.J. DaSilva, H. Taguchi, Mechanism for international co-operation in applied microbiology and biotechnology, *MIRCEN, J Appl Microbiol Biotechnol.* 2:27 (1986).
3. R.K. Saiki, T.L. Bugawan, G.T. Horn, K.B. Mullis, H.A. Erlich, Analysis of enzymatically amplified B-globin and HLA-DQ alpha DNA with allele-specific oligonucleotide probes, *Nature* 324:163 (1986).
4. T.L. Bugawan, R.K. Saiki, C.H. Levensson, R.M. Watson, H.A. Erlich, The use of non-radioactive oligonucleotide probes to analyze enzymatically amplified DNA for prenatal diagnosis and forensic typing, *BioTechnol.* 6:943 (1988).
5. C. Gorodezky, G. Sierp, E.D. Albert. "Proceedings of the V Latin American Workshop of Histocompatibility," Laboratory of Immunogenetics, University of Munich, Munich (1992).
6. M.J. Blackwell, C. Hale, B.M. Roberts, An H-11 linked gene has parallel effect on L. major and L. donovani infection in mice, *Immunogenetics* 21:385 (1985).
7. F.Y. Liew, Functional heterogeneity of CD4+ cells in leishmaniasis, *Immunol Today* 10:40 (1989).
8. F. Modabber. "Tropical Disease Research, a Global Partnership," Eighth Programme Report, TDR, J. Maurice, A.M. Pearce, eds., WHO, Geneva (1987).
9. WHO Technical Report series 701. "The Leishmaniases", Report of a WHO expert Committee, *Geneva* (1984).
10. C. Gorodezky, J. Guzman, G. De la Rosa, O. Hobart, L. Castro, O. Hernandez, J.M. Carranza, G. Granados, J. Flores, H. Perez, Class II antigens are associated with localized cutaneous leishmaniasis in Mexican Mestizos, *in*: "Visuals of the Clinical Histocompatibility Workshop," P.I. Terasaki, ed., One Lambda, Inc, Los Angeles (1992).
11. M.L. Lara, Z. Layrisse, J.V. Scorza, E. Garcia, Z. Stikow, J. Granados, W. Bias, Immunogenetics of human American cutaneous leishmaniasis: Study of HLA haplotypes in 24 families from Venezuela, *Hum Immunol.* 30:129 (1991).
12. M.L. Petzl-Erler, M.P. Bellich, F. Queiroz-Telles, Association of mucosal leishmaniasis with HLA, *Hum Immunol.* 32:254 (1991).
13. M.R.R. Amendoeira, L. Guilherme, M. Marin, I. Messias, M.C.A. Marzochi, H. Krieger, P. Cabello, J. Kalil, Linkage analysis: HLA and susceptibility to American visceral leishmaniasis, *in*: "Proceedings of the V Latin American Workshop of Histocompatibility," C. Gorodezky, G. Sierp, E.D. Albert, eds., Laboratory of Immunogenetics, University of Munich, Munich (1992).
14. H.A. Erlich, H. Debaz, N. Fildes, V. Trejo, C. Gorodezky, HLA-DQA1 PCR/oligonucleotide probe: Results of the Latin American workshop, *in*: "Proceedings of the V Latin American Workshop of Histocompatibility," C. Gorodezky, G. Sierp, E.D. Albert, eds., Laboratory of Immunogenetics, University of Munich, Munich (1992).
15. J.S. Dorman, C. Gorodezky, "Molecular Epidemiology: Advanced Biotechnology and Public Health Training for Disease Prevention," (submitted for publication) (1994).
16. H.A. Erlich. "PCR Technology. Principles and Applications for DNA Amplification," Stockton Press, New York (1989).

17. M.A. Innis, D.H. Gelfand, J.J. Sninsky, T.J. White. "PCR Protocols. A Guide to Methods and Applications," Academic Press, Inc., New York (1990).
18. L.T. Kirby. "DNA Fingerprinting: An Introduction," Stockton Press, New York (1990).
19. D.P. Drotman, J.W. Curran, Epidemiology and prevention of acquired immunodeficiency syndrome, *in*: "Public Health and Preventive Medicine," J.M. Last, R.B. Wallace, eds., Appleton and Lange, Norwalk (1992).
20. A. Eddleston, Hepatitis, *Lancet* 335:1142 (1990).
21. I. Baginski, A. Ferrie, R. Watson, D. Mack, Detection of Hepatitis B Virus, *in*: "PCR Protocols: A Guide to Methods and Applications," M.A. Innis, D.H. Gelfand, J.J. Snisky, T.J. White, eds., Academic Press, New York (1990).
22. H.S. Margolis, Viral hepatitis, *in*: "Public Health and Preventive Medicine," J.M. Last, R.B. Wallace, eds., Appleton and Lange, Norwalk, (1992).
23. WHO, Expanded program on immunization: global situation-poliomyelitis, *Geneva* (1989).
24. E.E. DaSilva, M.A. Pallansh, P.B. Holloway, M.J. Couto-Oliveira, H.G. Schatzmayr, O.M. Kew, Oligonucleotide probes for the specific detection of the wild poliovirus Types 1 and 3 endemic to Brazil, *Intervirology* 32:149 (1991).
25. J.M. Burstain, E. Grimprel, S.A. Lukehart, M.V. Norgard, J.D. Radolf, Sensitive detection of treponema pallidum by using the polymerase chain reaction, *J Clin Microbiol.* 29:62 (1991).
26. H. Shirai, M. Nishibuchi, T. Ramamurthy, S.K. Bhattacharya, S.O. Pal, Y. Takeda, Polymerase chain reaction for detection of the cholera enterotoxin operon of vibrio cholera, *J Clin Microbiol.* 29:2517 (1991).
27. D. Walliker, B. Fenton, T.E. Wellems, T.F. McCutchan, K.D. Vernick, Genetic recombination in plasmodium falciparum, *in*: "Current Communications in Molecular Biology: Molecular Genetics of Parasitic Protozoa," M.J. Turner and D. Arnot, eds., Cold Spring Harbor Laboratory, New York (1988).
28. M.J. Lockyer, S.C. Nicholls, Variation in plasmodium falciparum gene structure, *in*: "Current Communications in Molecular Biology: Molecular Genetics of Parasitic Protozoa," M.J. Turner, D. Arnot, eds., Cold Spring Harbor Laboratory, New York, (1988).
29. J. Ellis, T. Knapp, J. Crampton, Cloning of a polymorphic DNA fragment from the genome of leishmania donovani, *Mol Biochem Parasitol.* 34:261 (1989).
30. R.M.R. Coulson, D.F. Smith, Isolation of genes showing increased or unique expression in the infective promastigotes of leishmania major, *Mol Biochem Parasitol.* 40:63 (1990).
31. W.O. Rogers, D.F. Wirth, Generation of sequence diversity in the kinetoplast DNA minicircles of leishmania mexicana amazonensis, *Mol Biochem Parasitol.* 30:1 (1988).
32. A. Olivo, A. Sosa, J. Vilchis, C. Gorodezky, Clonacion de minicirculos de Leishmania mexicana mexicana IX Congreso Nacional de Inmunologia, Monterrey, NL, Mexico (Abstract) (1991).

INTERNATIONAL RESEARCH AND TRAINING THROUGH THE FOGARTY INTERNATIONAL CENTER

Josefa Ippolito-Shepherd

National Institutes of Health
Fogarty International Center
Building 31, Room B2C08
Bethesda, Maryland 20892
USA

INTRODUCTION

I am delighted to be here today to represent the Fogarty International Center (FIC) for Advanced Study in the Health Sciences, National Institutes of Health (NIH), and to participate in this very important Workshop. Dr. Philip Schambra, Director of the FIC, sends his best wishes for the success of this meeting.

As we approach the twenty-first century, it is impossible to ignore the vital importance of international biomedical and behavioral research collaboration, due to the speed of infections that cross national boundaries and the increasingly high cost of research. Fortunately, this collaboration is facilitated by the speed of communication and the availability of unique human resources, populations, and environments throughout the world. Biomedical science is definitely an international enterprise, as noted by the percentage of published papers reporting research advances achieved through international cooperation. In 1955, 1% of publications in Science and the New England Journal of Medicine showed international collaboration while in 1990 this percentage increased to 17% and 15%, respectively.

The United States Public Health Service has a long-standing commitment to international health through collaboration with national and international organizations. The NIH supports research both in the United States and overseas and collaborates with scientists worldwide on most diseases, primarily through foreign and U.S. biomedical scientists exchanges that make biomedical and behavioral research collaboration possible.

The FIC is the focal point for coordinating and promoting biomedical and behavioral research between the NIH and other U.S. government and private agencies involved in international programs, and foreign nations and scientists. In this role, FIC facilitates the assembly of scientists and leaders in the health field; provides policy guidance and direction; promotes international collaboration in the health sciences

Standardization of Epidemiologic Studies of Host Susceptibility
Edited by J. Dorman, Plenum Press, New York, 1994

through its research programs, conferences, and seminars; provides postdoctoral fellowships for research in the United States and abroad; promotes biomedical scientists exchanges between the U.S. and other countries; coordinates NIH Health Sciences activities at the international level; serves as a central resource for foreign scientists and visitors to the NIH; and analyzes international health issues.

The Center was established in 1968 and created in law in 1985 as a tribute to Congressman John E. Fogarty's commitment to international cooperation in all aspects of the health sciences. The Center is one of the 21 Institutes, Centers, and Divisions at the NIH. FIC has both staff and program functions. FIC responsibilities are carried out through four program activities: International Advanced Studies, International Research and Awards, International Coordination and Liaison, and International Services and Communication. FIC programs include International Research Fellowships, Senior International Fellowships, Small Research Grants to Support U.S.-Foreign Research Collaboration, and Special Joint Funds for U.S.-Foreign Research Collaboration. We also provide administrative support for foreign funded fellowships for U.S. scientists. Thus, to support international research and training we have activities in targeted health areas such as AIDS, neurosciences, and environmental diseases, as well as targeted geographic areas, such as Latin America and the Caribbean, and Central and Eastern Europe. We have 77 full-time staff and a 1992 budget of almost 20 million dollars.

The FIC scientific programs are grouped into three major areas. The first area includes fellowships, exchanges and training programs, such as the International Research Fellowship, Senior International Fellowship, Scholars-in-Residence awards, and scientists exchanges. The second area includes special scientific support programs, such as the AIDS Research and Training Awards, the neurosciences and environmental sciences awards, programs related to emerging microbes, Tuberculosis and other infectious diseases, and biodiversity activities. The third area includes the regional scientific collaboration support programs, such as the Regional Initiatives with Central and Eastern Europe, Latin America and the Caribbean, and Africa, including the small grant program. I will briefly describe these programs in terms of their purpose, provisions, and basic eligibility requirements.

INTERNATIONAL RESEARCH FELLOWSHIPS

The purpose of the International Research Fellowship is to stimulate international health-related research by enabling foreign postdoctoral biomedical or behavioral scientists to extend their research experience by working with U.S. scientists on health issues of mutual concern. This fellowship provides support for 12-24 months, including stipend, round trip travel, and supply and materials allowance to be used at the U.S. host institution. A scientist who is interested in being selected for this fellowship must be nominated for candidacy by a Nominating Committee in his/her own country, must have earned a doctoral degree in a health-related field, must have demonstrated the ability to engage in independent research, and must formulate a research proposal that is of mutual interest to the applicant and his/her sponsor. In addition, the applicant must have an agreement with a U.S. scientist who will accept the fellow and the proposed research project, have the assurance of a research position after completing the fellowship, and be able to speak and understand English. The Nominating Committee, established by each participating country maintains a liaison with the FIC, announces the availability of fellowships, receives and reviews applications according to our Center criteria, and, finally, submits the best applications to the Center.

SENIOR INTERNATIONAL FELLOWSHIPS

The Senior International Fellowship enables mid-career U.S. biomedical scientists to conduct research abroad and to share expertise with their foreign counterparts. Scientists interested in participating in this program must hold a doctoral degree in the health sciences, have 5 or more years of postdoctoral experience, have at least 2 years experience in the health sciences, hold a full-time appointment on the staff of a U.S. institution, be invited by a foreign institution, be a U.S. citizen or permanent U.S. resident, and may not have had more than one previous Senior International Fellowship.

FOREIGN FELLOWSHIPS FOR U.S. SCIENTISTS

Foreign fellowships for U.S. scientists support research in the foreign country/area by postdoctoral biomedical scientists who are in the formative phase of their career. Countries which provide fellowship support to U.S. scientists are Finland, France, Germany, Ireland, Israel, Japan, Norway, Sweden, Switzerland, United Kingdom, and Taiwan.

EXCHANGE OF U.S. AND FOREIGN SCIENTISTS

FIC's exchange program fosters biomedical collaboration between scientists on topics of mutual interest. Participating countries are Austria, Bulgaria, France, Hungary, Poland, Romania, Countries of the CIS, and Yugoslavia. Generally, exchanges are joint efforts between the sending and the host countries. The sending country typically pays for travel expenses and the host country for in-country costs. To participate in this program, a U.S. candidate must be a U.S. citizen or permanent resident with a doctoral degree, be affiliated with an institution, and have appropriate professional experience. The applicant's scientific merit is reviewed by an NIH committee and it is followed by a secondary review of the program relevance, and host country consideration and approval.

SPECIAL JOINT FUND FOR U.S. - FOREIGN RESEARCH COLLABORATION

The special joint fund for U.S.-Foreign Research Collaboration utilizes funds contributed by the United States and foreign countries under formal agreements to support collaborative research of mutual interest and benefit. Funds are awarded to a foreign participant if the proposed collaborative research is of interest to both the U.S. and foreign researchers, if benefits from the research will advance the NIH mission, and if applicants have passed scientific and policy reviews in the United States and the foreign country.

FOGARTY INTERNATIONAL RESEARCH COLLABORATION AWARDS (FIRCA)

This award facilitates collaborative research between scientists in the United States and Central and Eastern European countries and between the United States and Latin American and Caribbean countries. Small grants up to $20,000 per year, for up to 3 years, are available for the purchase of supplies and equipment in the foreign laboratory and to provide travel for U.S. and foreign scientists. No salaries or stipends for personnel are available through this award. To be eligible for this award, the U.S.

participant must be the Principal Investigator of an NIH research project grant. The foreign scientist must be from a country within the targeted region, and must hold a position at a public or private nonprofit institution. Application deadlines are on the first day of February, June, and October of each year. Awards are made within 6 months of receipt of the application and are based on scientific merit and the adequacy of the project with respect to the Center's mission.

The FIC also functions as a WHO collaborating Center for Research and Training in Biomedicine, to stimulate contacts between the U.S. and foreign scientists and institutions, to provide research training through fellowships, to organize symposia, to transfer information, and to disseminate scientific publications. You may be particularly interested to know that the WHO has established a Collaborating Center for Diabetes Research, Information and Education and a Collaborating Center for the Methodology, Design and Analysis of Epidemiological and Clinical Research in Diabetes with the National Institute of Diabetes and Digestive and Kidney Diseases (NIDDK). The NIDDK has both formal and informal collaborative research efforts with numerous institutions throughout the world. Country-to-country agreements include the U.S.-Japan Malnutrition Panel, the U.S.-Italy Science and Technology Agreement; the U.S.-Israeli Binational Science Foundation; and agreements under the Special Foreign Currency Program with Poland and Yugoslavia.

NIH VISITING PROGRAM

The NIH initiated the Visiting Program for foreign scientists in 1950. The FIC assumed the administrative responsibilities for this program in 1968. Under this program scientists are eligible for three levels of appointments: Visiting Fellows, Visiting Associates, or Visiting Scientists. Visiting Fellows receive an initial award for 2 years and may extend their stay for a maximum length of 3 years. The stipend for this program varies according to professional experience. The Visiting Associates and Visiting Scientists are appointed as NIH employees for a maximum of 4 years. Salary is based on scientist training and experience. You may be interested in knowing that on any one day there are approximately 1800 foreign scientists working in NIH laboratories--approximately 1/3 of the NIH scientific workforce.

NIH GUEST RESEARCHER PROGRAM

Since the 1930s, NIH has made research and study facilities available to qualified scientists who are supported by their home institutions and/or foreign and domestic foundations. The FIC assumed responsibility for foreign scientists sponsored by this program in 1968.

NIH SPECIAL VOLUNTEER PROGRAM

The NIH Special Volunteer Program, established in 1987, is financially supported by home institutions and/or foreign and domestic organizations. Scientists participate in directed programs of ongoing research.

SCHOLARS-IN-RESIDENCE PROGRAM

The Scholars-in-Residence Program enables eminent scientists and scholars to exchange information and ideas with the NIH scientific community and conduct studies of importance to contemporary biomedicine and international health. Candidates are nominated by either senior NIH staff members or former Scholars. These nominations are reviewed by the Scholars Advisory Panel and the FIC Advisory Board. The final selection is made by the Director of the Fogarty Center. About one third of the research staff in NIH laboratories are visitors from other countries supported by either NIH, their host governments or universities, or other sources.

INTERNATIONAL STUDIES PROGRAM

The International Studies Program is an expression of Congressman Fogarty's concept in 1963 that led to the creation of the FIC: "I should like to see a center dedicated to international cooperation and collaboration in the interest of the health of mankind. This center would encompass...facilities...to permit the assembly for discussion, STUDY, and research of the outstanding health scientists of the world..."

INTERNATIONAL POSTDOCTORAL RESEARCH AND TRAINING IN AIDS

The International Postdoctoral Research and Training in AIDS program supports collaborative research and training for U.S. and foreign scientists, to expand their capabilities in the epidemiology, diagnosis, preventions, and treatment of AIDS. AIDS researchers from developing countries receive the highest priority for acceptance into this program. As a result of this program, strong partnerships have evolved between U.S. schools of medicine and public health and their counterparts in other countries that provide training to both scientists and health technicians in clinical and epidemiological skills related to HIV/AIDS. The skills acquired through this training apply not only to AIDS but to other communicable diseases as well.

INTERNATIONAL TRAINING IN EPIDEMIOLOGY RELATED TO AIDS

Long-term training takes place in the U.S., while short-term training occurs in both the U.S. and other countries. Training-related research occurs mostly in other countries. Five types of training are offered to foreign scientists from developing countries. The training is designed to increase the scientists' capability to conduct epidemiological studies, clinical trials, and prevention measures. The international postdoctoral research and training in AIDS is our largest program in terms of funding.

In addition, FIC administers international agreements on behalf of NIH, such as the "umbrella agreements" in science and technology generated by the Department of State, to advance bilateral relations and scientific cooperation. Agreements are both at the institutional and at the ministerial level. In most cases, these agreements do not provide a source of dedicated funding, but they do provide a useful framework for cooperation.

Finally, the Center has established a Central and Eastern European Initiative and a Latin American and Caribbean Initiative. These initiatives facilitate cooperation between NIH scientists and biomedical and behavioral scientists in these regions. Through these initiatives FIC provides support for **new or expanded research cooperation,** to preserve centers of excellence during a period of economic and political transition, and to lay the basis for continuing research partnerships with U.S. universities.

CONCLUSIONS

In conclusion, I share the enthusiasm of my colleagues that the increased collaboration between scientists throughout the world will facilitate the control and, hopefully, prevention of major diseases and epidemics of the future. The health of the people of the world will depend increasingly on this international cooperation in biomedical and behavioral research. Fostering this international cooperation remains our major challenge for the future.

INCIDENCE, SEASONAL PATTERNS AND CLINICAL CHARACTERISTICS OF IDDM IN THE "THREE CITIES STUDY" IN POLAND FOR THE YEARS 1986 - 1990

Z. Szybinski,[1] A. Czyzyk,[2] R. Wasik,[3] H. Dziatkowiak,[4] M. Szurkowska,[1] D. Pach,[1] H. Stanuch,[5] L. Walczycka,[5] A. Zarnecki,[5] L. Szopinska-Ciba[6]

[1]Department of Endocrinology
Collegium Medicum
Jagiellonian University in Krakow
POLAND

[2]Department of Gastroenterology and Metabolic Diseases
Medical Academy in Warszawa
POLAND

[3]Department of Endocrinology for Children and Adolescents
Medical Academy in Wroclaw
POLAND

[4]Department of Endocrinology for Children and Adolescents
Polish-American Institute of Pediatrics
Collegium Medicum
Jagiellonian University of Krakow
POLAND

[5]Department of Informatics
Collegium Medicum
Jagiellonian University in Krakow
POLAND

[6]Regional Diabetologic Outpatient Clinic in Krakow
POLAND

INTRODUCTION

From the early 1980's,[1-3] interest in epidemiology of IDDM has increased over the world. In Poland, the first register was set up by Rewers et al. for the midwest part of the country.[4] In 1986, the "Three Cities Study" in Krakow, Warszawa and Wroclaw was initiated on a population basis, including more than 1.6 million people.[5] The register was included in the EURODIAB program in 1989[6] and the WHO DIAMOND Project in 1990.[7]

The aim of our study was:

1. To map the incidence of IDDM nationwide in those regions in Poland where the disease has never been studied by standardized epidemiologic methods,
2. To test the hypothesis of an increasing trend in incidence of IDDM in the Polish population,
3. To evaluate the features of IDDM in the Polish population in terms of age distribution, sex ratio, seasonal pattern and clinical course of the disease,
4. To create a necessary basis for studies on risk factors and trends in the incidence of IDDM in Poland.

SUBJECTS AND METHODS

Population Base and Area

The population base was calculated from the data of the Polish Demographic Annual. There were 1.64 million subjects aged 0-29 years, representing approximately 10% of the Polish population of the same age group. In Warszawa and Wroclaw, it was an urban population; in Krakow, the area covered was 57% urban and 43% rural. Krakow and Wroclaw represent the southern, and Warszawa the central part of Poland. The population at-risk is ethnically homogenous, with over 99% European Caucasians.

Case Definition

All registered subjects had to fulfill the following criteria to be included: 1) IDDM diagnosed in accordance with the National Diabetes Data Group recommendation, 2) insulin-dependent diabetes at the time of diagnosis, 3) onset of diabetes between age 0-29 years after December 31, 1985, and 4) resident of the areas selected at the time of diagnosis. All three cities are university centers and most patients were hospitalized at diagnosis. Other types of diabetes were excluded.

Case Ascertainment

The registry originates from two sources. The primary data source consists of the Provincial Pediatric Outpatient Clinics and the Specialized Diabetic Outpatient Clinics for adults in Krakow, Warszawa and Wroclaw. In these clinics, a specialized physician was required to complete a standard questionnaire for each newly diagnosed patient who met the criteria. The secondary data source was provided by pediatric and internal medicine departments in the respective medical school and pediatric internal medicine wards of the hospitals. At the end of each year, hospital records were reviewed and all newly diagnosed cases which met all the criteria entered the register. The degree of ascertainment was over 90%.

Statistical Analysis

The incidence rates have been age-adjusted and expressed as the number of newly diagnosed cases of IDDM per year and per 100,000 residents at-risk in the age groups studied. The 95% confidence interval for the incidence rates were also calculated.

The seasonality pattern was assessed by the Cosinor method used by Halberg and Fourier and regression analysis. A statistical significant seasonal variation was assessed by the amplitude test ($p < 0.05$).

RESULTS

From January 1, 1988 to December 31, 1990, 400 newly diagnosed patients (males/females ~ 1.0) with IDDM were registered. The mean annual incidence rate was 5.2 per 100,000 (3.2 - 7.4) (Table 1), and no increasing temporal trend was observed. Our data are comparable to the Midwest Poland Register for the same period of time. According to the data from the EURODIAB program, the incidence rate of IDDM in Poland represents one of the lowest observed in Europe compared to the highest found in Finland (42.9/100,00 per year) and in Sardinia, Italy (30.28/100,000per year).[6]

Table 1. The incidence rates of IDDM per 100,000 per year in Krakow, Warszawa and Wroclaw in 1986 - 1990.

	1986	1987	1988	1989	1990	Total
Krakow	6.9	4.8	7.6	3.6	7.3	6.0
Krakow voivod.	-	5.7	4.8	6.6	4.8	5.5
Warszawa	4.9	3.5	3.2	3.2	3.2	3.6
Wroclaw	6.1	4.7	4.7	5.8	5.1	5.3
Wroclaw voivod.	7.6	3.9	4.6	5.8	5.0	5.4
Total	6.3	4.5	4.9	5.0	5.0	5.2

Figure 1 represents the incidence of IDDM according to age, year and area investigated. Our register covers the whole sub-population of IDDM for ages 0-29 years. The incidence rate for the 0-14 year age group is comparable to the 15-29 year age group, with some variation in the given calendar year. Extension of the register to older age groups is especially important when data are analyzed in terms of economics of diabetes. The adult sub-population of IDDM may also be valuable when genetic and immunologic markers are evaluated.

Figure 2 represents the age at onset distribution for IDDM males and females. An increased incidence was observed for age groups 10-14 years in girls and 15-24 years in boys. The duration of symptoms to first insulin injection in 84% of the cases was less than two months (Table 2).

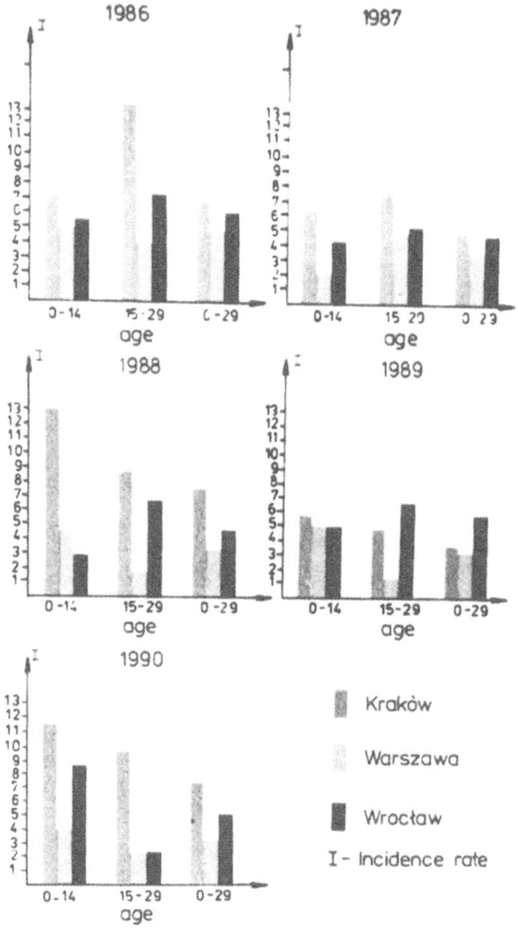

Figure 1. Incidence rate of IDDM according to age, year and area investigated.

Table 2. Duration of symptoms from the onset of diabetes to diagnosis and first insulin injection.

Duration of Symptoms (months)	Number of cases	%
<2	280	84.3
2-4	33	10.0
>4	19	5.7
Without Symptoms	0	0
Total	332	100.0
No Information	68	-

Table 3. Laboratory data at diagnosis.

	Age 0-14 Years		Age 15-29 Years	
Clinical Data	Males	Females	Males	Females
Glycaemia $\pm$ SEM	23.6 $\pm$ 2.03	22.6 $\pm$ 2.14	22.1 $\pm$ 1.49	17.68 $\pm$ 1.16
Range	8.4 - 96.2	7.7 - 68.4	8.3 - 56.1	6.9 - 36.3
Glycosuria % of cases	96.0%	95.4%	94.6%	95.3%
Ketonuria % of cases	66.3%	73.6%	64.8%	60.9%

Table 3 represents laboratory findings at diagnosis of the disease. Regarding some risk factors, we found approximately 30% of cases had a familial history of diabetes initially. However, when data were verified according to EURODIAB, the frequency diminished to 16.4%. Among other factors, infections were present in 37.4% of cases. Ketonuria was observed, depending on age and sex, among 60.9% - 73.6% of cases. Among older females, lower values of glycaemia were observed.

In the Polish population, a seasonal pattern of IDDM incidence was observed quite clearly (Figure 3, Figure 4).

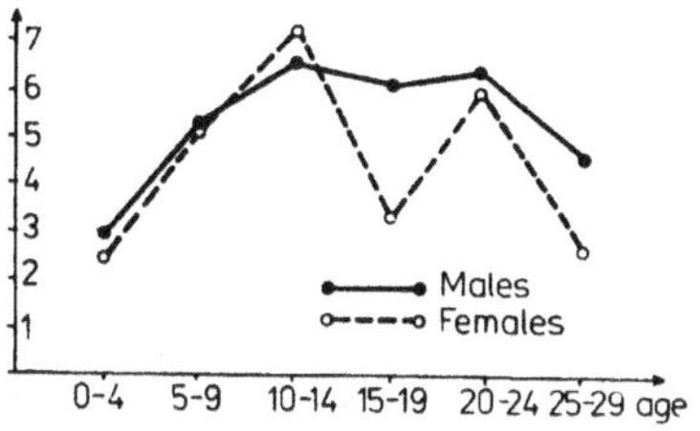

Figure 2. Age at onset by sex in area investigated 1986-1990.

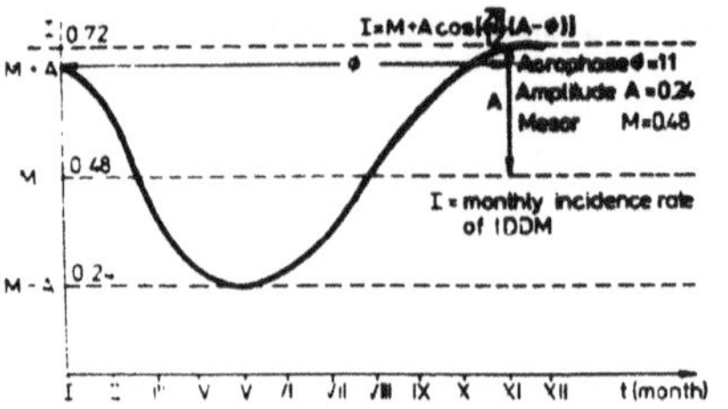

Figure 3. Annual biorhythm of the incidence of IDDM in Krakow.

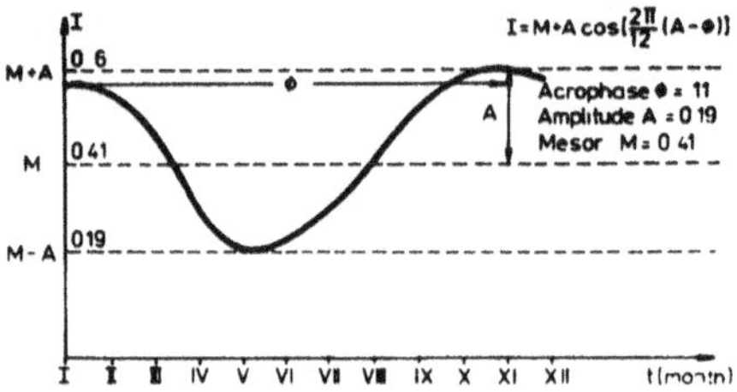

Figure 4. Annual biorhythm of the incidence of IDDM in Wroclaw.

To evaluate the data, we used the Cosinor method of analysis by Halberg and Fourier.[8] It allows verifying a true annual biorhythm. We found a very well defined biorhythm in the Krakow and Wroclaw region, with a similar acrophase, amplitude and mesor in November. Selection of cases with IDDM onset placed around acrophase creates a very interesting sub-population for the future studies on genetic and immunologic markers, especially with respect to the observed gradient in incidence in Europe and possibly over the world.

CONCLUSIONS

In 1986, our group within the "Three Cities Study" Program set up a register of IDDM for a population base of over 1.6 million people. In 1988, we joined the EURODIAB Study, and in 1990, the WHO DIAMOND Project. All registered subjects had to have IDDM diagnosed according to National Diabetes Data Group recommendations, and have the onset of diabetes between 0-29 years of age after December 31, 1985. The register originated from three independent sources: primary data sources were provided by Provincial Outpatient Clinics, by respective Medical Academies, Pediatric and Internal Medicine Departments, or parallel town hospital records. From 1986-1991, 205 males and 195 females with IDDM aged 0-29 years had been registered and the mean annual incidence rate was 5.2 per 100,000 (95% confidence intervals: 3.3-7.7). For age groups 0-14 and 15-29, the mean incidence rates were 5.29/100,000 and 5.4/100,000, respectively. To evaluate seasonal patterns of IDDM, we used the Fourier analysis method applied to biorhythm evaluation. Using the criteria of biorhythm, we found the true biorhythm (seasonality) in Krakow and Wroclaw regions only, with maximal incidence (amplitude) in autumn and winter months. Approximately 85% of newly diagnosed cases with IDDM were registered within two months, and 6% within four months of the onset of clinical symptoms and insulin treatment. Approximately 65% of cases demonstrated ketonuria at the onset of clinical symptoms. Sixteen percent of patients had a family history of diabetes and approximately 50% had some infections or injuries before clinical symptoms.

In summary,

1. The incidence rate of IDDM in Poland (5.2/100,000) is one of the lowest in Europe.
2. We did not observe a difference in incidence of IDDM between the 0-14 year

and 15-29 year age groups. However, the extension of the register to older age groups is important for future studies on economics of diabetes, mortality, genetic and immunologic markers.

3. We observe in Poland true annual biorhythm of incidence of IDDM. The Cosinor method and Fourier's analysis may be recommended to evaluate and verify the biorhythm of IDDM incidence rate.

REFERENCES

1. R.J. LaPorte, N. Tajima, H.K. Akerblom, et al., Geographic differences in the risk of insulin-dependent diabetes mellitus: The importance of registries, *Diabetes Care* 8 (suppl. 1):101 (1985).
2. A. Green, H. King, R.E. LaPorte, The role of IDDM registries in diabetes research and care, *in:* "Diabetes", M. Serrano-Rios, P.J. Lefebvre, eds., New York, Excerpta Med (1986).
3. Diabetes Epidemiology Research International Group: Geographic patterns of childhood diabetes, *Diabetes* 8:1113 (1986).
4. M. Rewers, R.E. LaPorte, M. Walczak, et al., Apparent epidemic of insulin-dependent diabetes in Midwestern Poland, *Diabetes* 36:106 (1987).
5. Z. Szybinski, M. Wysocki, H. Dziatkowiak, R. Wasik, et al., Incidence of insulin-dependent diabetes mellitus in three cities (Warszawa-Krakow-Wroclaw, "Three cities study") during the years 1986-1988, *Endokrynologia Polska* 41:1 (1990).
6. A. Green, E.A.M. Galle, C.C. Patterson, Incidence of childhood-onset insulin-dependent diabetes mellitus: The Eurodiab ACE study, *Lancet* 339:905 (1992).
7. WHO Multinational Project for Childhood Diabetes, *Diabetes Care* 13:10 (1990).
8. C. Bingham, B. Arbogast, G.C. Guillaume, et al., Interferential statistical methods for estimating and comparing cosinor parameters, *Chronobiologia* 9 m397 (1975).

HIGH INCIDENCE OF INSULIN-DEPENDENT DIABETES MELLITUS AMONG AFRICAN AMERICAN AND HISPANIC CHILDREN IN CHICAGO

Rebecca B. Lipton,[1] Erica L. Gibson-Rosado,[1] and Arthur Kohrman[2]

[1]Loyola University of Chicago Stritch School of Medicine
2160 S. First Avenue
Maywood, IL 60153
USA

[2]University of Chicago Pritzker School of Medicine
LaRabida Children's Hospital and Research Center
65th Street at Lake Michigan
Chicago, IL 60649
USA

INTRODUCTION

Virtually all studies of insulin-dependent diabetes mellitus (IDDM) have focused on European-origin populations in the developed world, and data regarding the incidence and prevalence of IDDM among non-Caucasian groups in the U.S. and elsewhere are extremely limited. Data from Birmingham, Alabama and from Pennsylvania indicate that the incidence among African Americans is about 2/3rd to 3/4th that of Caucasians.[1,2] However, these estimates are based on fewer than 150 incident cases. Data from the Caribbean show an extremely low incidence of 5.9/100,000 among African Americans in the US Virgin Islands,[3] with a significant increase in Caucasian admixture among African-origin IDDM children when compared with healthy controls. While there are very few data on the incidence of diabetes among Hispanic youth, the Colorado registry demonstrated an incidence rate of 9.5/100,000 per year, approximately 2/3rd that of the non-Hispanic Caucasian rate in Colorado.[4] Data from Cuba show an extremely low incidence of 1.8/100,000 in 1988.[5]

The degree of genetic admixture, and presumably susceptibility to IDDM, varies greatly among migrant groups in the New World. In view of the north-south gradient which has been observed for IDDM incidence across the world,[6] it is of great interest to determine the incidence of IDDM among Hispanic groups in a northern urban setting, and to contrast this with data from more equatorial and more rural settings. Clearly, it is equally important to determine if there is disparity in IDDM risk experienced by Hispanics of differing national origins residing in a similar environment. Among both African Americans and Hispanics in

Standardization of Epidemiologic Studies of Host Susceptibility
Edited by J. Dorman, Plenum Press, New York, 1994

the U.S., an epidemic of noninsulin-dependent diabetes mellitus is in progress. The relationship within families and ethnic groups of the two major types of diabetes is an additional important question which will be addressed in this study cohort.

A registry of Type I diabetes cases among African American and Hispanic children in the city of Chicago provides unique and essential data. The city of Chicago is home to large numbers of African Americans and Hispanics. There is diversity within these groups with respect to socioeconomic status and the length of time that they have lived in the urban setting.[7] African Americans began migrating to Chicago in large numbers in the 1920's, and waves of migration have occurred subsequently. Chicago is the third largest center of Hispanic settlement, after New York and Los Angeles, and its Hispanic community exhibits more cultural diversity. Mexican immigrants comprise the majority of Chicago's Hispanic community, but substantial numbers of Puerto Rican, Central American, and South American migrants are also represented. These individuals began coming to Chicago during the Great Depression in search of work, with several subsequent waves of immigration distinguished by place of origin and political and economic characteristics. Thus, second- and third-generation descendants, as well as newcomers, are found both in the African American and Hispanic populations. These communities are well established and have progressed economically and now include members from a spectrum of social class backgrounds. It is unusual to have access to such large and eclectic minority populations, and it is therefore especially attractive as a study design to investigate the genetic and environmental components of IDDM etiology which may be observed in this setting.

METHODS

The Chicago metropolitan area encompasses 1,911 square miles. There were approximately 3,033,000 residents of the city of Chicago at the beginning of 1990, of whom about 1,610,000 were of African American or Hispanic origin.[8] Children under the age of 18 years comprise approximately 413,000 African American and 206,000 Hispanic Chicagoans.

Procedure for Identifying Cases

The registry is based on retrospective surveillance of hospital records since 1985. Primary identification is through hospital discharge indices, with verification by contacts with pediatricians and neighborhood clinics, and diabetes camp lists.

There are 40 general hospitals in the city of Chicago and nearby locations which have more than 200 beds and are likely to admit children. As of 1 October 1992, medical record review was complete at 30 institutions (75%) and applications for access to the records were pending at seven institutions.

Inclusion/Exclusion Criteria

Cases are included in the registry according to the following criteria, which are in accordance with the standards published by the WHO DIAMOND Project Group:

1. Diagnosis of IDDM
2. Age at onset less than 18 years
3. Insulin-using at diagnosis
4. Resident of Chicago, Illinois, at the time of diagnosis
5. Diagnosis on or after 1 January 85

6. Diabetes not secondary to another medical condition
7. African American ethnicity as stated on discharge summary or medical record, or Hispanic surname as documented by the U.S. Census Bureau, 1980.[9]

Secondary case ascertainment is planned for the coming 12 months and will be accomplished through pediatrician and clinic records. These data are to be compared with those derived from the primary source (the hospital records), and an estimate of the overall number of cases will be made using the capture/recapture method of Sekar and Deming.[10]

Statistical Methods

Rates are calculated using estimates of the population at-risk determined by the City of Chicago Planning Department.[8] The Poisson distribution was used to calculate 95% confidence intervals around the rates. Other univariate analyses were conducted with the SPSSX statistical package.[11]

RESULTS

As of 1 October 1992, 272 African American children and 94 Hispanic patients were registered. It is possible to calculate "minimum" incidence rates using these preliminary data, based on the assumption that no more patients will be ascertained, which is clearly not true. This exercise provides a useful picture of the findings we may ultimately expect to derive; it also provides an estimate of the lower bounds of the incidence rates. Table 1 demonstrates that the minimum yearly incidence of IDDM among African American males was 9.8/100,000, while among females it was on average 10.7/100,000. The estimated risk per 100,000 for African American males ranged from 5.8 in 1986 to 12.1 in 1988, while among African American females the rates ranged from 8.2 in 1986 to 14.1 in 1990. Thus, we are not observing the lower risk for African American males compared to females which was previously reported by the smaller registries in Birmingham, Alabama and Pittsburgh, Pennsylvania.[1,2] Although the data are incomplete, it appears that the risk of diabetes is equally distributed among males and females in both ethnic groups here in Chicago, a pattern that has been generally found across the world. These minimum rates clearly equal or exceed those reported for African American populations elsewhere, and may reflect either the greater risk associated with residing in a large, Northern urban setting, greater precision due to a larger number of cases ascertained in Chicago, or both. These findings, while intriguing, are preliminary, since the registry database is as yet incomplete.

The minimum incidence among Hispanic females was on average 8.3/100,000 (range 3.1-10.3) and among males it was 6.6/100,000 (range 5.2-7.7). In comparison to the data from Colorado, the risk for IDDM among Hispanic Chicagoans is similar.

A note is in order about denominator data. One explanation for the greater than expected number of diabetic children among African Americans is that the overall number of children at-risk may have been underestimated. The U.S. Department of Commerce has recently admitted to an undercount in the 1990 census which disproportionately affected urban minorities.[12] The incidence rates shown here are based on the population projections made by the City of Chicago Planning Department, which are substantially higher than the U.S. Census figures.[8,13] For example, the City of Chicago demographer estimated that there were 412,900 African American Chicagoans <18 years of age in 1990, while the U.S. Census counted 342,659, or 20% fewer. The discrepancy among Hispanics appears to be less serious, with the City of Chicago estimate exceeding the U.S. Census count by only 5500, or 3% (205,982 vs. 200,545 respectively). Therefore, we plan to continue to rely on the City of Chicago figures for calculating rates, although it is a persistent concern that these, too, may be underestimating the true population at-risk.

Mean age at onset (±SD) was 10.8 years (±4.5) for African Americans and 9.9 years (±4.3) for Hispanics (Table 2). This relatively young age at onset could be an artifact of the ascertainment process, since we examined records at the major tertiary centers first. This might be expected if older children were more likely to have been diagnosed and admitted at community hospitals, while younger children were transferred to the tertiary hospitals. It is of interest to note that diabetes onset was on average almost one year earlier among Hispanics than among African Americans.

Table 1. Minimum IDDM incidence rates, ages 0-17, per 100,000 as of October 1, 1992.

		Year						
Group		85	86	87	88	89	90	Total*
African American	N	22	17	22	22	20	29	132
Females	Rate	10.2	8.2	10.7	10.7	9.7	14.1	10.7
African American	N	24	12	23	25	18	20	122
Males	Rate	11.5	5.8	11.1	12.1	8.7	9.7	9.8
Hispanic Females	N	8	9	3	10	8	10	48
	Rate	8.7	9.6	3.1	10.3	8.1	10.0	8.3
Hispanic Males	N	5	7	6	6	8	8	40
	Rate	5.2	7.1	6.0	5.9	7.7	7.6	6.6

Average Minimum Yearly Risk, 1985 - 1990, per 100,00 (95% Confidence Limits):

African American Children	7.4 (6.0 - 9.2)
Hispanic Children	10.2 (9.0 - 11.6)

*Excludes patients registered but diagnosed after 1990 (n = 24)

Of considerable interest is a potential association with NIDDM in these minority populations where the disease is extremely prevalent. Currently available data are based only on medical records, so that the occurrence of diabetes among family members may or may not be noted on the chart, and information on the onset or severity of the disease among relatives is extremely unreliable. However, only 22% of the currently registered probands have a negative family history for diabetes mellitus, 41% have at least one diabetic second-degree relative recorded on the chart, and a full 21% have at least one first-degree relative (sibling or parent) with diabetes. These families provide an invaluable resource for studying the inheritance patterns of diabetes in populations highly susceptible to both IDDM and NIDDM.

CONCLUSIONS

In examining these preliminary results, several matters are of interest. First, it is clear that the risk for insulin-dependent diabetes mellitus among minority children in Chicago is potentially greater that reported elsewhere. "Minimum" incidence rates calculated from this incomplete database equal or exceed those from completed registries in Pennsylvania, Alabama, and Colorado. Although this may in part reflect underestimation of the population at-risk, it may well be due to environmental factors present in the Northern urban setting.

Onset of disease was relatively early in life among these minority children, and those of Hispanic ethnicity were particularly at-risk. In contrast to data compiled elsewhere, there was no sex difference in age at onset, perhaps because on average, disease onset occurred before puberty (10.6 years in African Americans, and 9.4 years in Hispanics). There was a high prevalence of diabetes in the families of these patients, reflecting the increased prevalence of diabetes mellitus in U.S. minority populations compared with non-Hispanic Caucasians.

Table 2. Mean age at onset, Chicago IDDM patients ages 0-17 registered as of October 1, 1992.

Group	N	Mean Age, Years, ± SD	
African American Females	144	10.8	± 4.4
African American Males	128	10.8	± 4.6
Hispanic Females	43	9.7	± 4.3
Hispanic Males	51	10.1	± 4.2

The African American and Hispanic populations in Chicago are uniquely heterogeneous with respect to the duration of residence in the urban setting, socioeconomic development, and the range of countries of origin among Hispanics. By participating in the WHO DIAMOND Project, these data will be directly comparable with those from other populations. Secondary phases of the project will collect information on family history, socioeconomic factors, and genetic markers for IDDM.[14] Thus, within-ethnic group, as well as cross-national comparisons of risk will be possible. There is a high prevalence of both IDDM and NIDDM in the families of the registered patients. For these reasons, the potential information derived from the Chicago registry will provide critical data on the etiology of insulin-dependent diabetes mellitus in African American and Hispanic children.

REFERENCES

1. L.E. Wagenknecht, J.M. Roseman, W.J. Alexander, Epidemiology of IDDM in African-American and White children in Jefferson County, Alabama, 1979-85, *Diabetes* 38:629 (1989).
2. R.E. LaPorte, N. Tajima, J.S. Dorman, et al., Differences between Blacks and Whites in the epidemiology of insulin-dependent diabetes mellitus in Allegheny County, Pennsylvania, *Am J Epidemiol.* 123: 592 (1986).
3. E.S. Tull, J.M. Roseman, C.L.E. Christian, Epidemiology of childhood IDDM in US Virgin Islands from 1979 to 1988: Evidence of an epidemic in early 1980s and variation by degree of racial admixture, *Diabetes Care* 14:558 (1991).
4. E.C. Gay, R.F. Hamman, P.J. Carosone-Link, et al., Colorado IDDM registry: Lower incidence of IDDM in Hispanics: Comparison of disease characteristics and care patterns in a biethnic community, *Diabetes Care* 12:701 (1989).
5. O. Diaz, T. Norat, Estudio epidemiologico de la diabetes Tipo I, Cuba 1988, *Bol Col Nac Invest Epidemiol Diabete* 1:8 (1990).
6. M. Rewers, R.E. LaPorte, H. King, et al., for the Diabetes Epidemiology Research International Study Group, Trends in the prevalence and incidence of diabetes: Insulin-dependent diabetes mellitus in childhood, *World Health Stat Q.* 4:179 (1988).
7. S. Bernardo, "The Ethnic Almanac," Doubleday, New York (1981).
8. M.V. Bousefield, "Population forecast for the city of Chicago, 1980-2010," City of Chicago Dept. of Planning, Chicago (1988).
9. United States Bureau of the Census, "Spanish surname list: Census of population and housing," Dept. of Commerce (1980).
10. C.C. Sekar, W.E. Deming, On a method of estimating birth and death rates and the extent of registration, *Am Stat Assoc J.* 1949:101 (1949).
11. SPSSX User's Guide, 2nd Ed. Scientific Software, Inc., Chicago, IL (1986).
12. D.A. Freedman, Adjusting the 1990 census, *Science* 252:1233 (1991).
13. United States Bureau of the Census, Census of Population and Housing, Dept. of Commerce (1990).

INDEX

GPSR Compliance

The European Union's (EU) General Product Safety Regulation (GPSR) is a set of rules that requires consumer products to be safe and our obligations to ensure this.

If you have any concerns about our products, you can contact us on ProductSafety@springernature.com

In case Publisher is established outside the EU, the EU authorized representative is:

Springer Nature Customer Service Center GmbH
Europaplatz 3
69115 Heidelberg, Germany

Batch number: 10761599

Printed by Printforce, the Netherlands